CONTINENTAL TRADING BLOCS

CONTINENTAL TRADING BLOCS

The Growth of Regionalism in the World Economy

Edited by
Richard Gibb
Department of Geographical Sciences
University of Plymouth, UK

Wieslaw Michalak
School of Applied Geography
Ryerson Polytechnic University, Toronto, Canada

JOHN WILEY & SONS
Chichester • New York • Brisbane • Toronto • Singapore

Published 1994 by John Wiley & Sons Ltd,
Baffins Lane, Chichester,
West Sussex PO19 1UD, England
Telephone National Chichester (0243) 779777
International (+44) (243) 779777

Reprinted August 1996

Other Wiley Editorial Offices

John Wiley & Sons, Inc., 605 Third Avenue,
New York, NY 10158–0012, USA

Jacaranda Wiley Ltd, 33 Park Road, Milton,
Queensland 4064, Australia

John Wiley & Sons (Canada) Ltd, 22 Worcester Road,
Rexdale, Ontario M9W 1L1, Canada

John Wiley & Sons (SEA) Pte Ltd, 37 Jalan Pemimpin #05–04,
Block B, Union Industrial Building, Singapore 2057

Library of Congress Cataloging-in-Publication Data

Continental trading blocs: the growth of regionalism in the world
economy/edited by Richard Gibb and Wieslaw Michalak.
p. cm. – (Belhaven series in economy, society, and space)
Includes bibliographical references and index.
ISBN 0-471-94909-4
1. Trading blocs. 2. International trade. 3. International
economic integration. 4. Regionalism. 5. General Agreement on
Tariffs and Trade (Organization) I. Gibb, Richard, 1961–
II. Michalak, Wieslaw. III. Series.
HF1418.7.C66 1994
382—dc20

93–47089
CIP

British Library Cataloguing in Publication Data

A catalogue record for this book is available from the British Library

ISBN 0-471-94909-4

Typeset in 10/12pt Times by Saxon Graphics Ltd, Derby
Printed and bound in Great Britain by Bookcraft (Bath) Ltd

Contents

Figures

Tables

Notes on contributors

Michael J. Bradshaw

Dr Bradshaw is a Lecturer in the School of Geography, and an Associate Member of the Centre for Russian and East European Studies, University of Birmingham, England. He received his Ph.D. in Geography at the University of British Columbia, Canada. His research interests include foreign trade and regional development in the post-Soviet Republics and the economic development of Siberia and the Russian Far East. He is editor of *The Soviet Union: A New Regional Geography?*, published in 1991 by Belhaven Press. Most recently, he contributed a report entitled *The economic effects of Soviet dissolution* to the Royal Institute of International Affairs.

The University of Birmingham, School of Geography, Edgbaston, Birmingham B15 2TT, United Kingdom

Richard A. Gibb

Dr Gibb is a Senior Lecturer in the Department of Geographical Sciences at the University of Plymouth, England. In 1986 he received his D.Phil., which examined transfrontier regions within the European Community, from the School of Geography, University of Oxford. His research interests focus on economic cooperation and integration, with a particular emphasis on the European Community and Southern Africa. He has published widely on various aspects of European integration and, more recently, on the consequences of the democratisation of east–central Europe for western Europe as a whole. In 1992, he published a book with Mark Wise – *Single Market to Social Europe* – examining the economic and social consequences of Community integration. He has also developed a keen interest in aspects relating to the Channel Tunnel and is at present working on a book – *The Channel Tunnel: A Geographical Perspective* – for John Wiley. In 1992 he spent a sabbatical year at the University of Cape Town, South Africa, and has since published on the prospects and potential for regional economic cooperation in post-apartheid southern Africa.

Department of Geographical Sciences, University of Plymouth, Drake Circus, Plymouth PL4 8AA, United Kingdom

Robert N. Gwynne

Dr Gwynne is Reader in Latin American Development at the School of Geography, University of Birmingham, England. He is the author of a book examining the relationship between industrial and urban growth in Latin America and, more recently, he published a book comparing the process of industrialisation in Latin America and East Asia. In addition, he has written two books on the economic and political development of Chile and numerous academic articles and reports on the economic development of Latin America and Chile. He is well known as a leading Chilean specialist and has advised the British Foreign Office and British companies on Chilean economic affairs.

School of Geography, The University of Birmingham, Edgbaston, Birmingham B15 2TT, United Kingdom

Rupert Hodder

Dr Hodder read Geography at the University of Sussex (School of African and Asian Studies). He studied Chinese at SOAS (London) and at The Chinese University of Hong Kong; he followed this with a doctorate at the University of Leeds. He was appointed Lecturer in Human Geography at the London School of Economics in 1990, and in 1993 he accepted an appointment at the Chinese University of Hong Kong. Among his many academic publications are two books – *The West Pacific Rim* (1992) and *The Creation of Wealth in China* (1993) – both published by Belhaven Press. He is at present working on a new book, *Merchant Princes of the East*, which examines the Chinese diaspora in the Far East.

Department of Geography, Chinese University of Hong Kong, Shatin N.T., Hong Kong

Alan D. MacPherson

Dr MacPherson is an Associate Professor of Geography and Associate Director of the Canada–United States Trade Center at the State University of New York at Buffalo. His research interests are in small company export development, international investment, and industrial promotion. He is the author of several research reports on producer services in Ontario and New York State. He has published widely on the provision and trade of services and the impact of the Canada–US Free Trade Agreement on the American economy.

Department of Geography, University of Buffalo, State University of New York, Box 610023, Buffalo, New York, USA

James E. McConnell

Dr McConnell is a Professor of Geography and Director of the Canada–United States Trade Center at the State University of New York at Buffalo. His research interests are in international business, regional growth and development, and corporate decision making. He is the author of numerous articles on trade between Canada and the United States, foreign direct investment in the United States and the regional impact of trade on manufacturing and services.

Department of Geography, University of Buffalo, State University of New York, Box 610023, Buffalo, New York, USA

Wieslaw Z. Michalak
Dr Michalak was educated in Poland, France and Canada. Currently he is an Assistant Professor in Geography at the Ryerson Polytechnic University, Toronto, Canada, having worked previously in England and Australia. His main research interests are economic and social geography, industrial society, philosophy and post-totalitarian transition in eastern Europe. He has published widely on the role of producer services in industrial society, social and economic change in eastern Europe, and economic integration in Europe and North America. His current research includes work on the effects of free trade in western Canada, foreign direct investment and trade in eastern Europe, and trading blocs.

School of Applied Geography, Ryerson Polytechnic University, 350 Victoria Street, Toronto, Ontario, Canada M5B 2K3

Mark Wise
Dr Wise is Principal Lecturer in Geographical Sciences and European Studies at the University of Plymouth, England. He has studied at universities in England, Canada, Belgium and France and has long been interested in the problems of integrating different nations and states into coherent international regional entities. Following research in Centres for European Studies at the Universities of Sussex and Brussels, he wrote his doctoral thesis on policy making within the European Community. Since then he has written on a vareity of European topics including a book on the EC's Common Fisheries Policy as well as articles ranging from Community regional policy to security issues in Europe. In collaboration with Richard Gibb, he recently published a book entitled *Single Market to Social Europe* which examines the economic, social and political dimensions of the EC.

Department of Geographical Sciences, University of Plymouth, Drake Circus, Plymouth PL4 8AA, United Kingdom

Preface

One striking feature of the international economic order unfolding in the 1990s is the growth of international regionalism. There is a growing consensus that the process of multilateral trade liberalisation is on the decline and regionalism is on the ascendency. This book therefore examines, in both a theoretical and an empirical manner, the nature and evolution of international regionalism. The aim of the book is twofold: first, to evaluate some of the arguments and issues surrounding trading bloc formation, and in particular whether regionalism is a result of, or a response to, a major transformation in the nature of contemporary capitalism; and second, to examine the character, structure and components of existing trading bloc arrangements. As far as the editors are aware, this is the first book to evaluate systematically the issue of regionalism alongside its wider theoretical and empirical perspectives.

The idea of editing a book on continental regionalism arose from an ESRC-sponsored research project into regional economic integration in post-apartheid southern Africa. Richard Gibb and Wieslaw Michalak, who have shared equally the task of editing this book, then produced the first sketches of such a volume in 1993. Having identified those examples of continental regionalism best suited for such a text, the editors invited a number of regional specialists to contribute chapters. All the chapters of this book have been written by geographers who are experts in their respective geographical regions. The editors would like to thank those authors for their support and enthusiastic participation. Not only did all those contacted contribute, but, quite unexpectedly, the tight production schedule was adhered to by all!

The first two chapters, written by the editors, examine the rationale and philosophy underpinning the often conflicting principles of free trade and regionalism. Although there was a general collaboration on these chapters, Richard Gibb assumed primary responsibility for Chapter 1 which deals with the historical perspective and neo-liberal arguments. This chapter examines the contribution made by the creation of the United Nations and the Bretton Woods Agreement, and in particular the General Agreement on Tariffs and Trade (GATT), to the process of trade liberalisation. Focus here is placed upon the strengths and failings of the GATT to manage the rise of regionalism effectively. Wieslaw Michalak assumed primary responsibility for Chapter 2 which examines why regional trading blocs emerged in the first place and the relationship that exists between these regional arrangements and more fundamental economic and social trends. This chapter highlights the relationship between regionalism and theories concerned with the economic and social transformation of contemporary industrial society.

This introductory section is followed by three chapters examining international regionalism in different parts of Europe: the European Community in the west; the demise of CMEA and the desirability of 'large integration' in the east; and finally, the disintegration of the Soviet Union and the potential for reintegration amongst the Commonwealth of Independent States. Chapters 6 and 7 focus on the Americas. The prospects and potential for the North American Free Trade Area (NAFTA) are reviewed in Chapter 6 followed by an analysis of regional integration in Latin America in Chapter 7. Chapters 8 and 9 extend the geographical scope of study to examine regional economic cooperation and integration in southern Africa and the West Pacific Rim. Both editors worked on the concluding chapter to the book which tries to draw together the many conflicting trends evident in the regionalism versus multilateralism debate.

The editors of this book would like to express once again their thanks to the contributing authors, Mark Wise, Mike Bradshaw, James McConnel, Alan MacPherson, Robert Gwynne and Rupert Hodder, for their support. All the illustrations in this book have been drawn by Tim Absalom of the Cartographic Resources Unit of the Department of Geographical Sciences at the University of Plymouth. The editors would like to express their sincere thanks to Tim for his excellent work and ability to meet wholly unreasonable deadlines. Finally, thanks are also due to Carys and Kathie for their tolerance, days missed on the coast and some no doubt tedious but heated conversations in the Dolphin.

Richard Gibb and Wieslaw Michalak
Plymouth
November, 1993

Acronyms

ACP	Africa, the Caribbean and the Pacific states
AFTA	ASEAN Free Trade Area
ALADI	Asociacion Latinoamericana de Integracion Economica (same as LAIA)
ANC	African National Congress
APEC	Asia Pacific Economic Cooperation
ASEAN	Association of Southeast Asian Nations
BLS	Botswana, Lesotho and Swaziland
CACM	Central American Common Market
CAP	Common Agricultural Policy (of the European Community)
CCP	Common Commercial Policy (of the European Community)
CCT	Common Customs Tariff
CEFTA	Central European Free Trade Agreement
CERTA	Closer Economic Relations Trade Agreement
CET	Common External Tariff
CFP	Common Fisheries Policy (of the European Community)
CIS	Commonwealth of Independent States
CMEA	Council for Mutual Economic Assistance
CONSAS	Constellation of Southern African States
CPSI	Comprehensive Programme for Socialist Integration
CPSU	Communist Party of the Soviet Union
CSIS	Centre for Strategic and International Studies
EAEC	East Asia Economic Council
EAEG	East Asia Economic Grouping
EC	European Community
ECSC	European Coal and Steel Community
ECU	European Currency Unit
EEA	European Economic Area
EEC	European Economic Community
EFTA	European Free Trade Association
EMS	European Monetary System
EMU	European Monetary Union
ERM	Exchange Rate Mechanism
ETM	Elaborately Transformed Manufacturing
EURATOM	European Atomic Energy Community
FDI	Foreign Direct Investment
FTA	Free Trade Agreement
G7	Group of Seven

GATS	General Agreement on Trade in Services
GATT	General Agreement on Tariffs and Trade
GDP	Gross Domestic Product
GDTA	Geographically Discriminatory Trading Arrangement
GEIS	General Export Incentive Scheme
GNP	Gross National Product
GSP	Generalised System of Preferences
IMF	International Monetary Fund
ISI	Import Substitution Industrialisation
ITO	International Trade Organisation
LAFTA	Latin American Free Trade Area
LAIA	Latin American Integration Association
LDC	Less Developed Country
MERCOSUR	Mercado Comun del Cono Sur (same as SCCM)
MFA	Multi-Fibre Agreement
MFN	Most Favoured Nation
MNC	Multinational Corporation
MNE	Multinational Enterprise
NAFTA	North American Free Trade Agreement
NATO	North Atlantic Treaty Organisation
NIC	Newly Industrialised Country
NIS	Newly Independent States (of the ex-Soviet Union)
NTB	Non-Tariff Barrier
OAU	Organisation of African Unity
ODECA	Organisacion of Estados Centroamericanos
OECD	Organisation for Economic Cooperation and Development
OEEC	Organisation for European Economic Cooperation
OMA	Orderly Market Arrangement
OPEC	Organisation of Petroleum Exporting Countries
PAFTDA	Pacific Free Trade and Development Area
PAR	Preferencia Arancelaria Regional (same as RTP)
PBEC	Pacific Basin Economic Council
PECC	Pacific Economic Cooperation Conference
PPS	Purchasing Power Standard
RTP	Regional Tariff Preferences
SACM	South African Common Market
SACU	Southern African Customs Union
SADC	Southern African Development Community
SADCC	Southern African Development Coordination Conference
SCCM	Southern Cone Common Market (same as MERCOSUR)
SEA	Single European Act
SEM	Single European Market
SICA	Sistema de la Integracion Centroamericana

SII	Structural Impediments Initiative
SSR	Soviet Socialist Republic
TNC	Trans-National Corporation
TRIM	Trade-Related Investment Measures
TRIP	Trade-Related Intellectual Property
UN	United Nations
UNCTAD	United Nations Commission on Trade and Development
UNCTC	United Nations Centre on Transnational Corporations
UNTCMD	United Nations office for Transnational Corporations and Management Division
VER	Voluntary Export Restraint
VIE	Voluntary Import Expansion
WEU	Western European Union
WTO	World Trade Organisation

1 Regionalism in the world economy

Richard Gibb
University of Plymouth

The 1980s witnessed a growth of international regionalism in western Europe, North America, Latin America, Africa and Asia. This resurgence of regionalism will significantly influence the nature and evolution of the world economy in the 1990s and beyond. However, the future of regionalism depends also on developments taking place within the world economy, particularly the multilateral trading system. Regionalism and multilateralism are therefore inevitably and inextricably interlinked. One of the principal objectives of this introductory chapter is to explore the complex relationship that exists between regional trading blocs and the multilateral trading system.

The globalisation of business through the spread of multinational companies, changing the fundamental relationship between markets and states, has led a number of scholars (O'Brien, 1992; Hopkinson, 1992) to predict 'the end of geography'; in other words, a diminishing importance of the spatial dimension in the organisation of the world economy. However, somewhat paradoxically, the globalisation of business appears to have promoted international regionalism as states try to control at the regional level what they have increasingly failed to manage at the national and multilateral levels. At the same time as international economic relations become truly global in character, the multilateral trading system is in decline and regionalism is on the ascendency. This apparent shift away from globalism to interstate regionalism raises a number of important questions. Does regionalism constitute a threat to multilateralism and undermine the principles and obligations of the General Agreement on Tariffs and Trade (GATT)? If it does, what is the 'true' nature of the threat? Or should the creation of regional trading blocs be considered as a desirable evolution, likely to reinforce global free trade? Or, perhaps, should regionalism be perceived as a good second best option compared to the unattainable first option of multilateralism? There are certainly no clear-cut answers to these very difficult questions. Two well-respected economists, Preeg (1989) and Schott (1989), find them-

Continental Trading Blocs: The Growth of Regionalism in the World Economy
Edited by R. Gibb and W. Michalak

selves in sharp disagreement over the desirability of regionalism. Preeg interprets trading blocs, particularly free trade areas, as having the potential to support a more liberal multilateral trading system. Schott, on the other hand, regards regionalism as a threat to free trade and the GATT process of liberalisation without discrimination. Politics, politicians, policy-makers and business are no less divided over the merits of regionalism. Much depends on whether regionalism is perceived to be a response to the detrimental impacts arising from the unqualified application of the principle of free trade or a strategy used to promote multilateral liberalisation.

Before proceeding to evaluate some of the key issues in the regionalism debate, it is first worth emphasising that regional cooperation is as much an economic process as it is a political one. Economic cooperation and political sovereignty are at the very heart of any regional initiative, in Europe or elsewhere. The temptation to examine regionalism as a purely economic and technical phenomenon must therefore be avoided (Gibb, 1993). This chapter seeks to evaluate the principal arguments in the regionalism versus multilateralism debate and to clarify some aspects of the integration process. It is divided into four main parts. The first part reviews briefly the post-1945 movement to free trade and its associated economic rationale. The second part examines the multilateral trading system and the GATT principles of non-discrimination and market-orientated trading. It focuses on Article XXIV of the GATT which sanctions free trade areas and customs unions. The third part clarifies the multitude of complex arrangements covered by the term 'regionalism', ranging from sectoral agreements to political–economic unions. As observed by Wise and Gibb (1993), semantic muddle and conceptual confusion over the different forms of regionalism often make communication of ideas extremely difficult. This section therefore clarifies the concepts and terminology. Finally, the fourth part examines the theory of economic integration in order to explore whether regional trading blocs are consistent with the GATT. Do they really threaten multilateralism and will they fragment the world economy?

THE HISTORICAL PERSPECTIVE

According to the United Nation's Department of Economic and Social Affairs:

> The maintenance of an open global trading framework is of the utmost importance to all nations ... All countries have an interest in ensuring that regional trade arrangements are part of the global trend towards openness and trade liberalisation, since global welfare will be increased more through non-discriminatory than discriminatory reductions in protectionism. (United Nations, 1990, 3)

Theoretical and practical policy arguments over the relative merits of free trade and protectionism date back to the mid-nineteenth century, when the United Kingdom repealed the Corn Laws and unilaterally embraced a policy of free trade. Reflecting the United Kingdom's comparative advantage in the world economy as it then existed, a move away from mercantilism towards free trade was considered to be the best policy option regardless of whether trading partners were for free trade or protectionism. Tension between free traders and protectionists has since been a dominant feature of national and international trade negotiations. The cause of free trade received a major boost following the devastating impacts of the 1930s recession when 'beggar my neighbour' policies imposed high tariff barriers and strict quota regimes to curtail free trade. Indeed a prominent economist and adviser to the GATT, Bhagwati, goes as far as to argue that post World War II trade liberalisation was driven by a pro-trade bias generated by the 1930s 'experience with tariffs that was widely perceived to have been a spectacular failure' (Bhagwati, 1988, 20).

In an excellent analysis of the 1930s World Depression, Kindleberger (1973) examines the reasons that lay behind the Depression. There were of course many contributory factors, such as the misuse of the gold standard, deflation, America's restrictive monetary policy and structural disequilibrium in the world economy. Last, but by no means least, there were the detrimental effects associated with exchange rate depreciation and tariff escalation. Kindleberger dispels the view of the World Depression resulting from economic nationalism, bilateralism and tariff rises. Few academics would disagree with his analysis that 'beggar my neighbour' policies were not the sole cause of the Great Depression. However, to what extent tariffs were a product or a cause of the depression is clearly debatable. As Bhagwati (1988, 21) observes, 'it is certainly arguable that the tariff escalations deepened the Depression'.

Throughout the 1920s, the liberal international trading system came under mounting pressure from an increase in tariffs. In 1927, the League of Nations responded to this threat to free trade by holding a World Economic Conference in Geneva. Its aim was to negotiate a reduction in tariffs or, at the very minimum, a tariff truce. The Conference was largely unsuccessful. In fact, Irwin (1992, 30) argues that international gatherings of this type, aimed at restoring multilateralism in Europe, were at the time misplaced and probably counterproductive. Regional liberalisation, enabling at least some countries to reduce tariffs, was hampered by intransigence over the interpretation of the 'most-favoured-nation' clause. In June 1930, the United States of America (USA) passed the infamous Smoot–Hawley Tariff Act, resulting in the average level of tariff increasing from 38 to 53 per cent (Oxley, 1990). The Smoot–Hawley Act led to a wave of retaliatory action, with virtually every country, especially the United Kingdom, increasing average tariff levels significantly. The United Kingdom raised its tariff levels in November and December 1931 and, in addition, depreciated sterling. Most of continental Europe followed a similar course of action, leading Switzerland, for example, to abandon a German–Swiss trade agreement. Throughout Europe, tariffs were reinforced by quantitative restrictions and exchange controls were applied in a discriminatory fashion. Between 1929 and 1932, the volume of trade declined by over 40 per cent while world output fell only 20 per cent (Irwin, 1992).

The contracting spiral of world trade between 1929 and 1933 is illustrated graphically in Figure 1.1. Kindleberger (1973) concludes his analysis into the causes that lay behind the Great Depression by highlighting the destructive consequences arising from a lack of leadership in the world economy:

> The world economic system was unstable unless some country stabilised it, as Britain had done in the nineteenth century and up to 1913. In 1929, the British couldn't and the United States wouldn't. When every country turned to protect its national private interest, the world public interest went down the drain, and with it the private interests of all. (Kindleberger, 1973, 292)

The inability or unwillingness of countries to assume responsibility for stabilising the world economy led to, amongst other things, the chaos of protectionism in the 1930s and the Great Depression. To what extent there are similarities between protectionism in the 1930s and regionalism in the 1990s, which may be attributed to a lack of leadership in the world economy with perhaps the US unable and Japan unwilling to assume responsibility, is an issue that will be examined further in Chapter 10.

The 1930s World Depression was much deeper and more widespread than any other in modern history. Protectionism undermined living standards throughout the world and 'played a major role in giving rise to the militarism which precipitated the Second World War' (UN, 1990, 11). Whatever the exact cause of the Great Depression, it left experts and politicians convinced of the need to impose limits on the extent of governmental interference in international trade (Culbert, 1987; GATT, 1992). A perception that protectionism led to a fragmentation of the world economic system, depression and ultimately disaster became widespread. Post-war trade liberalisation was therefore initiated by a free trade bias in the governing elites arising from the destructive consequences of the 1930s Depression. There is no doubt that the growth of international trade, instigated by the process of multilateral trade liberalisation, has been one of the most outstanding characteristics of the economic changes of the past 50 years. But how robust is the theory supporting free trade and multilateralism as the organising principle for the world economy?

THE CHALLENGE OF FREE TRADE

For many economists, the arguments supporting multilateralism and trade liberalisation are unquestionably strong:

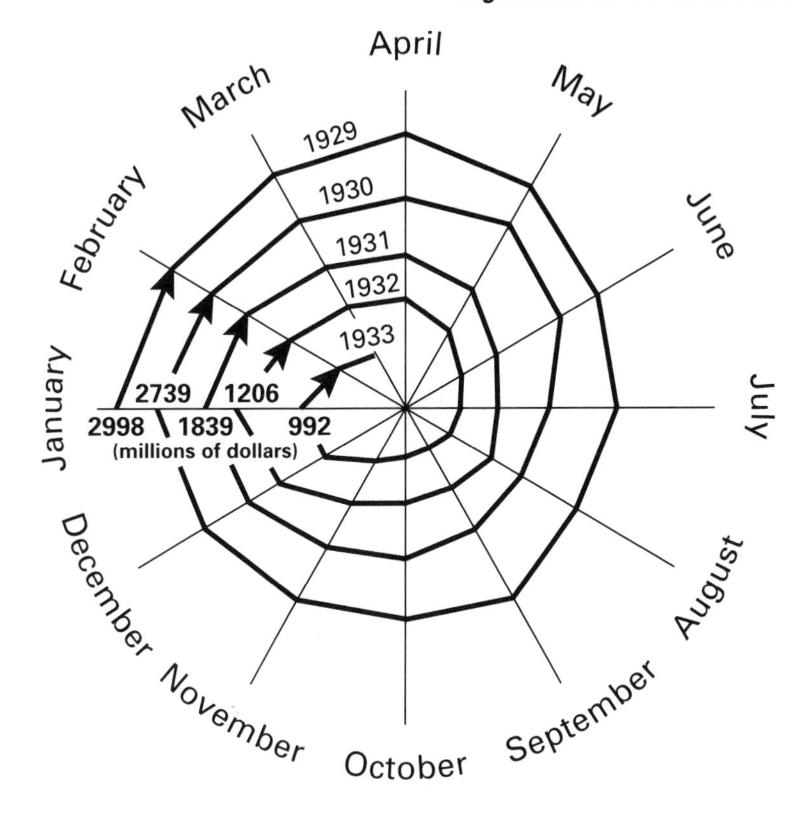

	1929	1930	1931	1932	1933
January	2997.7	2738.9	1838.9	1206.0	992.4
February	2630.3	2454.6	1700.5	1186.7	944.0
March	2814.4	2563.9	1889.1	1230.4	1056.9
April	3039.1	2449.9	1796.4	1212.8	
May	2967.6	2447.0	1764.3	1150.5	
June	2791.0	2325.7	1732.3	1144.7	
July	2813.9	2189.5	1679.6	993.7	
August	2818.5	2137.7	1585.9	1004.6	
September	2773.9	2164.8	1572.1	1029.6	
October	2966.8	2300.8	1556.3	1090.4	
November	2888.8	2051.3	1470.0	1093.3	
December	2793.9	2095.9	1426.9	1121.2	
Average	2858.0	2326.7	1667.7	1122.0	

Source: Kindleberger, 1973

Figure 1.1 World trade, 1929–1933

> the general presumption in favour of freer trade on a most-favoured-nation basis is one of the most robust conclusions to come out of the study of economics in the past two hundred years. (Wonnacott and Lutz, 1989, 60)

Multilateralism and the GATT are based on an essentially laissez-faire approach to international trade. Furthermore, international trade theory is founded on the fundamental principles of the free market system as formulated by the classic liberal thinkers Adam Smith and David Ricardo. By removing impediments to the free movement of the factors of production, international trade will, it is argued, be determined by the optimal utilisation of production factors structured according to comparative advantage. In theory, comparative advantage improves resource allocation and utilisation. Theoretically then, 'any country will ultimately be better off by participating in foreign trade than would otherwise have been the case' (Scheepers, 1979, 83). Detail aside, the starting point for this process is the removal of impediments to trade. The initial impact of removing barriers is felt at the most disaggregated microeconomic level. Companies and firms are forced to rethink their development strategies as their costs of production decrease and overall competitive pressures increase. Domestic markets are then exposed to international standards of productivity. This should, in turn, lead to enhanced economies of scale as the competitive pressures of the market-place restructure industries. Significantly, such a conceptualisation fails to incorporate the impact of external economies, viewing them as either unimportant or fundamentally unquantifiable (see Chapter 2).

The removal of barriers to trade will therefore reduce the costs for goods and services and, as a consequence, promote growth in demand. This new demand strengthens competitive pressures and leads to a restructuring of industry based upon newly enhanced economies of scale. Inefficient companies are forced to close down and new investment is directed to those plants and industries which possess a beneficial comparative advantage within the international economy. However, the removal of barriers and enhanced competition should, in theory, lead to gains far beyond those associated with economies of scale. For example, market fragmentation permits a whole series of inefficient business practices to operate. Overmanning, excess overheads and overstocking result in what the European Commission describes as 'x-inefficiency' (Wise and Gibb, 1993). X-inefficiency promotes the wasteful allocation of human, physical and financial resources and permits excess and monopoly profits to develop as a result of weak competition. The removal of barriers to trade and a more competitive economy helps to eradicate these inefficiencies and encourages a rationalisation of business activity. The combined effects of economies of scale, enhanced productivity, and income and consumption implications, promotes economic development in those countries able to participate in free trade. Thus, despite the sometimes painful short-term costs associated with economic restructuring, free trade induces a more efficient and productive, and ultimately more competitive, domestic economy.

However, the formation of regional trading blocs casts doubt on the unqualified application of the theory of free trade. International trade theory is founded upon the Ricardian principle of comparative advantage which, because it assumes perfect competition, full employment and constant returns to scale, has been widely criticised, particularly by the advocates of new trade theory (see Chapter 2). The problems associated with orthodox trade theory promoted several new theories of international trade (see for example Barker, 1977). The creation of regional trading blocs can be attributed, in part, to the realisation that international trade, based on the fundamental principles of the free market, will not have the desired economic results. Commenting on the formation of the European Economic Community in 1957, Scheepers observed that:

> The ... experience showed that the practice of free trade in the pure sense in a world perspective did not result in an acceptable geographical distribution of economic growth, and that some intervention was necessary to achieve it. (Scheepers, 1979, 83)

Regionalism can be interpreted as an attempt to promote the theory of free trade on a more restricted geographical basis. Simply put, free trade may be a good thing for the performance of the world economy as a whole but, inevitably, there will be winners and losers. Whilst the GATT Uruguay Round may have the potential to raise global welfare permanently by more than $100 billion a year, spur economic growth and extend competition to hitherto sheltered parts of the world economy (The Economist, 1992), the benefits of such a process will be spread unevenly. Those French farmers who riot in order to maintain trade protection are in no doubt about the merits of the GATT multilateral trading structure being applied to European agriculture! This whole debate relates to a central theme underpinning the current chapter: whether regionalism is a response to the detrimental consequences arising from free trade or an approach designed to promote multilateral liberalisation. Since the end of the Second World War, the GATT has been dedicated to the latter cause: non-discrimination among trading partners and a mutual lowering of barriers to trade. Over the past 45 years the GATT has significantly influenced the nature and evolution of the world trading system. It is for this reason that the current chapter turns to an examination of the GATT Treaty and its provisions for managing regional trading blocs.

THE GENERAL AGREEMENT ON TARIFFS AND TRADE

The severity of the Great Depression and the disastrous consequences of discriminatory bilateralism during the inter-war period strengthened the resolve of policy makers to limit the extent of governmental interference in international trade. Anglo-American deliberations during the Second World War were instrumental in establishing the post-war economic order. Negotiations were, however, protracted, with serious disagreement over the desirability of trade preferences. The United Kingdom's insistence on maintaining the Imperial Preference was in direct conflict with America's desire to eliminate all forms of discriminatory practice in international commerce. According to Culbert (1987), in the negotiations to establish an Atlantic Charter in 1941, Churchill 'strenuously objected' to a clause calling for 'access without discrimination and on equal terms to world markets' (Culbert, 1987, 386). In the end, the Atlantic Charter was a compromise, allowing states to respect their existing obligations (i.e. imperial preferences) whilst at the same time promoting the cause of non-discrimination and the most-favoured-nation principle. The issue of non-discrimination was a hotly contested one, as reflected in the following letter from Maynard Keynes to Dean Acheson, then Assistant Secretary of State in the USA:

> [the] most-favoured-nation clause and all the rest which was a notorious failure and made such a notorious hash of the old world. We know it won't work. It is the clutch of the dead, or at least the moribund, hand. (quoted in Harrod, 1951, 512)

America's insistence on non-discrimination and Britain's determination and tenacity to maintain the Imperial Preference were reflected in all trade agreements during and immediately after the Second World War, the GATT being no exception.

Anglo-American cooperation during the Second World War led to the 1944 Bretton Woods Conference that established the institutional framework for the post-war economic order. The primary goal of Bretton Woods was macroeconomic stability, seen as an important prerequisite to facilitating multilateral free trade and the adoption of liberal trade policies. In order to establish a stable international system, Bretton Woods designed an institutional infrastructure formulated on three international organisations: the International Monetary Fund (IMF), the World Bank and the International Trade Organisation (ITO). The ITO was to be for trade what the World Bank is for developing nations and the IMF is for macroeconomic policy. Under the auspices of the United Nations, over 50 countries worked on a draft charter for the ITO. The original draft included provisions on employment, international investment, services and commodity agreements (GATT, 1992).

The negotiations to establish a world trading organisation were both time-consuming and acrimonious, and ultimately unsuccessful. Again, the source of the friction lay with the disputed merits of the most-favoured-nation principle, multilateralism and preferential market access. As the negotiations dragged on, a requirement to implement some tariff reductions in the short term was recognised. As a temporary measure only, the USA proposed that the commercial policy provision of the Draft ITO Agreement be incorporated in a General Agreement on Tariffs and Trade. In 1947, 23 countries met in Geneva and agreed to a mutual set of tariff reductions codified in the GATT (Figure 1.2). The original intention was for the GATT 'to be a provisional agency that would go out of existence on the establishment of the ITO' (Culbert, 1987, 396). Although the ITO Charter was agreed at a UN conference in Havana in March 1948, its ratification by national legislatures proved difficult (GATT, 1992). In 1950, when the Havana Charter failed to be presented for US Congressional ratification, the ITO was effectively dead and the GATT became the principal instrument for commercial policy and international trade regulation. The GATT therefore had an inauspicious start, based on a compromise between America's desire to lower tariffs on a selective basis without discrimination and a British aspiration for across-the-board tariff reductions with an element of discrimination allowing for the Imperial Preference. These tensions are reflected in the GATT's Articles of Agreement.

THE GATT TRADING SYSTEM

By 1992, the GATT had 105 Contracting Parties (as the GATT member states are formally known) with a further 27 countries applying the rules of the Agreement on a *de facto* basis (Figure 1.3). According to the GATT secretariat, the 105 Contracting Parties account for over 90 per cent of world merchandise trade (GATT, 1992). The GATT aims to promote a non-discriminatory open market in which only prices and tariffs determine comparative advantage. As such, the GATT is not a classic free trade organisation in that it permits the imposition of tariffs. Its primary goal is to promote 'open, fair and undistorted competition' (GATT, 1992, 8). However, the Agreement does aim to liberalise trade and open up markets through an overall reduction of trade barriers. As the preamble to the Agreement makes explicitly clear:

> relations in the field of trade and economic endeavour should be conducted with a view to raising standards of living, ensuring full employment and a large and steady growing volume of real income and effective demand ... by entering into reciprocal and mutually advantageous agreements directed to the substantial reduction of tariffs and other barriers to trade and to the elimination of discriminatory treatment in international commerce. (GATT, 1992, 1)

Instrumental to the GATT process of reducing tariffs was the establishment of major conferences, which later became known as the 'rounds'. Since 1948 there have been seven rounds of trade negotiations. The Uruguay Round, completed in December 1993, was the eighth (Figure 1.4). The procedures adopted by the

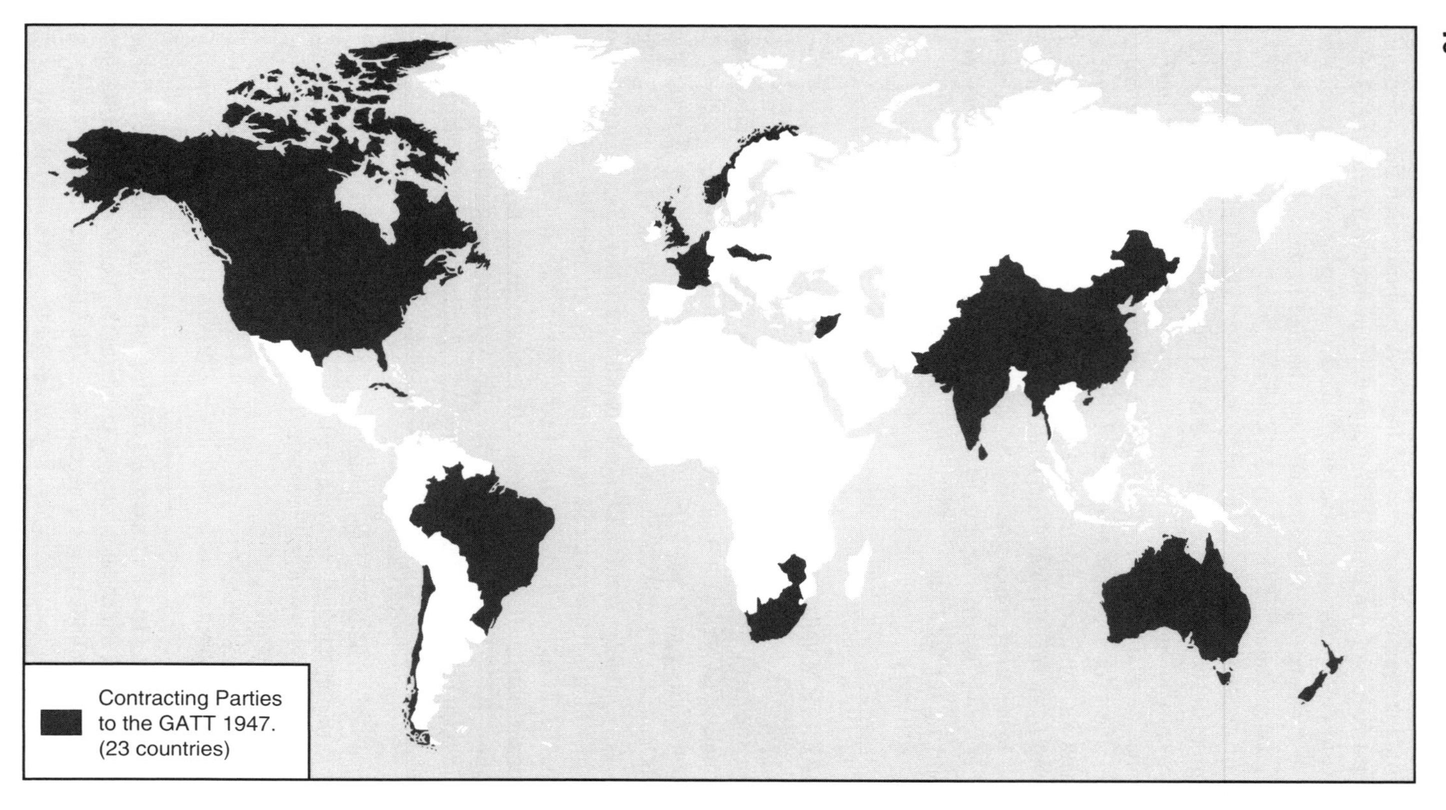

Figure 1.2 Members of the General agreement on Tariffs and Trade, 1947

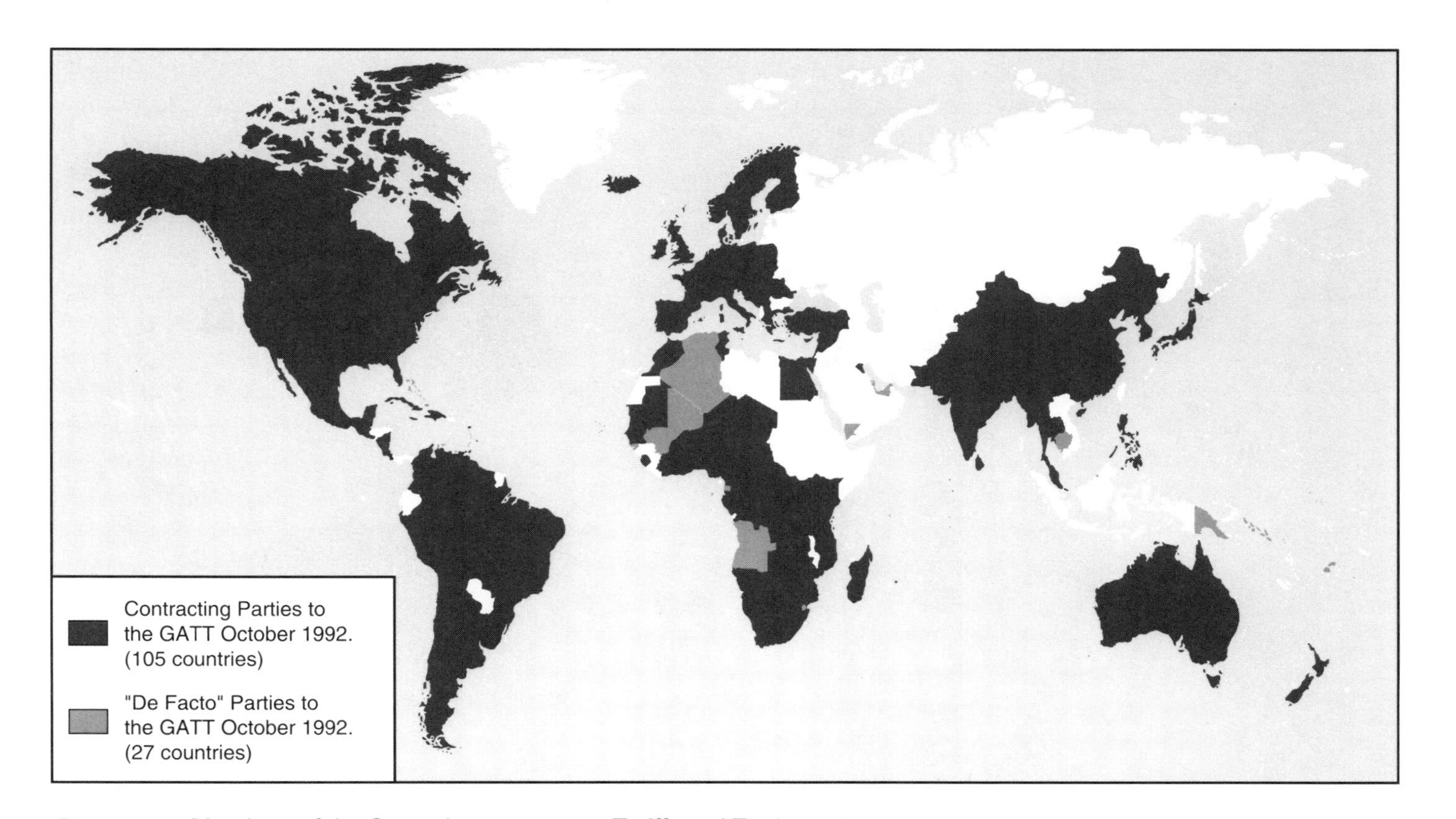

Figure 1.3 Members of the General agreement on Tariffs and Trade, 1992

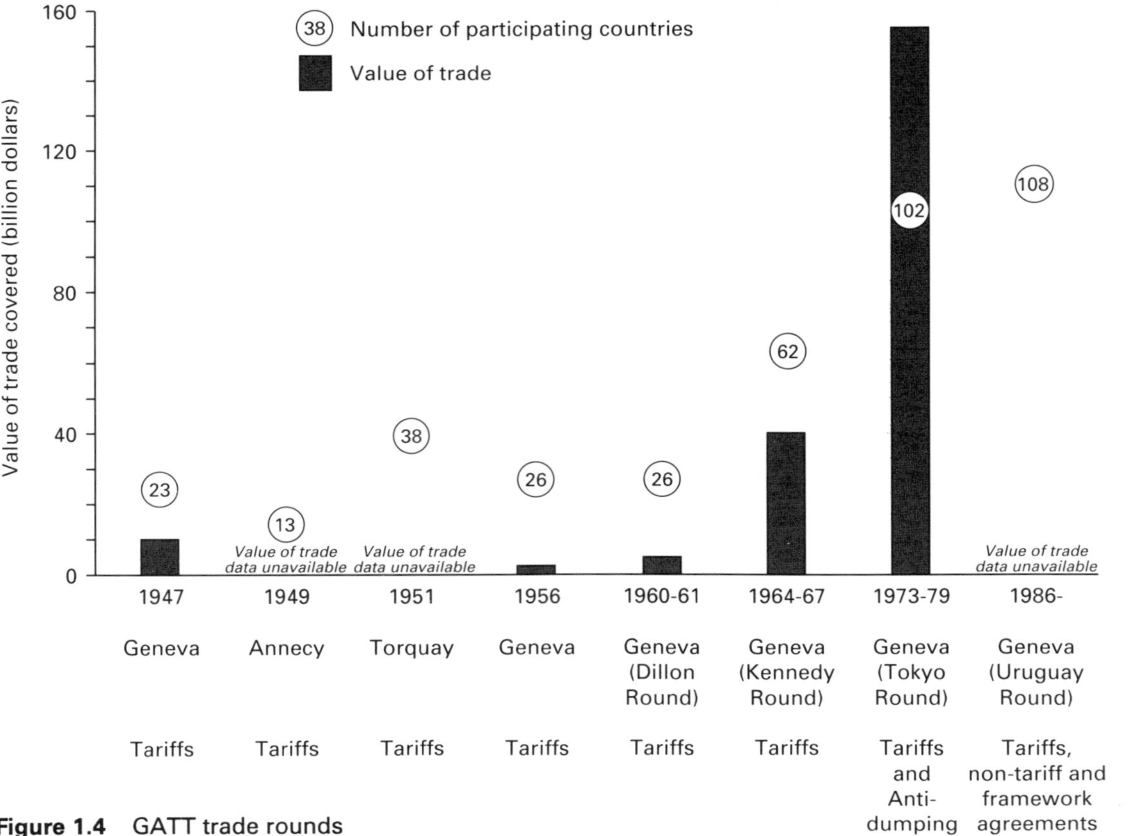

Figure 1.4 GATT trade rounds

GATT for negotiating multilateral tariff reductions have seen tariff levels fall from approximately 40 per cent in 1948 to 5 per cent in 1987 (Figure 1.5). After the 1974–1979 Tokyo Round, average tariffs on manufactured goods were 4.9 per cent in the USA, 5.4 per cent in Japan and 6 per cent in the European Community. This reduction in tariffs was accompanied by an unprecedented growth of free trade and income that favoured mainly, but not exclusively, the developed countries (Preeg, 1989). Between 1946 and 1985, the volume of world trade multiplied nine times. In the post-war period, trade grew significantly more rapidly than world output and income. During the 1950s and 1960s, the value of trade increased at a rate 50 per cent greater than the growth of output and income (Figure 1.6). Most analysts are in agreement with Preeg that:

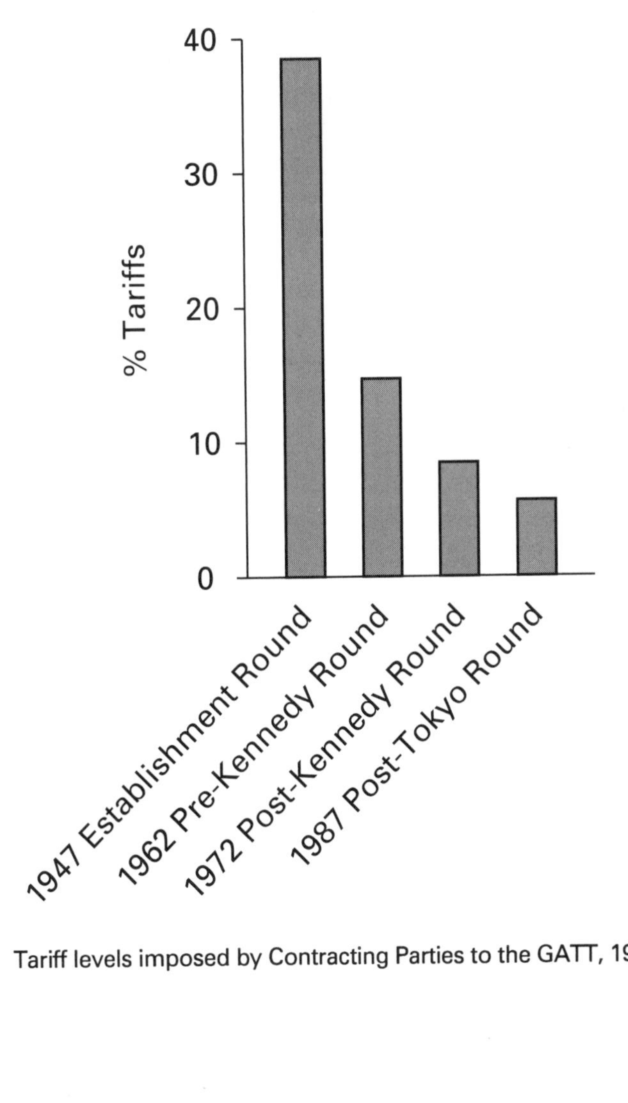

Figure 1.5 Tariff levels imposed by Contracting Parties to the GATT, 1947–1987

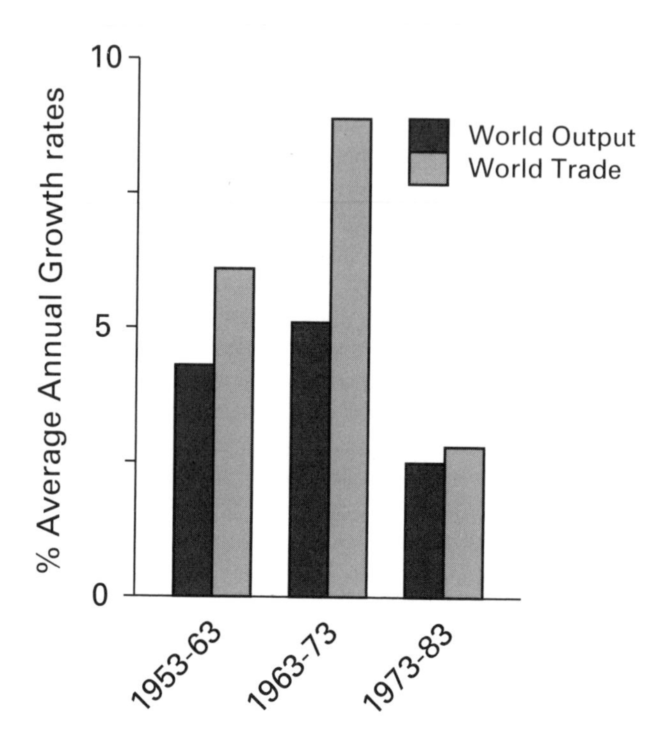

Figure 1.6 Average annual growth rates of world output and world trade, 1953–1983

> The unprecedented growth in trade and national income in the 1950s and 1960s is rightly credited, in large part, to this [the GATT] process of multilateral trade liberalisation. (Preeg, 1989, 201)

The preamble to the Agreement identifies the two indispensable principles of the GATT: non-discrimination, or most-favoured-nation status, and reciprocity. The very first Article of the Agreement outlines the meaning behind the most-favoured-nation principle. It is the backbone to the GATT Agreement and worth quoting at some length:

> any advantage, favour, privilege or immunity granted by any contracting party to any product originating in or destined for any other country shall be accorded immediately and unconditionally to the like product originating in or destined for the territories of all other contracting parties. (GATT, 1986, 2)

Thus, according to Article 1, no country shall confer any special privilege to another or discriminate against it: all are on an equal basis and share the benefits and costs of any change in trade barriers. International trade must therefore be conducted on the basis of non-discrimination. The most-favoured-nation principle is supported by Article III, which requires 'national treatment' of all foreign goods once they have passed through customs. Foreign goods from Contracting Parties should not therefore be liable 'to internal taxes or other internal charges of any kind in excess of those applied, directly or indirectly, to like domestic products'.

The principle that a reduction in one tariff barrier will be extended to all other Contracting Parties contradicts, or at the very least causes tension with, the second most important principle of the GATT, reciprocity. As Winters (1990) observes, reciprocity is never defined formally by the GATT and is open to interpretation. Reciprocity, in terms of the Agreement, can be seen as reinforcing the conventional approach to reducing tariff barriers in that the benefits and costs of changes are balanced through negotiation. In the first instance, tariff reductions were negotiated bilaterally item by item. However, in the Kennedy Round of multilateral trade negotiations from 1963 to 1967, across-the-board tariff reductions were negotiated. After a package has been agreed, any party wanting to raise the tariff is obliged to negotiate with the party with which the original reduction was agreed in order to compensate for lost trade. Reciprocity therefore provides the 'means for prizing open markets' (Winters, 1990, 1289) and exposing them to the forces of free trade and competition. However, after a tariff reduction has been negotiated, it is then generalised by the most-favoured-nation clause, thus preventing the swapping of favours between Contracting Parties and the growth of bilateralism.

The American insistence on non-discrimination in international trade is therefore forcefully supported by Articles I and III of the Agreement which prescribe unconditional most-favoured-nation treatment.

THE GATT'S INFLUENCE ON REGIONAL TRADING ARRANGEMENTS

Although the principle of non-discrimination is central to the GATT, preferential trade arrangements in effect at the time of the Agreement were allowed to remain. More importantly from the viewpoint of regional trading blocs, Article XXIV of the Agreement permits Contracting Parties to establish free trade areas and customs unions. Thus regional trading blocs, despite epitomising the very principle of discrimination, are indeed GATT-consistent. As observed by Patterson (1989), regional trading blocs are 'GATTable'. It would therefore appear that regional and multilateral trading arrangements can coexist, with regionalism complementing multilateralism and promoting free trade. The General Agreement explicitly recognises the value of closer integration through regionalism:

> the provisions of this Agreement shall not prevent, as between the territories of contracting parties, the formation of a customs union or of a free trade area or the adoption of an interim agreement for the formulation of a customs union or of a free trade area. (GATT, 1986, Article XXIV, Paragraph 5)

Thus the General Agreement permits regional trading blocs as an *exception* to the general rule of most-favoured-nation treatment on the condition that certain criteria are met. Three principal criteria must be satisfied in order for trading blocs to conform with, and be sanctioned by, the GATT:

1. With respect to a customs union (essentially the same criteria are applied to a free trade area), 'duties and other regulations of commerce imposed at the institution of any such union ... shall not on the whole be higher or more restrictive than the general incidence of duties ... prior to the formation of such union.' (Article XXIV, paragraph 5a)
2. The Agreement also specifies that 'duties and other restrictive regulations of commerce are eliminated with respect to substantially all the trade between the constituent territories of the union or at least with respect to substantially all the trade in products originating in such territories.' (Article XXIV, paragraph 8a)
3. Any interim agreement leading to a free trade area or a customs union 'shall include a plan for the formation of such a[n] ... area within a reasonable length of time.' (Article XXIV, paragraph 5c)

In theory, if a review by the Contracting Parties found that a proposed customs union or free trade area did not satisfy the above criteria, the countries proposing the arrangement would have to modify the proposal, withdraw from the GATT or convince the Contracting Parties of the merits of such an arrangement. Paragraph ten of Article XXIV states that Contracting Parties may, in exceptional circumstances and by a two-thirds majority vote, approve proposals that do not comply fully with the Agreement's conditions.

Article XXIV, by sanctioning discriminatory behaviour, has been much criticised for allowing violations of the most-favoured-nation principle. Furthermore, although the GATT insists that regional trading blocs are permitted 'provided that certain *strict* criteria are met' (GATT, 1992, 10), the criteria established are ambiguous and easily satisfied. The conditions established by the GATT for the creation of regional trading blocs, as outlined above, are very much open to interpretation. Perhaps the most important qualification, that trade barriers remaining against non-members are not more restrictive than those previously in effect, is dependent on the ambiguous phrasing '*shall not on the whole be any higher*' There are many ways to interpret and calculate tariff levels before and after the creation of a customs union or free trade area. Tariffs, perhaps weighted to the population size of the country, can be calculated individually, collectively or as discrete items. If, for example, a customs union increases average tariff levels imposed against non-member steel imports but reduces them on agriculture and coal, are tariffs '*on the whole*' then lower? The six Contracting Parties of the GATT that established the European Economic Community in 1957 calculated tariff levels according to an arithmetic mean and refused to negotiate on other possible methods of calculation (Dam, 1970). If the US-Canadian free trade area were extended to include countries from central and south America, with higher tariff barriers, it could then significantly increase tariffs on world trade as a whole and, whilst conforming to the GATT, hinder multilateral agreements and trade liberalisation.

The second most important condition established by the GATT, that regional trading blocs should eliminate duties and tariffs on '*substantially all trade*' among members, is no less ambiguous. There is clearly enough scope in the phrasing of this amendment for the Contracting Parties to avoid creating a 100 per cent tariff-free trading bloc. '*Substantially all trade*' leaves much latitude for interpretation. If, for example, 100 per cent tariff cuts are introduced on 70 per cent of all trade, perhaps leaving significant and strategic sectors untouched by tariff cuts, does that then represent '*substantially all trade*'? Alternatively, will a 70 per cent reduction in tariffs on 100 per cent of trading transactions, although obviously at variance with the spirit of Article XXIV, be accepted as meeting the basic requirements of this condition?

At first sight, the 100 per cent tariff-free requirement may appear to contradict the principle of non-discrimination. Article XXIV states that countries which discriminate should eliminate ***all*** restrictions on *substantially all trade*. Only 'big-time discrimination' (Finger, 1992, 8) will therefore be tolerated by the GATT. However, the rationale behind this requirement is to help prevent the explosion of partial discriminatory arrangements that would inhibit the development of a liberal trading system and promote bilateralism. Furthermore, the complete abolition of barriers, even on a regionally discriminatory basis, may promote more open trade and investment. Preferential trading arrangements may provide the most practical route to multilateral free trade and therefore positively support the GATT's ultimate objective. This is the basic argument for the convergence between the objectives and practice of multilateral and regional trade policy initiatives. It is the GATT's contention that regional trading blocs, if properly managed and policed, can support multilateral free trade. However, there is much evidence to suggest that in practice, the GATT has failed in its endeavour to control the nature and behaviour of regionalism. In particular, many preferential trading blocs, whilst not raising barriers to external trade from outside the region, have manifestly failed to create a 100 per cent tariff-free trading area. Article XXIV of the GATT Treaty is therefore widely criticised as being incapable of managing simultaneously the growth of regionalism and promoting multilateralism (see for example Schott, 1989; Bhagwati, 1992a; Finger, 1992; Hopkinson, 1992). Whilst there is considerable diversity of opinion over whether regional arrangements are building blocks or stumbling blocks to world trade, there is rare unanimity in criticising the inability of Article XXIV to control effectively the character of such arrangements. Patterson notes that:

> of all the GATT articles, this [Article XXIV] is one of the most abused, and these abuses are amongst the least noted ... those framing any new free trade area need have little to fear that they will be embarrassed by some GATT body finding them in violation of their international obligations. The effective destruction of Article XXIV as a serious restraint on free trade areas and customs unions began in earnest when the European Community was examined ... (Patterson, 1989, 36)

This criticism of Article XXIV is supported by the fact that of the more than 70 preferential trading arrangements of which the GATT has been notified, not one has ever been formally disapproved of (OECD, 1990, 18). This brings forward a much wider issue over the inability of the GATT to manage not only the spread of regionalism but also multilateralism and free trade in the world economy of the 1990s. The weakness and lack of competence of the GATT, examined briefly below, to influence significantly the behaviour of a truly globalised world economy have in part, been responsible for the resurgence of regionalism in the 1990s.

FAILINGS IN THE GATT

The circumstances of world trade in the 1980s, particularly the growth of protectionism, the diminishing importance of manufacturing in many old industrial regions and the increasing levels of international trade outside the aegis of the GATT, have focused attention on the shortcomings and suitability of the Agreement. The inability of the GATT to manage the world trading system effectively is an important contributory factor behind the development of preferential trading arrangements. The pedestrian pace of trade liberalisation negotiations has led to the criticism that the GATT is in fact the 'General Agreement to Talk and Talk'. The Agreement is frequently portrayed as being 'dead', 'moribund', 'in disarray' and 'out-dated', and is being viewed increasingly as inadequate and irrelevant. The rise of protectionism from the 1970s onwards has been the single most important factor discrediting the GATT's accomplishments.

As outlined earlier in this chapter, the process of multilateral trade liberalisation contributed significantly to the outstanding levels of trade, growth and income experienced in the 1950s and 1960s. However, the 1973 oil price shock instigated by the Organisation of Petroleum Exporting Countries (OPEC) fuelled protectionist instincts amongst Contracting Parties. As a consequence, the unprecedented post-war growth in trade and national incomes started to waiver as multilateral trade liberalisation became hampered by the imposition of various non-tariff barriers designed to circumvent the GATT. During the 1970s and 1980s, the drive to dismantle barriers to international trade slowed down in the face of world wide recession. Contracting Parties tried to protect their own domestic industries by lessening the competitive pressures from elsewhere, particularly Japan and the newly industrialising countries of east and south-east Asia. National policies and national solutions were sought to resolve the problems created by the international economic recession. Protectionist measures designed to favour domestic suppliers were the preferred policy. However, in order to avoid confronting the GATT and the international community as a whole, states used non-tariff barriers to restrict imports, which often took the form of national regulations and safety standards (Wise and Gibb, 1993). For example, the removal of such barriers within the European Community was the essential starting point for the '1992' programme to create a single market. Non-tariff barriers helped fragment the Community economy and their elimination was accorded high priority. The Cecchini Report (Cecchini, 1988) into the 'cost of non-Europe' estimated that the removal of non-tariff barriers to trade would result in a 7 per cent increase in GDP and an employment gain of over 5 million. Clearly non-tariff barriers can have substantial and detrimental repercussions for trade.

Most damaging of all has been the willingness of some states and trading blocs, particularly the USA and the EC, to embrace a variety of regional and sectoral arrangements. These so-called 'grey area measures' are in fact bilateral trading agreements between the importing and exporting states. They are designed purposelly to manage the market and control flows of international trade. Since the early 1970s, the number and breadth of grey area measures has increased dramatically, to include 'Voluntary Export Restraints' (VERs). 'Orderly Marketing Arrangements' (OMAs), 'Anti-dumping Provisions', 'Gentlemen's Agreements' and 'Countervailing Duties'. The European car industry provides a good example of how grey area measures have been employed by European Community Governments to implement protectionist policies designed to support domestic industries. In 1991, four Community states negotiated 'voluntary agreements' with Japan that limited considerably the number of vehicles permitted to enter their domestic markets: Britain (10 per cent), France (2.9 per cent), Spain (2000 cars per annum) and Portugal (10 000 cars per annum). In addition, Italy had a Community-sanctioned quota restriction limiting sales of Japanese cars to 3200 per annum (Wise and Gibb, 1993). These restrictions were successful in reducing the level of Japanese automobile imports into the European Community. According to the Nomura Institute (Europe 2000, 1989), these grey area measures cost the Japanese motor industry approximately £1.2 billion in lost sales during 1987. Another grey area measure is anti-dumping legislation, which has been described by Oxley as a 'new European speciality' (Oxley, 1990, 200). Although the GATT permits Contracting Parties to impose tariffs on imports that are being dumped, there is little doubt that countries use such measures as an instrument of protection rather than a check to unfair competition. There are no established or widely recognised criteria for calculating what constitutes dumping and the whole procedure is wide open to abuse. The Americans have developed their own speciality in the form of countervailing duties. These duties focus on the level of 'unfair' subsidy an exporting company receives from its national government. An appropriate tariff is then imposed on 'offending' imports in order to negate what would otherwise be distorting competition. Similar to the judgements made about dumping, the determination of 'fair' subsidy and 'fair value' is vulnerable to abuse. Bhagwati (1991) cites several examples where countervailing duties have been exploited in order to secure a protectionist advantage.

The growth of market management measures through the use of various grey area measures has both reduced free trade and undermined the authority of the GATT. Such measures unquestionably violate the fundamental principles of the GATT; they are discriminatory, hidden not transparent, and often quantitative and bilateral in nature. So-called 'voluntary' agreements, whereby an exporting country agrees to restrict the level of exported goods to an agreed figure, undermines the philosophy and principles underpinning the GATT-sponsored multilateral approach to free trade. As a consequence, such agreements are often designed explicitly to circumvent multilaterally agreed safeguards (Patterson, 1989). The growth of export-restraining arrangements led Bhagwati to remark that:

> The growth of this set of non-tariff barriers not only halted but partially reversed the process of trade liberalisation. It also signified to many the possibility of a menacing shift in the nature of the international trade regimes. (Bhagwati, 1988, 47)

The inability of the GATT to control the rise of non-tariff barriers effectively and the consequent spread of bilateral negotiations has helped to emphasise the benefits of regionalism. As the European Community's Single European Market (SEM) programme has illustrated, it may be more efficient and certainly quicker to eliminate non-tariff barriers to trade from within a regional trading bloc of 'like-minded' countries.

Another important motivating factor supporting the resurgence of regionalism is the growing level of trade conducted outside the aegis of the GATT. The Uruguay Round was the first to start negotiations on a whole host of activities central to the world economy; services, trade related investment measures and intellectual property guarantees. There are, in addition, a multitude of more traditional industries, particularly agriculture and textiles, excluded from the GATT multilateral system. To this list can be added steel, electronics, automobiles and electrical goods. The GATT's rules and principles are therefore being flagrantly ignored across an increasingly wide range of products. This is undoubtedly a most serious criticism of the GATT-sponsored process of multilateral liberalisation. Estimates vary as to exactly what percentage of trade takes place outside the GATT. Oxley (1990) cites figures to suggest that only 7 per cent of world trade is governed by the General Agreement. In 1988, Choate and Linger calculated that for all global transactions, trade as well as capital flows, the figure may be as low as 5 per cent. However, most analysts agree on a figure much higher, estimating that between 50 and 60 per cent of world trade in goods and services is subject to the GATT disciplines. Significantly, even these higher estimates leave almost 50 per cent of world trade GATT-free. Whatever the precise figure, a substantial amount of trade is conducted outside the GATT rules. This has helped to undermine further the General Agreement's authority and push states towards some form of regional activity that is geographically more constrained but, at the same time, more comprehensive in both coverage and depth.

Yet another factor motivating states to move towards some form of regionalism is the GATT's poor record of enforcement and the problem of the 'free-rider'. Multilateral trade negotiations, where concessions and obligations are extended to all parties, raise the politically difficult issue of what action to take when a state accepts the concessions but refuses to honour the obligations. The GATT's record of enforcement has been, at best, mixed. Consultation, conciliation and settlement of disputes are fundamental to the GATT's enforcement process (GATT, 1992). The vagueness of the GATT obligations, particularly in agriculture over which the European Community has been criticised many times, makes decision making difficult and vulnerable to substantial delaying tactics. The European Community's Common Agricultural Policy (CAP) is the very antithesis of an open and free market based on multilateral principles. However, the GATT has failed spectacularly to change the fundamentally protectionist nature of this policy. The USA challenged the CAP in the Uruguay Round, arguing that the Community's internal support mechanism for agriculture, subsidising both domestic prices and exports, distorts world markets to an unacceptable degree. The quantity of export subsidy is a particular focus of attention. However, agricultural subsidies are not the exclusive preserve of the European Community, as this quote from Preeg makes abundantly clear:

> an escalating export subsidy war between the European Community and the United States beginning in 1985 ... caught up other grain exporting countries in a ruinous cross fire. An unbridled export subsidy war between the two largest traders is antithetical to everything the GATT stands for. (Preeg, 1992, 85)

Clearly, the GATT has many weaknesses, both economic and political, that have the potential to motivate states to seek preferential trading arrangements. Such arrangements are increasingly perceived to be a more expeditious mechanism of achieving trade liberalisation. The European Community's successful

1992 programme to remove non-tariff barriers to trade has helped to reinforce this perception further. Trading blocs therefore offer the possibility of closer policy integration than that achievable at the multilateral level. The trend towards regionalism is also being strengthened by a desire not to be left out of trading agreements. The emergence of new trading blocs, coupled with a deepening of economic integration in the European Community, provides a strong stimulus for other countries to intensify collaboration (Gibb, 1993). There is a real prospect that the world economy will be dominated by a small group of rival trading blocs. The fear of being isolated and disadvantaged, in terms of unfavourable trading arrangements, trade diversion and economies of scale, is a stong motivating force behind the resurgence of regionalism in the 1980s and 1990s.

Having outlined briefly some of the motivations supporting the growth of preferential trading arrangements, attention now turns to a focus on the nature and character of regionalism.

DIFFERENT FORMS OF REGIONALISM: FROM SECTORAL AGREEMENTS TO POLITICAL–ECONOMIC UNIONS

Thus far, international regionalism has been referred to as a single homogenous entity. However, regional trading arrangements take several forms and cover a multitude of different schemes. By definition, all trading arrangements involve an element of geographic discrimination in that various factors of production are subject to differing tariff barriers according to country of origin/destination. It is for this reason that the United Nations (1990) refers to such arrangements as 'geographically discriminatory trading arrangements' (GDTAs). All such arrangements have one unifying characteristic: the preferential terms of trade participants have over non-participating countries. According to Schott (1991, 2), there is amongst all the different forms of regionalism a threefold uniformity of purpose: to promote welfare gains through income and efficiency; to enhance negotiating capacity with third countries; and, more often than not, to augment regional political cooperation. Whilst at this most basic of levels there is undoubtedly a common set of objectives, it is important to distinguish between sectoral cooperation, free trade areas, customs unions, common markets, and economic and political unions. A great deal of literature fails to recognise the core characteristics that distinguish these different forms of regionalism, incorrectly referring to these quite specific terms as if they were somehow interchangeable. A clarification of the concepts of regionalism avoids conceptual confusion.

In his pioneering work on the theory of economic integration, Balassa (1962) interprets regional economic integration as either a static state of affairs or a process. As a state of affairs, regionalism involves the elimination of various forms of discrimination between participating countries; in effect, a sort of 'once and for all' agreement. Regionalism as a process 'encompasses measures designed to abolish discrimination between economic units belonging to different national states' (Balassa, 1962, 1). As a process, regionalism occurs stage by stage as states contemplate higher and more complex levels of integration. The stages outlined by Balassa have been adopted and adapted by numerous authors (Lipsey, 1957; Robson, 1980; Hansen et al., 1992, Wise and Gibb, 1993).

Specific levels of regional trading arrangements will now be analysed, from the lowest forms of cooperation to the highest (Figure 1.7). Although this framework provides a most useful tool for analysis, it is worth emphasising that regionalism is a complex phenomenon which is not easily understood by recourse to simple theoretical formulations of the type discussed below.

Sectoral cooperation

The degree of integration required in a sectoral trading agreement is very limited. Balassa (1962) classifies this form of agreement as reflecting cooperation as opposed to integration. The latter is concerned with the complete elimination of some forms of discrimination whereas the former aims only to lessen discrimination in certain sectors of the economy. Sectoral cooperation limits the agreement to well-defined sectors, such as coal and steel. This form of cooperation is the preferred option for those countries unwilling or unable to encompass higher levels of integration within a mutually acceptable time frame. The sectoral approach may also be part of a much wider process leading to further integration, such as the European Coal and Steel Community (ECSC) established in 1952. The Preferential Trade Area for Eastern and Southern Africa, established on 21 December 1981, is a more recent example of a sectoral agreement which aims gradually to eliminate barriers to trade or, at the very least, grant preferential import and export status on a *restricted list of commodities*. The Preferential Trade Area for Eastern and Southern Africa is considered as the first step towards creating a common market and eventually an economic community similar to the European Community.

However, the sectoral approach may also be regarded as a desirable end product in itself. Such agreements are an attractive option for those states disinclined to expose most sectors of their economy to enhanced competition. It is also a form of regionalism that transfers economic benefits but imposes the least cost on national sovereignty. Sectoral cooperation protects states from having to relinquish the level of control often implicit in more wide-ranging regional agreements. Many economists perceive sectoral cooperation agreements as a second-best option to integration. However, if they are successful, as in the case of the ECSC, they may encourage participating states to seek more complex and higher levels of regionalism.

	Removal of internal quotas & tariffs	Common external customs tariff	Free movements of land, labour, capital and services	Harmonization of economic policies and development of supra national institiutions	Unification of and political and powerful super national institutions
Sectoral Cooperation	◖				
Free Trade Association	●				
Customs Union	●	●			
Common Market	●	●	●		
Economic Union	●	●	●	●	
Political Union	●	●	●	●	●

Figure 1.7 Varying levels of regional integration

Free trade areas

A free trade area requires participating countries to eliminate completely quantitative trade restrictions and customs tariffs against each other's goods. However, countries in a free trade area retain the sovereign power to decide individually whether and to what level trade restrictions should be imposed on non-members. The level of economic and political integration required to establish a free trade area is limited. Regionalism at this level does not require a high level of institutional agreement. There is no need to establish new regional decision-making and management organisations that restrict the sovereign decision-making authority of participating states (Scheepers, 1979). Free trade areas can therefore be used between countries without them necessarily having strong political ties.

One of the principal problems associated with free trade areas stems from the 'privilege' enjoyed by individual member states being able to adopt separate arrangements for imports from third countries. There is an obvious problem of goods being cheaply imported from a third country into a member state of the free trade area (with low tariffs and competitive industries), and then being transferred without tariffs to another member state attempting to resist such imports (with high tariffs and vulnerable industries). In order to avoid the problem of trade deflection, free trade is usually restricted to those products and commodities originating from participating states. Although complex 'rules of origin' regulations aim to define the status and 'nationality' of goods able to be traded freely, the globalisation of business and the accounting practices adopted by multinational companies seriously restrict the effectiveness of such legislation.

The advantages of a free trade area, particularly independence in determining and formulating trade policies and the fact that national sovereignty remains largely intact, attract those states preferring a loose-knit regional structure. However, the problems of policing trade deflection can make the next tier of regionalism, the customs union, appear attractive.

The customs union

A customs union is similar to a free trade area except that participating countries agree to adopt uniform import tariffs and common quota restrictions *vis-à-vis* countries outside the union (Balassa, 1962). Members of a customs union therefore have identical trade barriers, usually termed a common customs tariff (CCT) or common external tariff (CET), in order to negate the problems associated with trade deflection. A customs union agreement does not necessarily imply an overt surrender of national sovereignty, but to establish identical tariff barriers against imports from non-members requires a commitment to common decision-making. The ability of participating countries to determine national trade policies independently is therefore weakened.

Common markets

In a common market, the customs union arrangements are extended to include the free movement of the factors of production (goods, people, capital and services) within and across the borders of participating states. A common market is therefore concerned with not only the free trade of goods inside the CET, but the free movement of those things needed to produce those goods (Wise and Gibb, 1993). In addition, 'freedom of establishment' is also considered to be an essential component of a common market. Thus any citizen belonging to any one of the participating countries has the right to apply for employment and establish a business in other countries of the common market. The core characteristics of a common market are those outlined in the 1957 Treaty of Rome establishing the European Economic Community: the free movement of goods, people, capital and services, as well as the freedom of establishment. However, Brooks (1983) argues that rules on competition, although not explicitly mentioned in the Rome Treaty, are in fact essential to the proper functioning and maintenance of a common market. In order for states to compete on a level playing field, harmonisation may also be sought in health and safety, social security and education.

A common market therefore requires a much greater degree of political and economic cooperation than one finds in sectoral agreements, free trade areas or customs unions. The creation of a common market represents a significant reduction of national sovereignty in that cooperative decision-making processes, needed in order to develop common policies and harmonisation, become necessary to ensure the effective functioning of the market.

Economic unions

An economic union reflects an even higher and more complex level of regionalism, incorporating all the previous stages of integration and adding monetary and fiscal policy harmonisation. Economic union presupposes the unification of monetary, fiscal and social policies and requires the formation of supranational institutions and organisations whose decisions are binding on participating countries. A single common currency is an essential prerequisite to the effective functioning of a genuinely free common market (Hansen et al., 1992). Exchange rate fluctuations and government-induced currency devaluations aimed at managing balance of payments problems distort competition and fragment economic union. The formation and operation of relatively fixed exchange rates represent a weak form of monetary union. An economic union also presupposes uniform rules and procedures in regard to taxation and many other policies necessary to ensure that the basic conditions of industry are the same throughout all participating countries.

The establishment of supranational institutions is widely recognised to be a core characteristic of economic unions. Participating countries decide to pool parts of their sovereignty, those necessary to administer the market, within a supranational institution. Supranational institutions, that have legislative powers throughout the territories of all participating states, significantly erode the sovereign base of national states. Supranational laws take precedence over national laws. However, an economic union will restrict supranational decision making to what are conventionally perceived as economic matters, in particular monetary and fiscal policies.

Political unions

At the top end of the regional hierarchy is a political union which retains all the features of an economic union, but exerts supranational decision making beyond the purely economic. Political union removes the confusion generated by categorising activities as political or economic. In reality, of course, the two are inevitably and inextricably interlinked. Political union extends supranationalism to areas at the very heart of national sovereignty, such as defence and foreign policy. Member states of a political union deal with appropriate matters at the supranational level and abandon goals of preserving national sovereignty intact. As observed by Wise and Gibb (1993), participating states:

> are here moving towards federal concepts such as 'pooling sovereignty' and dividing responsibilities between different governments and legislatures in a complex intermeshing of institutions at national, supranational and, if desired, regional levels. At this state of integration there would be no more semantic splitting of hairs about where 'economic' matters ended and those of a political nature began. (Wise and Gibb, 1993, 31)

Summary

To summarise, the term regionalism covers many different types of trading bloc which can be identified by the extent and depth of the resulting integration. Regionalism can be a state of affairs or a process. Viewed as a process, it encompasses measures designed progressively to eliminate all forms of discrimination between participating states. The process is driven by the economic and political benefits of regionalism prompting member states to contemplate higher levels of integration. The six degrees of regional integration identified above can therefore be interpreted as defining six separate forms of regionalism (as a state of affairs), or a process that can promote more complex and deeper levels of integration amongst participating states. Taken as a process, and therefore a model, the six degrees of regional integration are neither discrete and separate nor inevitable stages leading to political union. Most existing trading blocs do not fit neatly into any one of the integration levels described. For example, the European Community is a fully developed customs union, a fairly well developed common market and has fragments of monetary and political union. There is also nothing inevitable about regionalism as a process precipitating more complex and in-depth integration. Indeed, trading blocs can and do disintegrate, as evidenced by the East African Community and the Central African Federation. Participating countries may also be eager to view regionalism as a static state of affairs; there is little evidence to suggest that the free trade area between the USA and Canada is seen as anything else. The participating countries do not envisage the free trade area developing into a customs union, common market or economic union.

Clearly, international regional integration is a complex phenomenon incorporating many different types of integration with an equally diverse number of aims and objectives. It does not readily lend itself to simple theoretical formulations (Gibb and Wise, 1993). Having outlined the different forms of regionalism, attention now turns to the difficult question of whether regional trading blocs promote trade liberalisation or hinder it. In other words, is regionalism a defence mechanism, which aims to protect the territorial integrity of markets, or a mechanism that promotes multilateral free trade?

THE ECONOMIC MERITS OF REGIONALISM

Since the early 1950s, the economic value of preferential trading arrangements has been the subject of vigorous debate. This debate has centred upon a basic question related to this chapter's pivotal theme: to what extent does regionalism promote trade liberalisation or represent a step towards protectionism? Are preferential trading arrangements simultaneously trade liberalising for participant countries and trade restricting for those states excluded? There is clearly no consensus over whether regionalism should be interpreted as a complement to, a substitute for, or a threat to the GATT-sponsored multilateral trading system. From a neo-classical economic perspective, all discriminatory trading arrangements are perceived to be a 'second-best' option compared to multilateral free trade. Although an extensive amount of literature is devoted to examining the theory of economic integration, it is 'not very good at answering questions about the welfare effects of moves from one distorted policy to another – that is, in dealing with the problem of the second best' (United Nations, 1990, 22).

Viner was one of the first scholars to evaluate the economic consequences of forming a customs union. In his pioneering and classic analysis, *The Customs Union Issue* (1950), he questioned the then accepted wisdom that customs unions implied a step towards free trade. Although Viner's concepts of 'trade creation' and 'trade diversion' were established more than 40 years ago, they have been accepted by most political economists and almost every scholar of international trade, and continue to form the basis of the current debate (Tovias, 1991). Viner established what Wonnacott and Lutz (1989, 59) describe as the 'most important single observation about free trade areas': that they do not necessarily represent a step towards free trade. Indeed, Viner was of the opinion that preferential trading arrangements are more likely to promote protectionism:

> It is a strange phenomenon which unites free-traders and protectionists in the field of commercial policy, and its strangeness suggests that there is something peculiar in the apparent economics of customs unions ... with respect to most customs union projects the protectionist is right and the free-trader is wrong in regarding the project as something, given his premises, which he can logically support. (Viner, 1950, 41)

The seminal distinction between the trade-creating and trade-diverting impacts of preferential trading arrangements remains at the very core of the theory of economic integration and is, for this very reason, examined in detail below.

A preferential trading arrangement promotes trade creation when a country's high-cost domestic production is replaced by a cheaper product originating from a participating country. This move towards lower-cost sources results in greater domestic consumption and generates additional trade and welfare in the process. Conversely, trade diversion occurs when imports of efficiently produced products from countries outside the trading arrangement are displaced by more expensive imports from partner countries. Increased intraregional trade therefore takes place at the direct expense of lower-cost imports from outside the trading agreement. Trade diversion does not therefore increase either total trade or consumption and, in theory, lowers real income and imposes global welfare losses. The costs of trade diversion are borne by both consumers and exporters.

The example in Figure 1.8 clarifies the concepts of trade creation and trade diversion. It envisages three states: H (the home country), M (a member state of the trading arrangement) and T (a third country outside the arrangement). Figure 1.8 also illustrates the unit cost of good X in the three countries followed by the price of good X when imported into H at 100 per cent (example A) and 50 per cent (example B) tariff rates respectively. Under free market conditions, the home country (H) would import good X from the cheapest source (T). If, however, H wanted to protect its domestic market, it could do this by imposing a 100 per cent tariff on M and T. If countries H and M then agree to participate in a free trading arrangement, M displaces H as the source of supply for good X (example A). At the same time, H gains an increase in welfare. Trade creation therefore occurs as M's exports displace H's higher-cost domestic production.

However, this situation changes radically if instead of having a 100 per cent tariff, H has a 50 per cent import tariff and purchases good X from T at 37.5 (instead of a world price of 25 or a domestic price in H at 40). If H and M then create a free trading agreement with a 50 per cent CET, H, instead of purchasing

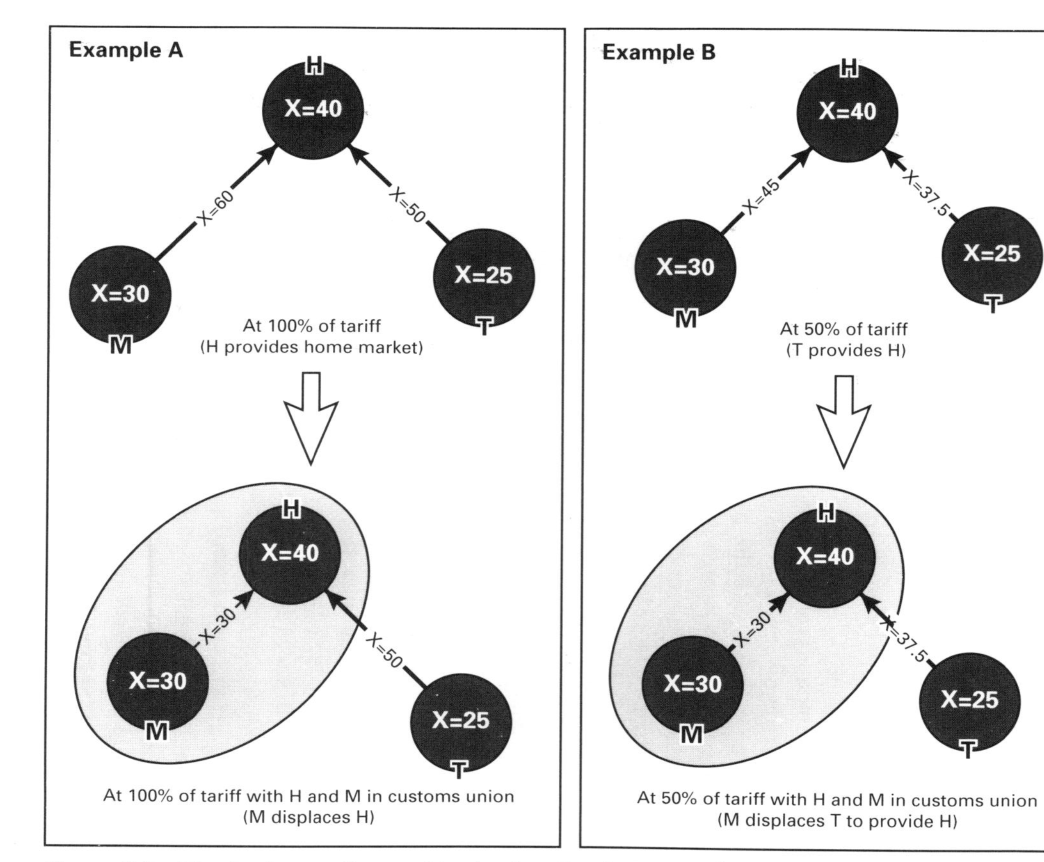

Figure 1.8 The trade creating and trade diverting impacts of a customs union

X from T at 37.5, purchases X from M at 30. Although consumers in H are purchasing the product at a rate cheaper than would have otherwise been the case (30 instead of 37.5), the trading arrangement has been trade diverting. Trade between H and M has increased at the direct expense of T, thus lowering overall welfare levels (example B).

Although Viner was of the opinion that preferential trading arrangements would cause either trade creation or trade diversion (Tovias, 1991), it is now generally accepted that trading blocs usually contain elements of both simultaneously. However this simple theory, which has since been reworked and criticised many times, provides a basic framework for the analysis of whether members and non-members of a trading bloc benefit from such an agreement. Viner's work remains at the 'very core of customs union theory' (Wonnacott and Lutz, 1989, 62).

Perhaps the most significant refinement made to Viner's original analysis was proposed by Meade (1955) and Lipsey (1957), who argued that a preferential trading agreement may be desirable for its members even if it is predominantly trade diverting. They criticised Viner's exclusive focus on resource allocation and production efficiency at the expense of consumption patterns, particularly the effect on consumer demand when prices decrease. Lipsey (1960, 501) was critical of 'Viner's implicit assumption that commodities are consumed in fixed proportions independent of the structure of relative prices.' The potential impact on consumption patterns can be illustrated in the example given above (see Figure 1.8), where trade diversion takes place (example B) at the same time as allowing consumers in country H to purchase good X more cheaply (at 30 from M instead of 37.5 from T). Meade and Lipsey argue that the pattern of consumption is therefore less distorted and that trade diversion has the potential to generate a consumer surplus which may compensate for the negative impact on production costs. Trade diversion can therefore generate additional trade and raise real national income and welfare.

Many others have since developed alternative approaches to customs union theorising. Most notable are Cooper and Massell (1965) who, in an analysis of small developing countries, argued that preferential trading arrangements should be viewed as an instrument of protection and not as a mechanism for promoting free trade. Nonetheless, they argued that trade integration based on the free movement of goods and a CET was an appropriate policy for the promotion of import-competing industrialisation (Bhagwati, 1991). This argument was based on the possibilities of increasing scale economies and specialisation. By increasing the market for suppliers within a region, trading blocs, although potentially trade diverting, can promote scale effects that reduce the costs of that diversion.

During the 1970s, an influential and alternative approach to customs theorising was developed by Kemp and Wan (1976). Their work focused on the 'terms of trade' effects between participating and non-participating states of a trading bloc. Contradicting Viner's analysis, Kemp and Wan argued that customs union formation made a positive contribution to world welfare. By means of a redistributive mechanism within the trading area, member states could compensate those members who benefited least. Equally, by adopting an appropriate CET, trading blocs could fully compensate those non-member states outside. It is therefore possible for any group of countries to create a customs union, with an appropriate external tariff, that will improve the welfare of members and non-members alike. Commenting on this observation, Tovias notes:

> The discovery that creating customs unions for economic reasons was rational was important in itself, since before it, most economists believed that the real reason for establishing unions were political, sentimental or cultural ... after the mid-1970s, theoretical optimism came back through the back door: customs unions were not irrational creatures after all. (Tovias, 1991, 10 and 12)

More recently, a spatial perspective has been added to the literature on customs union theory. Many analysts argue that discriminatory trading arrangements are often, but not always, more beneficial when regionally based. Krugman (1990a) argues that regionalism improves the likelihood of preferential trading arrangements having a welfare-generating impact. His argument is based on the assertion that countries in close geographic proximity usually have a disproportionately high level of trade, thus reducing the threat of trade diversion. This conclusion is supported by Balassa and Bauwens (1988), who in an analysis of regionalism note that geographic proximity is an important determining factor affecting the level of intra-European trade in manufactured goods.

CONCLUSIONS

International regionalism is a complex phenomenon covering a multitude of different schemes, from sectoral agreements to political unions, and can be seen as a static state of affairs or as a process. Since Viner's seminal work, the complexity and depth of economic analysis directed towards international regionalism has been widened to include economies of scale, specialisation, increased productive capacity, gains in export markets, improved terms of trade and a lot more besides. Throughout the post-war period, numerous attempts have been made to construct a general theory of economic integration. However, none of these theories has been fully successful. The analytical challenge posed by preferential trading arrangements is that they can be simultaneously trade creating and trade diverting for both participatory and non-participatory states. It is a subject of inherent complexity and ambiguity.

As this chapter illustrates, theoretical analysis based on liberal economic principles is ambivalent and inconclusive in evaluating the impact of trading blocs on living standards, redistribution of income and welfare. Furthermore, no overall consensus is likely to be reached in the foreseeable future. According to economic analysis, preferential trading arrangements may or may not bring welfare gains for participating countries. They may benefit or harm global welfare levels. Third countries outside the trading bloc may be better or worse off after the formation of a discriminatory arrangement. The academic community is equally divided over the question of whether regionalism, in its various forms, constitutes a menace or an opportunity to the GATT-sponsored process of multilateralism. Whilst Preeg (1989) considers west European and North American regionalism to be a 'desirable evolution potentially supporting a more multilateral liberal system', Schott (1989) perceives free trade areas as potentially damaging the GATT and undermining trade liberalisation. Clearly, there are no straightforward answers to the question of whether regionalism is a response to the detrimental consequences of free trade or a strategy used to promote multilateral liberalisation. Customs union theory can be equally convincing in both supporting and opposing both sides of the argument.

The current chapter focused on the principal arguments in the debate over the relative merits of international regionalism and multilateralism. The results of this analysis are somewhat ambiguous, reflecting perhaps both the theoretical and pragmatic difficulties in implementing and evaluating the practical outcomes of these divergent approaches to international trade. On the one hand, there are few doubts that the multilateral approach embraced since 1945 has contributed significantly to the increase experienced in the volume and frequency of international cross-border trade. The rise of the international, or even global, economy is a direct result of the liberal trade policies instituted by the GATT. On the other hand, the resurgence of international regionalism and protectionist measures signifies some important change in the relationship between the principal economic actors and the GATT, and perhaps ultimately the merits of regionalism.

As the practice of international trade drifts further and further away from the organising principles embodied in liberal economics, it is all too easy to focus on the operational shortcomings of the GATT. It is therefore important to remember that the GATT is a collection of sovereign states and can be no more than the sum of its constituent parts. To focus blame on the GATT is, in effect, to blame its member states, the most influential of whom are the principal economic actors. The practical difficulties of policy making in an increasingly volatile and fluid international environment reflect, in part, the analytical difficulties posed by preferential trading arrangements.

The failure of classical liberal economic theory to explain adequately the resurgence of international regionalism may be the result of what Lipietz (1992, IX) observes to be a tendency amongst economists to treat their subject 'as if it could be defined by immutable laws, behaviour and tendencies'. However, since the early 1970s attention has increasingly been focused on the broader phenomena of economic transformation and restructuring. In fact, as the next chapter will make clear, the issue of regional integration is part of what many commentators believe to be a transformation of 'industrial society' and the 'capitalist system of production'. The resulting mixture of economic uncertainty and political instability poses new challenges and problems, not least in the realm of regional integration and cooperation. The resurgence of international regionalism is therefore the result rather than the cause of growing tensions within the established economic and social order. The next chapter explores some of these themes in more detail.

References

Balassa, B. (1962) *The Theory of Economic Integration*, Allen & Unwin, London.

Balassa, B. and Bauwens, L. (1988) The determinants of intra-European trade in manufactured goods, *European Economic Review*, 32, 1421–1437.

Barker, T. (1977) International trade and economic growth: an alternative to the neoclassical approach, *Cambridge Journal of Economics*, 1, 153–172.

Bhagwati, J.N. (1988) *Protectionism, The 1987 Ohlin Lectures*, MIT Press, Cambridge, Mass.

Bhagwati, J.N. (1991) *The World Trading System at Risk*, Harvester-Wheatsheaf, Hemel Hempstead.

Bhagwati, J.N. (1992a) *Regionalism and Multilateralism: an Overview*, World Bank and CEPR conference on new dimensions in regional integration, April 2–3 1992, Session 1, paper No.1, Washington D.C.

Brooks, F.E.J. (1983) *The EEC and a southern African common market in legal perspective*, Unpublished Ph.D. thesis, University of Exeter.

Cecchini, P. (1988) *The European Challenge: 1992 – The Benefits of a Single Market*, Wildwood House, Aldershot.

Cooper, C. and Massell, B. (1965) A new look at customs unions theory, *The Economic Journal*, 75, 742–747.

Culbert, J. (1987) War-time Anglo-American talks and the making of the GATT, *The World Economy*, 10, 381–408.

Dam, K. (1970) *The GATT: Law and international economic organisation*, University of Chicago Press, Chicago.

The Economist (1992) Free trade with luck, *The Economist*, October 17, 15–16.

Europe 2000 (1989) Foreign direct investment in the UK, *Europe 2000*, 4, 35–42.

Finger, J.M. (1992) *Gatt's Influence on Regional Arrangements*, The World Bank and CEPR conference on new dimensions in regional integration, April 2–3, Session V, Paper No.12, Washington, D.C.

GATT (1986) *The Text of the General Agreement on Tariffs and Trade*, GATT, Geneva.

GATT (1992) *GATT: What It Is, What It Does*, GATT, Geneva.

Gibb, R.A. (1993) A common market for post-apartheid Southern Africa: prospects and problems, *South African Geographical Journal*, 75, 28–35.

Hansen, J.D., Heinrich, M. and Nielson, J.V. (1992) *An Economic Analysis of the EC*, McGraw-Hill, London.

Harrod, R.F. (1951) *The Life of Maynard Keynes*, Macmillan, London.

Hopkinson, N. (ed.) (1992) *Completing the Gatt Uruguay Round, Renewed Multilateralism or a World of Regional Trading Blocs?*, Wilton Park Paper No.61, HMSO, London.

Irwin, D.A. (1992) *Multilateral and Bilateral Trade Policies in the World Trading*

System: An Historical Perspective, The World Bank and CEPR conference on new dimensions in regional integration, April 2–3, Session IV, Paper No.9, Washington, D.C.

Kemp, M. and Wan, H. (1976) An elementary proposition concerning the formation of customs unions, *Journal of International Economics*, 61, 95–97.

Krugman, P. (1990a) *Increasing Returns and Economic Geography*, National Bureau of Economic Research, Working Paper series No.3245, Cambridge, Mass.

Kindleberger, C.P. (1973) *The World in Depression 1929–1939*, Allen Lane, The Penguin Press, London.

Lipietz, A. (1992) *Towards a New Economic Order: Postfordism, Ecology and Democracy*, Oxford University Press, Oxford.

Lipsey, R. (1957) The theory of customs unions: trade diversion and welfare, *Economica*, 24, 40–46.

Lipsey, R. (1960) The theory of customs unions: a general survey, *The Economic Journal*, 70, 496–513.

Meade, J. (1955) *The theory of customs unions*, North-Holland, Amsterdam.

O'Brien, R. (1992) *Global Financial Integration: The End of Geography*, Pinter, London.

OECD (1990) *Recent Developments in Regional Trading Arrangements Among OECD Countries: Main Implications for Third Countries and for the Multilateral Trading System*, Trade Committee, OECD, Paris.

Oxley, A. (1990) *The Challenge of Free Trade*, Harvester-Wheatsheaf, Worcester.

Patterson, G. (1989) Implications for the GATT and the world trading system, in: J. Schott (ed.) *Free Trade Areas and US Policy*, Institute for International Economics, Washington, D.C., 353–366.

Preeg, E.M. (1989) The GATT trading system in transition: an analytic survey of recent literature, *The Washington Quarterly*, 12, 201–213.

Preeg, E.M. (1992) The US leadership role in world trade: past, present and future, *The Washington Quarterly*, 15, 81–91.

Robson, P. (1980) *The Economics of International Integration*, Allen & Unwin, London.

Schott, J. (1989) More free trade areas, in: J. Schott (ed.) *Free Trade Areas and US Trade Policy*, Institute for International Economics, Washington, D.C., 1–59.

Schott, J.J. (1991) Trading blocs and the world trading system, *The World Economy*, 14, 1–17.

Sheepers, C.F. (1979) The possible role of a customs-union type model in promoting closer economic ties in Southern Africa, *Finance and Trade Review*, 13, 82–99.

Tovias, A. (1991) A survey of the theory of economic integration, *Journal of European Integration*, 15, 5–23.

United Nations (1990) *Regional Trading Blocs: A Threat to the Multilateral Trading System: Views and Recommendations of the Committee for Development Planning*, Department of International Economic and Social Affairs, United Nations, New York.

Viner, J. (1950) *The Customs Union Issue*, Carnegie Endowment for World Peace, New York.

Winters, L.A. (1990) The road to Uruguay, *The Economic Journal*, 100, 1288–1303.

Wise, M. and Gibb, R.A. (1993) *From Single Market to Social Europe: the European Community in the 1990s*, Longman, London.

Wonnacott, P. and Lutz, M. (1989) Is there a case for free trade areas? in: J. Schott (ed.) *Free Trade Areas and US Trade Policy*, Institute for International Economics, Washington, D.C., 59–65.

2 The political economy of trading blocs

Wieslaw Michalak
Ryerson Polytechnic University

INTRODUCTION

Trading blocs are not new, and some forms of regional integration and close economic cooperation between states have been present in the economy since the inception of industrial society (Machlup, 1977; Greenaway et al., 1989). The previous chapter examined the liberal economic reasoning behind the often conflicting principles of free trade and regionalism. However, the emergence of trading blocs cannot be understood in economic terms alone. An explanation of continental regionalism demands a more general theory of social and economic transformation. The present chapter therefore evaluates the relevance of a number of theoretical perspectives by reviewing and integrating several concepts borrowed from a number of diverse sources. To facilitate this objective the chapter is divided into three sections. The first section deals with social theory and examines the reasons why trading blocs emerged in the first place. The focus here is on establishing a relationship between the process of trading bloc formation and more fundamental economic and social trends which took place in industrialised economies since the early 1970s. In the second and third sections the scope shifts to an application of this social theory to an explanation of the significance of multinational enterprises and protectionist economic policies to regionalism.

There is a large and rapidly expanding literature in the social and geographical sciences dealing with the change and transformation of the modern economy (Lipietz, 1987; Harvey, 1989; Storper and Scott, 1989; Drache and Gertler, 1991). A large portion is inspired by European philosophical traditions concerning historical and societal change. Although the intellectual sources motivating the literature are broad and varied, the principal stimulus is the work of Karl Marx or, more specifically, selected themes of mostly western European marxism. It is not our intention here to argue that marxism is the only approach to the

Continental Trading Blocs: The Growth of Regionalism in the World Economy
Edited by R. Gibb and W. Michalak

study of economic and industrial change. Indeed, there is ample and suggestive evidence that the attractiveness of historical materialism as a political ideology is on the decline. However, for various reasons, mostly beyond the scope of the present text, marxism remains one of the more fashionable theoretical constructs in Anglo-American social science (Michalak and Gibb, 1992a; 1993).

The first section of this chapter, therefore, reviews the most popular theories of industrial change, often formulated with a strong marxist flavour, and examines the relevance of these theories to the phenomenon of international regionalism. Prominent among the many sources are regulation theory, post-Fordism (linked to a certain degree with postmodernism), flexible specialisation (the only non-marxist attempt in this group) and flexible accumulation. Although none of these approaches is a complete and internally coherent theory in any philosophical sense, they nevertheless represent a genuine and important attempt to codify and account for the rapidly evolving social and economic international environment. Curiously, none of these approaches attempt to explain, or even to incorporate, the issue of regionalism in the world economy. This is a serious omission which, at least in our opinion, significantly limits the theoretical and empirical relevance of these approaches.

The intent of this chapter, therefore, is to sketch a common ground between these theories and the issue of regionalism. Our contention is that multilateralism and regionalism are associated with and are the result of different modes of regulation. Fordism and multilateralism were the foundation on which the long post-war prosperity in the West was built. Paradoxically, the systemic rigidities built in to the 'Fordist consensus' ultimately contributed to its crisis. Increasing demands for economic deregulation and flexibility led to the emergence of multinational enterprises and the 'global' economy. The 'post-Fordist' mode of regulation coincided with the progressive erosion of the ability of public institutions to curtail the effects of increased competition and demands for greater flexibility of capital and factors of production. The combined effects of these changes led to a weakening of the commitment to the principles of multilateral trade and the rise of regionalism as an alternative path to achieving free trade.

In the second and third sections of the chapter, this theory is applied to an explanation of the role which multinational enterprises and protectionist policies play in regionalism. To many social theorists, the multinational enterprise exemplifies many negative political and social consequences of economic organisation founded on the principle of profit maximisation. Continental and national protectionism are sometimes advocated in response to the alleged omnipotence of the multinational enterprise. The largely abstract debate on the evolving nature of the contemporary international economy is supplemented in the third section by a necessarily brief review of 'new trade theory'. Although new trade theory does not have any obvious and direct link with social theory, it is in fact a major theoretical attempt in economics to incorporate the problem of 'externalities' in the theory of international trade and regional integration.

Traditionally, the term '*externalities*' signifies the readily observed fact that the principle of perfect competition is merely a theoretical concept, a very useful one to be sure but no more than that: a purely theoretical generalisation. In fact, all enterprises and economies operate within the constraints of imperfect competition. Since the effects of imperfect competition are extremely difficult to model formally, the core of modern economic theory does not deal with imperfect competition in any significant manner beyond acknowledging its existence. By contrast, new trade theory attempts to incorporate some of the more readily detectable effects of the decline of the multilateral trade system. As such, it represents an interesting alternative to the liberal trade theory outlined in the previous chapter. More importantly, new trade theorists have influenced the 'fair trade' or 'managed trade' approach which appears to be the preferred policy of the Democratic administration of the United States in the mid-1990s. As such it has obvious significance for trade policies between the United States and its trading partners (Bhagwati, 1988; Anderson, 1991; Balassa, 1991; Yamazawa, 1992; Globerman, 1991; see also Chapter 6).

SOCIAL THEORY, INDUSTRIAL CHANGE AND REGIONALISM

The starting point of nearly all theories advanced to explain social and economic trends since the early 1970s is that there has been a major change in the way in which highly developed capitalist economies operate. The cycle of economic growth followed by a series of crises throughout most of the 1970s, 1980s and early 1990s is a result of some significant changes in the way in which the capitalist system functions. At the heart of this transformation is the shift from the relatively rigid methods and technologies of a system of production which emerged in its fullest form after the last world war, to a much more accommodating or *flexible* system of production that includes new forms of organisation, labour and production technologies. Although there are several ways in which these changes have been approached by social scientists, a change in the way the capitalist system of production and consumption is organised lies at the centre of most conceptualisations. However, none of these theories and conceptualisations takes the issue of international trade and regional integration seriously. The following section, therefore, examines several influential theories of social and industrial change and evaluates their relevance to the issue of continental integration.

Gertler and Schoenberger (1992) recently incorporated some elements of the regulationist approach, post-Fordism and flexible specialisation into an analysis of the trading blocs phenomenon. According to them, the contemporary trend of continental market integration is a macroeconomic extension of increased flexibility in diverse spheres of economic activity, including the use of labour and the deployment of machinery on a continent-wide scale. The chief objective of continental integration, i.e. the reduction of barriers to trade and flows of capital from one location to another, complements the more publicised forms of industrial restructuring leading to flexibility on the level of the individual enterprise or industrial region. Gertler and Schoenberger's approach to the issue of continental integration rests on an interpretation of several concepts borrowed from a number of diverse sources. Each of these sources will be examined separately. Perhaps the most influential of them is the regulation theory.

Regulation theory

Chronologically, one of the earliest attempts to interpret the economic and societal changes experienced by the highly industrialised countries after 1973 was introduced by a group of French political economists collectively referred to as '*the regulation school*', or simply '*regulationists*' (Jessop, 1990; Drache and Gertler, 1991). The credit for the initial impulse behind the development of *regulation theory* goes to Michael Aglietta (1979) and a team of economists grouped around the Paris-based research centre CEPREMAP (Centres d'études prospectives d'économie mathématique appliquées à la planification). The original concepts of Aglietta were later developed by Robert Boyer (1990a; 1990b), Alain Lipietz (1977; 1986; 1987; 1992; 1993), Pascal Petit (1984), Benjamin Coriat (1979; 1990; 1991) and Danièle Leborgne (Leborgne and Lipietz, 1988, 1989), among others. Although regulation theory draws generously from marxism and the French school of historiography of the *longue durée*, it was explicitly developed as an alternative to more orthodox versions of marxism epitomised by the theory of the new international division of labour (Lipietz, 1993). This structuralist attempt to conceptualise the ongoing changes in the nature of the capitalist economy was reworked by regulationists who objected to, in their view, a hasty reduction of the unique characteristics exhibited by distinct societies and regions into a general concept of the division of labour. Even though regulation theory is gaining widespread support amongst those who seek to understand changes in contemporary industrial society, it is not a single coherent theory in any conventional sense but rather a series of concepts used in an ongoing research programme within the marxist political economy (Webber, 1991).

Of particular importance to regulationists is the role of the local and nation state in the process of capitalist production and reproduction. The central role of *governance* or government-led *regulation* of the economy underlines the importance of state and government in economic development (Lipietz, 1992). According to regulationists, the economic development experienced in many highly industrialised countries after the Second World War was made possible by reaching the broad consensus between labour and capital, referred to as '*the mode of regulation*'. The long post-war boom, during which western European, Japanese and North American economies prospered, was the result of this

'Fordist' mode of regulation. Fordism is an industrial system based on a mass production of standardised goods and a mass consumption of consumer commodities regulated according to principles of Taylorist and Keynesian management. During the Fordist period, nearly all the principal pan-national institutions regulating international trade were created. One of the most important institutions crucial to the Fordist compromise was the General Agreement on Tariffs and Trade (GATT). Fordism, with the support of the GATT, stimulated a reduction in the level of impediments to free trade by converting most trade barriers to tariffs and extending the most-favoured-nation principle to an ever-increasing inner circle of member countries (see Figure 1.3). This powerful combination of government-regulated multilateralism and Fordism formed the basis for the long post-war boom which lasted until the first 'oil shock' in 1973. The key to that success, according to regulationists, was a series of social compromises between labour, capital and the nation-state. Thus the Fordist mode of regulation became the foundation on which multilateral free trade between the OECD countries could prosper. Multilateralism and Fordism are therefore intrinsically linked and any change in the mode of regulation would necessarily put the principle of multilateralism under strain.

The Fordist mode of regulation started to dissolve in the early 1970s as a result of a number of associated events and processes. Although there is no consensus as to the precise reasons for its alleged collapse, several areas of agreement have emerged (Moulaert and Swyngedouw, 1989). Already in the second half of the 1960s, productivity gains had begun to fall in most developed capitalist countries. Despite that decline, real wages in most branches of industrial activity continued to grow. Moreover, cuts in fixed capital spending began to rise in an ever-increasing circle of '*stagflation*'. The basic cause of these developments, according to regulation theory, was the rigidity of the Taylorist 'scientific' management methods and inflexible methods of industrial production (Taylor, 1947). Most enterprises reacted to the fall in productivity by increasing prices, a move which eventually led to double-digit inflation. Since the Fordist compromise required a straightforward relationship between inflation and wages, each wave of inflation led to a corresponding wage increase. The purchasing power of core currencies fell, further reducing demand. As a result, the post-war 'economic miracle' was jeopardised. Furthermore, the multilateral liberalisation of trade coupled with the industrial success of Japan and several developing economies in a part of South-East Asia (the latter often referred to as the 'newly industrialised countries') introduced a new element in post-war Fordism, i.e. fierce foreign competition. The crisis of the Fordist model of industrial development, especially in western Europe and North America, created a new international climate whereby increased calls for the protection of domestic enterprises and, in certain cases, whole sectors of the economy resulted in new policies of *managed* or *fair* trade which started to erode the Fordist multilateral framework (Baldwin, 1992).

According to regulationists, the globalisation of multinational enterprises arose from the increasing pressure to recover profitability levels. In response to these market pressures, multinational enterprises spread their operations to take advantage of low-wage regions such as Hong Kong, South Korea and Taiwan which, in turn, eventually evolved into fully fledged industrial economies modelled closely on their western prototypes. The bulk of production was directed back toward the European and American markets. Consequently, for the first time in the history of industrial society, the growth of trade outstripped the growth of industrial output. Gradually nation states and their governments began to lose control over the macroeconomic management of their domestic economies. The Fordist mode of regulation was 'outmoded' by the internationalisation of production, consumption and the growing flexibility of capital. There were only two avenues left to counter this erosion of control and regulation: to increase exports by selling more competitively; or to adopt 'non-tariff' or 'voluntary' protectionist measures. As Lipietz comments;

> There was, of course, no multinational agreement to balance growth in different countries, no international collective agreements, no supranational welfare state, no international treaty on working hours. To the supply-side international crisis of Fordism was added a demand-side international crisis. (Lipietz, 1992, 19)

The consequences of globalisation and the decline of multilateralism extend well beyond the economic sphere. In essence, most of the theoretical explanations mentioned in this chapter attempt to make sense of these consequences. The precise nature of the new mode of regulation that follows Fordism is still uncertain. Unlike some more deterministic conceptualisations, regulation theory does not imply any inherent teleology or 'historical laws'; there is nothing automatic or inevitable about the nature of the new mode of industrial development. Nonetheless, a general outline of the new mode of regulation, post-Fordism, can already be perceived. The emergence of trading blocs is an integral part of this new economic and social order.

In summary, Fordism and multilateralism created a solid basis on which the post-war prosperity of the West was founded. Both processes stimulated the reduction in the level of impediments to free trade by applying the most-favoured-nation principle to an ever-increasing virtuous circle of members. The key to that success was a social compromise between the state, capital and labour. The powerful combination of this Fordist mode of regulation coupled with multilateral regulation moulded the post-war 'co-prosperity zone' until the first 'oil-shock' exposed the weaknesses and rigidities of this system. Multilateralism and Fordism are therefore intrinsically tied to the same period of post-war economic and social history.

Post-Fordism

Post-Fordism is not a single or a coherent theory. Instead it is a collective term describing several theories and ideas which have emerged in the social sciences in recent years. Although some argue that post-Fordism describes a fundamentally different mode of production and regulation from Fordism (Jessop et al., 1991; Lash and Urry, 1987; Jacques and Hall, 1989), most agree that the distinction is not entirely clear. Post-Fordism is a new mode of capitalist regulation which emerged from the ashes of Fordism. In contrast to mass production, mass consumption and the welfare state, post-Fordism involves flexible production techniques and the flexible deployment of labour, segmented consumption patterns, postmodernist cultural forms and a restructured welfare system. The breakdown of Fordism and emergence of post-Fordism are primarily the result of a search for greater flexibility in the economy. More importantly for the present discussion, this new mode of economic regulation coincided with the gradual emergence of continental trading blocs. In short, the economics of flexibility at the level of an individual enterprise or nation-state is being extended into the international sphere in the search for flexibility on a continent-wide scale.

Lipietz (1992) argues that post-Fordism emerged gradually in three reasonably distinct periods throughout the 1970s and 1980s. The first period, from 1973 to 1979, was characterised by the inflationary policies instituted by the United States Federal Reserve to stimulate domestic demand. While this tactic succeeded in shielding the United States economy from a supply-side crisis, in the end it proved to be no more than a delaying tactic. The second stage started in 1979, when the newly elected conservative governments in the United States and Britain attacked the problem from a radically different angle by focusing on hardening 'soft' credit regulations. The reorganisation of capital markets in which firms operate stressed a return to liberal economic policies and more flexible practices in the utilisation of the factors of production, especially labour, machinery, technology and fixed capital. This '*liberal productionism*' was a radical attempt to rejuvenate the capitalist factors of production in what was perceived as an overregulated and too rigid economic environment.

The third stage commenced around 1983, when the United States Federal Reserve decided to relax and partly reverse its policies of credit restraint. This move triggered a long period of credit-led expansion in the American economy whose growth rate, until then, was dragging behind most other capitalist economies. A new school of thought developed which advocated a solution to the supply-side crisis. The intellectual novelty of that school can be characterised by one term, already mentioned several times – *flexibility*. The popularity of that adjective, applied to many forms of economic, political and social activity, is enormous and had resulted in an ever-expanding literature (for a good survey of that literature see papers by Meric Gertler, 1988 and 1992).

However, the actual economic policies of leading industrialised countries were guided by two, often competing, schools of thought. The first, associated most strongly with the liberal economic policies of the United States and Britain (and also France after 1983), strongly advocated a reduction of labour costs, introducing flexible working and contract practices (for example part-time work, casual work, short-term contracts, etc.), flexible subcontracting and vertical disintegration in the broadest sense of this term. The second school of thought emphasised instead the need for a new social compromise: a new contract of *collective bargaining* between labour, capital, the state and much more orthodox fiscal and monetary policies. Those principles, according to regulationists, were guiding the economic policies in Japan, Italy and also in Germany. By sharp contrast, the former school which guided policies implemented in the United States and Britain during the same period stressed a radical economic and social agenda aimed at introducing a much greater degree of flexibility in industrial and labour relations. In short, although post-Fordism disrupted the established social and productive relations, it did not replace them with any coherent long-term solution to the new economic and social challenges (Boyer, 1990b).

To summarise, post-Fordism is a new mode of capitalist regulation characterised by flexible production techniques and the flexible deployment of labour and capital. The search for greater flexibility resulted in a breakdown of Fordism and the decline of commitment to unconditional multilateralism. However, the new requirements of flexibility on the continental scale determine the limits of individual nation-states to act unilaterally. Consequently the emergence of trading blocs is a step toward a gradual reconstruction of governmental regulation at the international level. Although, as elaborated in Chapter 1, regional trading arrangements are not necessarily contrary to the principle of free trade, there is no doubt that the protectionist policies associated with regionalism contributed to an erosion of the multilateral framework of international trade relations. The resulting tension between multilateralism and regionalism contributed to the international instability and long-term crisis of capitalism. However, the exact nature of the emerging new relations is not entirely clear. There are at least two competing theories, flexible specialisation and flexible accumulation, which attempt to conceptualise this new order.

Flexible specialisation

A novel theory of social and economic transformation was proposed by two American economists, Michael Piore and Charles Sabel (1984), who attempted to interrogate changes in some core industrial districts in the context of a more fundamental process of industrial restructuring. Using some ideas of the regulationist school, they put forward a thesis that Fordist mass production is going to be supplemented by a parallel but entirely new regime of production characterised by craft production or 'flexible specialisation'. The core of this approach asserts that the problem of economic growth can be solved through an interaction between technology, institutions and polity.

The key to flexible specialisation is the notion of industrial districts consisting of small, medium-sized and large decentralised enterprises co-operating in a series of networks. Thus flexible specialisation is at once a general theoretical approach to the analysis of change in industrial society and a specific model of productive organisation. Regional integration in this context could be interpreted as an integration of several industrial districts on a continental scale. However, according to Piore and Sabel, in no sense should flexible specialisation be understood as a necessary outcome of some inherent logic of economic or technological development. On the contrary, both mass and craft production methods are merely ideal types rather than empirically established types of industrial organisation and production. In fact, both models can and do coexist and probably neither will dominate the global, or even a national, economy. Central to flexible specialisation are strategies of product development that seek to shape markets as well as rapidly adapting production.

In the mass production economy perfected since the Second World War, enterprises seek to reduce the anarchic tendencies of unregulated markets by administering a system of collective wage bargaining. This served two functions. First, it provided a much needed element of stability and predictability to the otherwise volatile and, ultimately, destructive economic environment. Second, it guaranteed a market for the vast quantities of commodities and services produced in a modern economy by linking wage increases to productivity rather than subsistence level, as in the case of the pre-Fordist stage of capitalist development. Though Piore and Sabel never examined the issue of multilateralism in detail, it can be extrapolated from their work that the GATT and the post-war free trade policies triggered by the Bretton Woods Agreement contributed to the subsequent Fordist compromise. Unfortunately, the inflationary shocks of the 1970s threw that system off balance and exposed its weaknesses and rigidities. Multilateralism was no longer sufficient as a framework for international trade relations. Increasingly, regional arrangements between clusters of geographically proximate industries and countries became the basis of new flexible specialisation.

According to Piore and Sabel, the collapse of Fordism arose from the inflationary shocks of the 1970s and the economic mistakes of the leading industrialised countries, particularly the United States. Fordism exposed the inherent inability of regulated capitalism to deal with a large economic system and a complex division of labour. Moreover, the Fordist regime of production led to a saturation of domestic markets and consequently increased trade between major industrial economies. In short, the Fordist regime of production became truly international. Multinationals were the most tangible evidence of that trend. However, there was no international mechanism that regulated demand and supply relations between competing economies and multinational enterprises. International Fordism was bound to falter as a result of the shortfalls in demand after the inflationary shocks. In addition, the emergence of the newly industrialised countries, as serious competition to the established international industrial order, further exposed the inherent instability of international economic relations.

The introduction of flexible production techniques and flexible specialisation prompted a radical reorganisation of the entire production system in key commodity and service industries. The emergence of a trend toward greater continental integration of the principal industrial players, the emergence of 'the triad' of western European, North American and Asian trading blocs, is seen as the institutional response to the contradictory requirements of flexibility; that is, to preserve and enhance the possibilities of mobility while at the same time limiting the threat to territorially defined markets (Gertler and Schoenberger, 1992). Thus regionalism can be interpreted as the result of a set of policies that seek to regulate economic activity in a strictly defined geographical area in order simultaneously to reduce the threat of outside competition *and* preserve the advantages of economic flexibility and large markets. In other words, government regulation of the market plays an extremely important role in the process of economic development, including multilateralism and regionalism. Ultimately, the two competing processes are associated with the post-Fordist transformation of both the modern capitalist state and the framework for the reconstitution of the system of production.

To summarise, the crisis of Fordism created conditions for the crisis of the GATT and multilateralism. The change in the mode of regulation to flexible specialisation created a new international climate whereby increased calls for protectionism promoted policies of managed trade. This new mode of economic regulation coincided with a new mode of international economic regulation – trading blocs. Thus regionalism is the spatial expression of the new mode of international regulation and flexible specialisation. However, the flexible specialisation concept is quite different from regulation theory, post-Fordism and conventional marxist political economy, in that it does not assign any significant role to either class relations or class struggle (Hirst and Zeitlin, 1991). Possibly the most traditional in this respect is the theory of flexible accumulation put forward by those who try to reconcile the evidence of change in the nature of the industrial society with marxism. Flexible accumulation theory is, in many ways, similar to post-Fordism because it uses a traditional marxist method of assigning a determining role to the forces of production in societal and economic change. However unlike post-Fordism, which attempts to incorporate some elements of postmodernism in its interpretation of industrial and economic change, flexible accumulation theorists resist postmodern theories and are more interested in a marxist interpretation of society, politics and culture, than of the economy.

Flexible accumulation

Flexible accumulation conveys an even stronger sense of some kind of an ongoing transformation underlying the changes in international capital accumulation. Although none of its advocates could be described as 'marxist' in the orthodox sense of the term, all of them base their analytical approach firmly within the marxist heritage. According to Harvey and Scott (Harvey and Scott, 1988; Harvey, 1989) and Storper and Scott (Scott and Storper, 1986; Storper and Scott, 1989), who are the strongest supporters of this theory, production for profit remains the basic organising principle of all social and economic activity despite the momentous and significant changes experienced by modern economies. Logically then, beneath the confusing surface of transitory and ephemeral appearances lies the unchanged foundation of all societal and economic relations – *capitalist accumulation*. Fordism fused with the GATT, and the rest of the Bretton Woods' mechanisms, governed the reduction of impediments and barriers to trade and set a course of worldwide expansion that extended well beyond the old industrialised world. Meanwhile, national governments engineered in Fordism both stable economic growth and rising material living standards through a mix of welfare provision, progressive fiscal policies and, through negotiated collective bargaining, a social consensus between employers and employees. 'Post-war Fordism has to be seen, therefore, less as a mere system of mass production and more as a total way of life' writes Harvey (1989, 135).

Fordism was also dependent on the massive expansion of world trade and international investment flows. The institutions of the GATT, IMF and the World Bank all played their role in assuring the expansion of the capitalist mode of accumulation. Although slow to develop outside the United States before the 1940s, Fordism became fully implemented throughout western Europe and Japan after the Second World War. The 1970s and 1980s have been a period of intense economic restructuring and, subsequently, social transformation and reform. Post-Fordism did not emerge suddenly but rather evolved gradually through a series of political and social experiments. These changes in the mode of production and regulation represent early signs of a passage to an entirely new mode of capital accumulation.

The new 'flexible' mode of accumulation builds upon solutions developed in response to the inefficiencies and rigidities of Fordism. Once again, the key phrase here is 'flexibility' and its application to labour, the production process, specialisation, patterns of consumption, and so on. The rise of services and high technology industries as well as the globalisation of production and markets are all responses to this perceived crisis of the Fordist mode of production and outmoded accumulation practices (Amin and Robins, 1990). Unfortunately, the new regime of accumulation reinforces uneven development between principal economic regions and less developed countries. Other negative consequences of flexible accumulation are a reduction in labour union power which led to the introduction of much less secure, but also much less rigid, labour contracts and practices. Harvey argues that declining union power and increasing subcontracting of labour and other intermediate inputs effectively transformed 'the objective basis for class struggle' (Harvey, 1989, 153). From a marxist point of view, this is perhaps the most fundamental influence of post-Fordism on the mode of accumulation. Class consciousness is no longer derived from class relations but rather from the social division of labour and its increasing fragmentation and specialisation. In this way the supporters of flexible accumulation depart significantly from more orthodox versions of marxism.

The theory of flexible accumulation implies that post-Fordism reconstructs the capitalist system of economic and social relations in new and innovative ways. Flexible accumulation, in effect, is a response by the capitalist forces of production to a much more sophisticated and complex division of labour and demand both from industry (intermediate inputs) and consumers (final inputs). However, contrary to a variant of this conceptualisation proposed by Lash and Urry (1987), which stresses the anarchic tendencies of flexible accumulation (see also Offe, 1985), Harvey proposed that the capitalist system is becoming more rather than less tightly organised. This arises from the ever-increasing pace of technological innovation. Particularly important have been the increased availability of information and, consequently, a complete reorganisation of the financial environment. Deregulation and computerisation of the financial system since the early 1970s have changed the 'balance of forces at work in global capitalism' (Thrift and Leyshon, 1988; Thrift, 1990). The new global financial and banking system transcends the nation-state and its political interests. Although there has always been a degree of autonomy in the tightly linked financial markets of capitalist nations, the breakdown of Fordism and the collective bargaining system reinforced and reworked the relationship between the capitalist state and increasingly independent and mobile financial capital. This, of course, has had a direct influence on the development of regional trading arrangements as an institutional response by nation-states to the threat of foreign capital.

It is reasonably easy to detect that the flexible accumulation thesis has been heavily influenced by the regulationist school. Nevertheless, the two differ in some key points. The crucial difference between the two approaches appears to be the contrasting treatment of post-Fordism and its broader significance to the central question of capitalist transformation. Whereas the regulationist school explicitly proposes its interpretation as an *alternative* to marxist theory, Harvey and others seem to stress the fundamentally *unchanged* nature of capital. Thus flexible accumulation is the result of a search for solutions to the crisis tendencies of capitalism within the broadly defined capitalist framework of social and economic relations. This view of flexible accumulation and post-Fordism is shared by many other writers including Gertler (1989), Pollert (1991), Sayer and

Walker (1992), although each of them criticises it for different reasons.

Almost the exact opposite of this view is, of course, the flexible specialisation thesis of Piore and Sabel (1984) who stress that new technologies open up the possibility for a reconstruction of labour relations and a production system founded on an entirely different social, economic and spatial basis. This interpretation of the developments since 1973 impressed regulationists who suggested that flexible specialisation and the mode of production are radically transforming the very core of late twentieth century capitalism. However, the advocates of flexible accumulation criticised regulationists for not providing any holistic theory explaining the mechanism and logic of the post-Fordist transition. In short, Harvey and others maintain that Marx's invariant elements and relations of the capitalist mode of production are beneath all the superficial 'froth and evanescence' of post-Fordism. Consequently capital accumulation, or rather the way in which overaccumulation is absorbed, is the main mechanism which can explain the formation of trading blocs. Interestingly, the proponents of flexible accumulation theory seem to revert here to a much older version of the marxist theory of accumulation proposed by Rosa Luxemburg some 70 years earlier. This suggests that despite the ongoing transformation, changes in contemporary international economic relations can be explained by a slightly modified version of the old marxist theory of capitalist accumulation. The '*spatial fix*' implies that the capitalist overaccumulation of commodities is disposed of by the creation of new capitalist spaces within which production and consumption can proceed (Harvey, 1989). This process can account for the growth of trade and investment abroad as well as the exploitation of labour by multinational enterprises.

The marxist theory of capital accumulation builds on the schemata of reproduction expounded in the second volume of *Capital*, today probably the most neglected portion of Marx's work (Marx, 1967, first published 1893). Marx maintained that capitalism was bound to destroy itself by the logic of its own internal contradictions, especially those associated with the progressive concentration of capital and the increasing pauperisation of the working class. Unfortunately, Marx never specified precisely under what conditions capitalism would actually become an economic impossibility. Rosa Luxemburg, in *The Accumulation of Capital* (1951, first published 1913), supplemented and expanded Marx's views. She suggested that the capitalist system would only work as long as it had at its disposal a non-capitalist market, whether internal or external. In her view, accumulation depends on an increasing demand for the goods produced by industry. Since industry cannot go on creating its own market indefinitely, and the spatial expansion of capital cannot go on indefinitely, capitalism was inevitably preparing its own economic ruin. There could be no such thing as a capitalist world system on a global scale. If capitalism extended to all geographical regions and exhausted its room for expansion, it would cease to exist (Kowalik, 1971).

Although some Russian marxists, such as Struve (Pipes, 1970; 1980), Bulgakov (1982) and Tugan-Baranovsky (1913a, 1913b), later Kalecki (1990; 1991), demonstrated that, at least in principle, capitalist accumulation can go on increasing indefinitely (incidentally putting in question the whole idea of scientific socialism along the way), the theory of capitalist accumulation survives in its various guises. Indeed, the theory of flexible accumulation and geographical expansion of capital borrows directly from this tradition. Multilateralism and regionalism in this context are merely two contradictory manifestations of the process of capitalist overaccumulation. Both forms of international trading arrangements serve the same purpose, i.e. the expansionary reconstruction of the capitalist mode of production.

Critics of flexibility

Although regulation, post-Fordism and accumulation theories proved enormously popular, there are at least two serious areas of weaknesses in these perspectives. The first relates to the notion of a flexible or post-Fordist system of production which is allegedly replacing older and much more rigid methods and techniques of Fordism. The second focuses on the concept of accumulation as the source of tension in the capitalist system. Each of these critiques is discussed below.

According to some critics, the far-reaching generalisations of the regulationist school and, in particular, the notion of the flexible regime of accumulation are exaggerating the real extent of changes in the modern economy as well as the degree of discontinuity which this 'radical break with the past' involves. Gertler (1988, 1992), Pollert (1991) and Sayer (1989) argue that some forms of flexibility have been present for a long time. In fact, various forms of production and organisation, including mass production and more flexible forms of organisation, coexisted throughout most of the nineteenth and twentieth centuries, almost since the inception of the industrial revolution. As Gertler writes, 'even in the heyday of mass production assembly lines, batch production was still extremely widespread throughout the American economy and remains so today' (Gertler, 1988, 429). In other words, the transformation into flexible forms of production represents the outcome of a long and, for the most part, unbroken chain of mostly technological and organisational innovation rather than a radical break with the past. Therefore, the flexible regime of accumulation does not represent a qualitative shift in the fundamental forces underlying capitalist accumulation but rather an improvement of the older version. The emergence of regionalism and decline of multilateralism are the logical consequences of this evolution.

Pollert (1988, 1991), from a marxist perspective, goes even further in her critique. First, the idea that the history of twentieth-century capitalism can be split into two clear-cut periods, Fordist and post-Fordist, cannot be sustained in the light of the marxist theory of accumulation. There can be no clear divide, in her view, between the two periods because standardised mass production continues to flourish in many parts of the world. Moreover, the allegedly rigid Fordist production methods were characterised by a considerable degree of broadly defined flexibility. In other words, post-Fordism relies on a very restrictive set of definitions about the nature of flexibility in the production process. Pollert objects even more strongly to the normative vision of the new type of relations between employees and employers. In contrast to the too general, and in her opinion overoptimistic, scenarios suggested by many 'post-Fordists', the new labour practices introduce much degraded and deskilled forms of work. In fact in her view, flexible labour practices amount to nothing more than a new strategy of capital to manipulate skilled and unionised workers.

Yet another angle on flexible accumulation was presented by Sayer (1989) and Gertler (1992). They argue that both the regulationist school and flexible accumulation theory focus on too narrow a range of industries, most notably automobile, consumer electronics, and motion picture (Lash and Urry, 1987; Storper and Christopherson, 1987). More importantly, the vast generalisations of post-Fordist theory which suggest a significant degree of transformation compared to Fordism are based on a very limited and selective empirical evidence. As Sayer writes, 'it is simply not adequate to ignore Japan in discussing allegedly epochal changes in western capitalism, for it is quite likely that some of the changes in the West have not been independent of Japanese competition but have occurred because of it' (Sayer, 1989, 670). In fact, the emergence of new forms of protectionism, particularly in Europe and North America, can be attributed, in part, to the competition from Japan and South-East Asia. Many organisational and technological practices which made Japan and, more generally, Asian economies competitive were introduced in the 1940s and 1950s, long before the alleged crisis of Fordism (Gertler, 1992).

To summarise, perhaps the most significant weakness of the post-Fordist thesis is its stress on the emergence, on the one hand, of localised new industrial districts or regions whilst, on the other hand, predicting the deepening of the international social and spatial division of labour. The bulk of post-Fordist research focuses on the local ramifications of the new mode of accumulation. The most celebrated examples of it are the much publicised (and criticised) studies of the Emilia-Romana region of the 'Third Italy' (Brusco, 1982; Piore and Sabel, 1984; Amin, 1989), Silicon Valley and Route 128 (Scott, 1988), and Orange County (Storper and Walker, 1989). By comparison, very little attention has been devoted to the implications of the new regime of accumulation to the emerging global and international economy. A typical way to conceptualise these implications is by stressing the significance multinational enterprises have to the process of accumulation. According to critics of Fordism, the flexibility of production at the international level amounts to an exploitation (in a marxist sense) of spatial variations in international wage levels, skill differentiation, labour availability and the level of labour specialisation within and between countries. However, as Tickell and Peck pointed out, 'it was precisely these strategies which exacerbated demand problems during the Fordist regime of accumulation and theorisation of this apparent contradiction is necessary' (Tickell and Peck, 1992, 200). As a result, more recent 'radical' accounts of the multinational enterprise are less critical of their 'exploitation' since they not only facilitate the linkages between industrial regions (Dicken, 1992b), but are also constrained by the nation-state and national trade policies. Nevertheless for many, multinational enterprises continue to be a symbol of both the corporate power of capital and the consequences of the free trade principle enforced indiscriminantly throughout the post-war period. The next section of this chapter therefore examines the nature of the multinational enterprise in the context of the social theories outlined above. Particularly significant is the clarification of their role in the issue of multilateralism versus regionalism.

MULTINATIONAL ENTERPRISES AND NATIONAL REGULATORY POLICIES

According to recent studies (Dicken, 1992a, 1992b; Hirst and Thompson, 1992), one of the most important economic trends during the 1970s and 1980s has been the acceleration of interaction and interdependence between the economies of highly industrialised countries. Production of goods, services and their trade are increasingly transcending national and regional borders. The growing frequency of cross-border transactions has been accompanied by the ever-increasing volume of international trade, the growing strength of international investment and the enhanced complexity of the international division of labour. The exact nature and form of the evolving international dependence may vary according to *de jure* and *de facto* criteria applied to the evaluation of integration, as defined in the previous chapter. Nonetheless, the *internationalisation* or, as some would argue, *globalisation* of economic activity is evident not only in industrialised regions but also in less developed countries. This phenomenon is a manifestation of the increasing interdependence of capital, primarily though not exclusively in the form of multinational enterprises. There is a wealth of literature dealing with the role of multinational enterprises in the world economy (see Vernon, 1985 for a good review), the scope and significance of which is well beyond the objective of this chapter. The issues of the primary functions and forces which led to the formation of multinational enterprises are charged with methodological and ideological arguments. Even the precise meaning of the term 'multinational enterprise' is surrounded by controversy (Dicken, 1992a; Dunning, 1993).

For these reasons the largest multinational enterprises attract an enormous amount of attention and have lately been the focus of much academic debate. In the literature two terms, '*international economy*' and '*global economy*', are used almost as bywords in any analysis of contemporary economy, society and polity. However, there seems to be a great deal of confusion about the exact interpretation of what those two terms actually mean. Despite these problems, as Hirst and Thompson (1992) argue, the generally accepted assumption is that whatever *internationalisation* and *globalisation* of the economy precisely mean, the processes described by these terms are well under way and are introducing a qualitatively new stage in the development of industrial society and international capitalism. It is important, therefore, to look more closely at the precise meaning of the two terms and their relevance to the issue of regionalism.

International or global economy?

According to Hirst and Thompson (1992), both the *international economy* and the *global economy* (they call it '*a world-wide international economy*' and '*a globalised international economy*' respectively) are ideal types or conceptual tools rather than actually existing phenomena. Although these terms are, to a degree, similar because both imply cross-border economic exchange, they differ significantly in some crucial attributes. An international economy defines an economy in which the principal economic entities – nation-states – are involved in the process of increasing interaction. The principal mode of this interaction is trade which takes the form of national specialisation (or competitive advantage) and the division of labour. In spite of this process, however, the domestic and external economic policies of nation-states remain relatively autonomous. Interactions are not necessarily direct but, instead, are 'refracted through national policies and processes' (Hirst and Thompson, 1992). In other words, the international economy is determined largely at the level of national economies and international phenomena are outcomes that emerge from the distinct and different performance of national economies. An international economy is an aggregate of nationally determined functions and policies.

By contrast, *a global economy* consists of distinct national economies which are *subsumed* and *rearticulated* into the international system by international transactions and processes. Although nationally determined policies still operate, they are *subordinate* to wider international determining factors. In the global economy the international economy becomes autonomous and transcends attempts at individual national policies and regulation. As Hirst and Thompson write:

> As systemic interdependence grows, the national level is permeated by and transformed by the international. In such a globalised economy the problem this poses for public authorities is how to construct policies that co-ordinate and integrate their regulatory efforts in order to cope with the systematic interdependence between their economic actors (Hirst and Thompson, 1992, 361).

As a result, the global economy, markets and industries are extremely difficult to regulate within the framework of multilateralism. Thus, the process of integrating some key national economies into regional economic pacts and trading blocs can be seen as an international attempt to coordinate regulatory agencies with the institutions of national governments. Consequently, the fundamental problem of the global economy is the integration and coordination of regulatory policies between the different national authorities. The objective of such integration is to regain at least some of the lost ground over the governance of economic activity.

Another important consequence of globalisation is the transformation of the multinational enterprise into a truly global or transnational enterprise, detached from constraints of government regulation. The transnational enterprise would, in this scenario, represent genuine footloose capital, unconstrained by any specific national base. For these reasons, multinational enterprises became the very symbol of the corporate power of capitalism. Such enterprises have also been charged with economic powers extending well beyond the sovereign interests of nation-states and transcending the economic sphere into polity and society. It has become almost conventional wisdom to view the nation-state and its economic policies as irrelevant to the economic activities of multinational enterprises (Schoenberger, 1988). Moreover, many critics stress their negative influence, especially in the developing world, as agents of hidden neocolonialism, dependency and imperialism. Indeed, regardless of the varied ideological perspectives adopted in this literature, many have argued that the multinational enterprise is the single most important force in industrial restructuring between and within nation-states (Vernon, 1971, 1977; United Nations, 1990a). As Peter Dicken (1992a) observes, the largest multinational enterprises are particularly significant in the international economy because first, they control economic activities in more than one country; second, they can exploit geographical differences between countries and regions in terms of factor endowment; and third, they are able to shift resources and operations quickly between locations at a global scale. All of these factors give multinational enterprises an enormously privileged position to benefit from economies of scale and scope, utilising various endowment factors in host countries. This had led some academics to predict that the tendency of the capitalist economy to concentrate capital would result in 200 to 300 multinational companies controlling over 80 per cent of capital by the mid-1980s (Perlmutter, 1969). Although a casual inspection of the largest multinationals will refute such exaggerated claims, there is no doubt that multinational enterprises play, and will continue to play, an important role in shaping the patterns of trade, foreign direct investment and transfer of technology.

Limits of the multinational enterprise

Despite the obvious significance of multinationals to both the international economy and the role which nation-states play in the complex process of regulation, little work has been devoted to a conceptualisation of these phenomena. The significance of multinational enterprises to the process of continental and regional integration has to be assessed against this background. The conventional wisdom has it that multinational enterprises are the prime force behind the development of multilateralism and free trade policies, not only because they are bound to benefit most from the opening of markets, but also because they can influence national trade and investment policies. In fact, the very label 'globalisation' implies that multinational enterprises operate practically unrestricted in nearly all significant markets. As such, multinational enterprises transcend the national and regional interests of nation-states, in effect representing none else but themselves, their capital and investors. In short, they symbolise the very core of the capitalist system, the raw capitalist enterprise in its purest form in the contemporary economy. Their commercial interests, that require free access to an ever-increasing number of consumers, are the prime force behind the deregulation of national markets and the whole ideology of free trade.

According to the proponents of such an explanation, multinationals have exerted a coercive pressure on the national governments of not only major industrialised countries but also, increasingly, the developing world (Krueger, 1992). The prime objective of this pressure is to enable unlimited access to markets by lowering tariffs and other non-tariff impediments to foreign imports of commodities, services and capital. Since it is in the interest of multinational enterprises to be able to move capital as freely as possible in order to benefit from differences in tax regimes, interest rates, financial regulations, and to relocate fixed capital from high to low wage labour markets, these enterprises have become the prime stimuli behind the liberalisation of trade and the deregulation of national economies. In other words, multilateralism and liberal free trade policies result from the concentration of corporate power, capital and political influence in the hands of a relatively small number of multinational enterprises.

Consequently, any limitations imposed on multinational companies through regional trade agreements are contrary to their broadly defined interests. The regional integration of markets may benefit only the most vulnerable multinationals which attempt to shelter from competition behind protective regional tariffs and non-tariff barriers. In other words, regional trade agreements are a mirror image of state regulation on a broader international level. Regional trading blocs are, therefore, an attempt to reinstate national regulation on a continental scale. Not surprisingly, the weakest multinational enterprises suffering from strong foreign competition are the most ardent supporters of regional protectionist schemes, state intervention and 'fair' trade. Prominent examples of these protectionist tactics are displayed by many European-based multinationals, such as Philips, Thompson-Bull, Fiat, and by General Motors in North America. Some radical critics argue from a marxist position, that multinational enterprises are the prime example of the basic unity of capitalist finance, production, and commodity trade (Harvey, 1989; Storper, 1992). Thus the capitalist world economy and multinational enterprises controlling it are subjects to the 'laws' of capitalist accumulation. Since protectionist policies, however beneficial in the short term, are contrary to the very 'logic' of capitalist overaccumulation, in time the resistance against the capital will crumble, bringing down with it multinational enterprises and the world economy.

Although there is no doubt that such an interpretation of the role of multinational enterprises in the process of multilateralism provides a suggestive explanatory tool, several serious problems emerge under closer examination. This explanation has at least three major weaknesses. First, however powerful the forces of integration in national economies, the reality of trade practices is far from the ideal type of free trade; nation states continue to play a key role in the interaction of the international rather than the global economy. Second, multinational enterprises are much less global and far more *regional* in their spatial scope than is implied by certain commentators. Consequently, the ability of the multinational enterprise to seek profits in the global economy has been exaggerated. Third, marxist interpretation oversimplifies the role of nation-states and their policies in regulating the international environment for multinational enterprises. In the following section, each of these weaknesses will be examined in more detail.

The myth of the global economy

Despite claims about the omnipotent nature of trade and foreign investment in the global economy, in reality only the highly industrialised economies comprise this rather exclusive membership. In 1989, over 80 per cent of world trade was conducted between OECD countries, and this increased recently to 85 per cent if the ex-socialist countries of eastern Europe are included in the calculations (Hirst and Thompson, 1992). Thus, at least in strictly spatial terms, there is no such thing as the global economy. There is an international economy where the major players are an exclusive club of the most economically and politically powerful nation-states. A frequently used example of globalisation is the financial sector which, after significant deregulation throughout the 1980s, achieved the status of 'global financial capital'. Of course, there is a great deal of truth about the ability of banks and financial markets to move freely and quickly enormous amounts of capital from one location to another (Thrift, 1990). However, one could argue that there is scarcely anything new in this development. The financial penetration by foreign capital of the UK and other key economies was actually greater between 1905 and 1914 than throughout the 1980s (Tomlinson, 1988; Keenwood and Loughead, 1992; The Economist, 1993a). Thus it is imperative to keep contemporary economic developments in historical perspective. For example, it is important to remember that the international economy, at least that consisting of the leading economic powers, was hardly less integrated before August 1914 than it is today. In other words, today's open world economy is not unique and unprecedented and, more importantly, its degree of openness is most probably reversible.

The spatial extent of multinational enterprises

One of the problems encountered when examining the spatial extent of multinational enterprises is the arbitrary definition of what actually constitutes 'domestic'. The distinction between domestic operations and those attributed abroad is in fact difficult to justify since a large proportion of transactions are, in fact, intrafirm rather than through open markets (Taylor and Thrift, 1986; Dunning, 1993). Therefore a less arbitrary and more realistic way to examine the geographical pattern of economic activities is to include operations in neighbouring countries and domestic sales as part of regional operations. Dunning (1993) presents data showing the distribution of economic activities, including 98 000 foreign affiliates of the leading multinational enterprises from 23 developed countries in 1980 (Table 2.1). The overwhelming majority of these activities are clustered within regional markets, even among very large multinational enterprises. In other words, the majority of European-based multinational enterprises operate mainly in Europe (western Europe), American and Canadian firms operate principally in North America and Europe, while an overwhelming proportion of Japanese multinational enterprises operate in South-East Asia. These broad trends are confirmed by the data collected by the United Nations office for Transnational Corporations and Management Division (quoted in Dunning, 1993). Similar results were reported regarding the recent trends in foreign direct investment in eastern Europe by western European multinationals (Murphy, 1992; Michalak, 1993).

Because of the rather limited empirical evidence, the significance of these findings is difficult to assess and has to be treated cautiously. There is no doubt that large multinational companies have increased the volume of their long-distance trade and investment throughout the last two decades. However, there is also evidence indicating that multinational enterprises, far from becoming *global*, are actually becoming increasingly *regional* (Ostry, 1992; Hirst and Thompson, 1992; Michalak, 1993). In other words, the trend toward the formation of regional trading blocs may be influenced strongly by multinational companies which simultaneously search for the economic advantages and flexibility of large markets as well as the protection afforded by regional trading arrangements. The extent of the omnipotence of multinational enterprises suggested by some authors is therefore at best questionable and at worst misleading. Most multinational companies still only operate principally in their domestic market and in a limited number of countries or, at most, regionally.

Table 2.1 Regional distribution of foreign affiliates of multinational enterprises in 1980 (percentage)

	Region					
Country	*Europe*	*North America*	*Latin America*	*Africa*	*Pacific Rim*	*Total*
Australia	19.8	8.7	1.7	0.7	69.1	100.0
Austria	73.0	6.2	9.8	1.2	9.8	100.0
Belgium	74.7	5.2	6.1	9.7	4.3	100.0
Canada	39.9	34.9	12.6	1.3	11.3	100.0
Denmark	73.5	7.3	4.9	3.5	10.8	100.0
Finland	80.6	11.4	3.2	1.6	3.2	100.0
France	58.1	8.2	7.7	18.7	7.3	100.0
Germany	68.4	9.0	9.2	3.0	10.4	100.0
Hong Kong	16.4	0.0	4.3	19.5	59.8	100.0
Italy	63.7	7.4	15.0	6.7	7.2	100.0
Japan	19.5	17.1	13.4	2.2	47.8	100.0
Luxembourg	88.7	1.4	7.9	0.7	1.3	100.0
Malaysia	29.1	0.0	2.3	2.3	66.3	100.0
Netherlands	71.2	6.3	6.1	5.0	11.4	100.0
New Zealand	11.0	4.2	0.9	0.4	83.5	100.0
Norway	74.6	8.8	3.6	3.0	10.0	100.0
Portugal	72.5	0.0	21.4	6.1	0.0	100.0
Singapore	8.7	1.9	1.2	0.5	87.7	100.0
Spain	62.1	2.4	27.7	4.7	3.1	100.0
Sweden	73.2	9.4	7.1	1.3	9.0	100.0
Switzerland	72.9	8.1	6.9	1.7	10.4	100.0
United Kingdom	35.7	14.1	4.7	7.7	37.8	100.0
United States	42.7	20.7	21.4	2.3	12.9	100.0

Source: Dunning 1993 (modified).

The role of the nation-state

The marxist interpretation of multinational enterprises oversimplifies and underestimates the role of nation-states and regulatory international institutions created after the Second World War (Hanink, 1989; Gaile and Grant, 1989; Dicken, 1992a; 1992b; Glasmeier, et al., 1993). Marxists have always had difficulties conceptualising the nation-state. They have seen multinational enterprises as an especially powerful force behind the gradual erosion of nation-states' sovereignty and their reduced ability to shape industrial and macroeconomic policies. However, it could be argued that financial deregulation and the resulting flows of short-term investment capital in and out of national currencies and securities markets provide a far more powerful influence on nation-states than multinational enterprises (Thrift, 1990). The limits of national sovereignty were revealed powerfully for many member states of the European Community during the collapse of the Exchange Rate Mechanism (ERM) in August 1993 (the ERM is the regulatory regime of currency exchange in the European Community). Moreover, the political sovereignty of nation-states has been significantly eroded as a consequence of other factors including technological advancements in telecommunications, computer networks and transport. These technologies have weakened the control of nation-states over domestic economic and political affairs far more than multinational enterprises (Leyshon, 1992; Grant, 1993).

It could also be argued that the emergence and evolution of the modern multinational enterprise in the world market is the *result* and not the *cause* of the multilateral liberalisation of markets in which the GATT was instrumental. After all, without the initial impulse (thanks mostly to American political power and capital) leading to the opening of the European and Japanese economies, it is highly unlikely that multinational enterprises would have developed into such powerful commercial institutions. Thus, despite the significant level of market penetration achieved by multinational enterprises, the nation-state, as well as the multilateral trading arrangements between nation-states, remains the single most powerful force shaping the international economy. Most multinational enterprises adapt relatively passively to governmental policy rather than continually trying to undermine it. Therefore the principal factor leading to the multilateral liberalisation of trade is the national government and pan-national institutions designed by nation-states after the Second World War. Although multinational enterprises represent a very important force orchestrating the shift in global economic activity, nation-states continue to provide the institutional and legal framework for these enterprises. Multilateralism and regionalism are two, albeit contradictory, outcomes of these policies. Having examined the role of multinational enterprises in the process of multilateralism, the attention now shifts to the role of the nation-state in regulating the international economy.

Regulation of the international economy

The regulatory environment established by nation-states poses an exceptionally complex milieu for multinationals to operate within. This is perhaps the most powerful reason why the international Keynesianism advocated by regulationists is very unlikely. According to Dicken (1992b), there are three types of regulation that can impress a powerful leverage on the operations of multinational enterprises. They are regulation of access to the market, regulation governing foreign investment and regulatory industrial policies. Each type of regulatory policies will now be examined in more detail.

The first type of regulation that governs access to markets and resources constitutes perhaps the most visible and most publicised type of state intervention. *Trade policies* are probably the most powerful and effective weapon in competing for economic primacy among nation-states. The primary function of the GATT is to coordinate such trade policies through an application of the principles of free trade, most favoured nation and multilateralism. However, the difficulties experienced by the GATT in the early 1990s indicate a weakening commitment of the most important participants, especially the United States, to these principles. Many new trade regulations and techniques have been developed by nation-states, such as non-tariff barriers (NTB), voluntary export restraints (VER) and voluntary import expansion (VIE), all of which, in essence, are designed to circumvent the spirit of the GATT (see Chapter 1). Moreover, the Uruguay round of the GATT includes an entirely new agenda covering among other items intellectual property, services and agriculture. The impact of these policies on the operations of multinational enterprises and trade in general has produced an enormous literature in economics, social science and history. A review of this literature is well beyond the scope of this chapter. However, as was noted above, the combination of multilateralism with the regulationist policies of the state created an exceedingly complex environment in which multinational corporations operate. Unfortunately, the critics of these enterprises often ignore and oversimplify the sheer complexity of the issues involved.

The second area of government regulation, *foreign direct investment*, is concerned principally with both the access and operation of foreign capital in a domestic economy. Since, as Dicken (1992a) and Lipietz (1992) aptly note, there is no international regulatory institution compatible with the GATT designed to regulate the rules of foreign investment, most nation-states are free to impose whatever regulations they perceive to be in their best interests (although the World Trade Organisation should address this issue). Since most foreign investment originates from highly industrialised economies, many developing countries have in the past introduced a whole host of deterrents and measures designed to restrict market access to foreign capital. The benefits of such policies, however, are increasingly questioned. It is argued by some economists (de Melo and Panagariya, 1992; Schott, 1991) that foreign investment contributed to the spectacular rise of the newly industrialised countries, particularly in South-East Asia. Conversely, the policies that severely restricted or banned foreign investment did not produce the hoped-for results. Probably the most spectacular example of the negative effects of such policies is that of the socialist economies of eastern Europe before 1989 (Michalak, 1993). As a result of the collapse of socialist policies, foreign investment is now not only allowed but actively encouraged by many east European and developing countries.

Industrial policy, the third type or regulation, aims to govern or manage industrial activities within the borders of a nation-state for, usually, a clearly defined long-term goal such as economic growth, increasing living standards, growth of exports, and so on. This type of policy is the most widespread and common way to regulate national economies. All nation-states, even those declaring to the contrary, have industrial policies that affect domestic economies to a very considerable degree. The specific policy mix tends to vary greatly from one country to another. However, almost all of them tend to oscillate between some form of developmental activism and economic liberalism. The ultimate objective of these policies is to improve or maintain the comparative advantage of a national economy over its competitors and assure a high standard of living and welfare for the majority of inhabitants. Despite some views to the contrary (for example Johnston, 1989), there is little doubt that nation-states continue to play a central role in moulding the economic performance of most countries.

All of these national policies have an enormous influence on the operations and locational behaviour of multinational enterprises. There is an ample evidence (Dunning, 1993; Yoffie, 1990; Yoffie and Milner, 1989) to suggest that without such regulatory differences, there would be no need for multinational enterprises, or indeed there would be no economic rationale for such enterprises. According to neoliberal economic theory, if international trade was genuinely free and there were no restrictions of any kind on the movement of capital and goods from one location to another, the reasons for a multilocational organisation of multinational firms would essentially cease to exist (Winters, 1990; Rugman and Verbeke, 1990). The continuing existence of differing government regulations on trade, foreign investment and industrial policy is the most obvious reason for the persistent growth of multinational companies.

The last section of this chapter focuses on the relationship between social theory and the so-called new trade theory. There is no obvious link between the two interpretations of the international economic environment and the associated functions of regionalism. However, new trade theory deserves attention because it is one of the most tangible results of the largely abstract debate on the relative merits of multilateralism and regionalism. It is a major theoretical attempt in economics to examine the problem of 'externalities' to the theory of interregional trade and regional integration.

MANAGED TRADE AND THE NEW TRADE THEORY

One of the most interesting developments in recent economic theory, *new trade theory*, examines the apparent discrepancy between the principles of free trade and the actuality of varying degrees of state regulation. In a broad sense, this theory attempts to legitimise the belief of many politicians and economists, especially in the United States and the EC, that the multilateral commitment to liberal trade policies is outdated. The advocates of this theory argue that given widespread government regulation, the free trade system is a myth or at best a distant objective. Consequently, trading blocs are the logical result of a set of policies focused at reconciling two contradictory aims: extending the benefits of the international division of labour and flexible specialisation, and protecting domestic and regionally based enterprises from outside competition. In effect, regional integration could be interpreted as a renewed attempt to address the classic question: is free trade merely an ideal type or a guiding principle for a pragmatic and realistic economic policy?

New trade theory responds to these concerns and throughout the 1980s its advocates produced a body of economic literature challenging the traditional arguments for liberal trade. This theory stresses the importance of the principle of increasing returns as a source of trade, as well as emphasising the significance of imperfect competition in international markets. The new trade theory was originally expounded by Dixit and Norman (1980), Lancaster (1980), Ethier (1982), and prominently by Krugman (1980; 1983; 1986; 1990a; 1990b; 1991; 1992), among others. The significance of this literature in the 1990s is that, unlike most other economic theories, it actually has influenced the trade and industrial policies of the American Democratic Administration (Tyson, 1987; 1992; Tyson, et al., 1988).

The advocates of this theory had originally attempted to explain the patterns of international trade. Their core argument asserts that countries do not necessarily specialise in trade solely because of their competitive advantage but also because of increasing returns. Trade is driven to a much greater degree than is commonly recognised by the economies of scale rather than the orthodox Ricardian principles reviewed in the previous chapter. In other words, international specialisation is the result of history, accident and governmental policies as well as national differences in resources and aptitudes. Krugman focuses on two areas: first, the introduction of industrial organisation theory at the expense of the principle of perfect competition; and second, he introduces the concept of external economies into the formal modelling of trade. In fact, possibly the most important contribution of this theory is that it incorporates the idea of external economies as crucial to the development of international specialisation. While economies of scale and imperfect competition are, of course, not new ideas, Krugman (1980) models them formally. His starting point is the contention that there has been a progressive dismantling of many trade barriers since 1945. However, the GATT lacks a theoretical basis and, in fact, is not built upon any solid economic theory at all. The liberal free trade theory, Krugman argues, is a low-level ideology that amounts to little more than 'enlightened mercantilism'. Enlightened mercantilism is a working theory that has nothing to do with the actual beliefs of economists and politicians and does not withstand confrontation with the practice of world trade. This liberal free trade ideology builds upon the following internal antinomy: each country would like to subsidise exports and restrict imports. As such, it captures some basic realities of the political significance of exports and import/competition practices. As Krugman writes: 'In effect, GATT-think sees the trade policy problem as a Prisoners' Dilemma: individually, countries have an incentive to be protectionist, yet collectively they benefit from trade' (Krugman, 1992, 429).

New trade advocates admit, however, that despite these criticisms the GATT Treaty has a great deal of practical relevance. This is because it guides the actual trade policies of the leading industrial states. Paradoxically, this function of the GATT is at the centre of the problem because the actual trade policies are already based on mercantilist principles and a degree of protectionism, *not* the principles of free trade. In other words, the GATT negotiations continue as if free trade principles were the actual principles of international trade, while in fact exactly the opposite is true. Thus the free trade theory does not offer an accurate representation of real world trade policies.

One of the most important assumptions of new trade theory is that most goods and services that are traded internationally are not produced by perfectly competitive industries. Indeed, there are a number of constraints on perfect competition that can, on aggregate, be described as economies of scale. The conventional liberal argument has it that imperfect competition reinforces the need for free trade precisely because trade increases competition and thus offsets some of the negative results of imperfect competition. This has been one of the principles used by Canadian officials to justify the free trade agreement negotiated between Canada and the United States. However, according to the new trade theory, this reasoning is unconvincing. It is not always the best policy to try to remove the negative consequences of imperfect competition by reverting to *laissez-faire*. Government regulation and intervention can achieve better results. A *strategic trade policy* or temporary protection and regulation of certain key sectors and industries can promote exports and, consequently, improve the competitive advantage. The key to this strategic trade policy is the concept of external economies. For example, if one can create through temporary protection an industry that would otherwise not exist, the higher earnings of the factors in that industry would compensate for any short-term distortions. Moreover, resulting agglomeration economies can add significantly to these gains because a new industry can lead to a whole series of spillover effects throughout the economy. In short, it is possible that a country which systematically tries to promote certain industries will raise its standard of living at the expense of other countries.

The new trade theory makes it plain that this is exactly what happens in trade between highly industrialised countries. Trading blocs are therefore a logical extension of the same basic principle. A government or a group of governments unified by a common external threat may make a credible commitment to expand domestic exports and enhance competitive advantage by means of subsidy. In this way a country or trading bloc can succeed in shifting profits from the competitors whilst at the same time raising the welfare of its citizens. Consequently, external economies provide a new rationale for *strategic* trade intervention (also called 'managed trade' or 'fair trade' or 'results-oriented trade', all of which are variations on the same theme) and regional economic integration (Milner, 1988; Pearson and Riedal, 1990; Richardson, 1990; Schott, 1990; Prestowitz, et al., 1991; Yoffie, 1993). Although new trade theory does not require a new trade policy, some form of industrial policy based on the desire to promote externalities, possibly on a regional basis, is inevitable in the view of its supporters.

New trade theory has legitimised government intervention against more radical liberal ideas, particularly in the United States. As Krugman writes:

> The new trade theory is probably the piece of that trend that has received the most attention, but intellectually it goes along with the revival of the ideas that linkages play a key role in development, that increasing returns play a key role in growth, and that co-ordination failures play a key role in business cycle (Krugman, 1992, 438).

Consequently, new trade theorists advocate an industrial policy involving a modest amount of annual subsidies for industries with a strong case for external economies. However, the main support for managed trade (strategic trade policy) and trading blocs is already the existing interventionist measures practised by the EC, Japan, the United States and other industrial competitors (Bhagwati, 1990a; 1990b; 1991; 1992b; 1992c; Bhagwati and Patrick, 1990; Woolcock, 1991; Winters, 1992). Although the GATT is not dead, it seems unlikely that it will regain its one-time vitality. The United States and the EC are bound to look for answers in closer regional integration and trade cooperation. The new regional approach to trade and industrial policy will almost certainly involve a substantial degree of industrial activism. It is hard to imagine that other economic blocs are far behind considering the importance of these markets to the rest of the world.

In a sense, new trade theory, for different reasons, advocates similar solutions to those laid out in the regulationist approach. Regardless of their theoretical underpinnings, both advocate a much more active role for the nation state at the regional level than the liberal policies of multilateralism embodied in the GATT would allow for. Although a clash between those two fundamentally opposing ideologies is not inevitable, there are clear signs that the world stage is set for confrontation.

CONCLUSIONS

In this chapter several theoretical approaches have been reviewed which attempt to interpret the nature and significance of social and economic changes in industrial societies since the early 1970s. These theories are necessarily fragmentary and incomplete since they attempt to conceptualise an ongoing and very contemporary process of transformation. No matter how diverse their philosophical and methodological foundations, however, they all have one thing in common: they acknowledge that some kind of fundamental change in the nature of the contemporary economy and society is under way. The nature of this change could best be described as a transformation in the form of regulation from Fordism to a more flexible system of production, organisation and consumption. One of the most enduring long-term consequences of this change is the emergence of trading blocs.

Unfortunately, despite the obvious importance of progressing continental integration, none of these influential theories took the issue of regionalisation seriously. This is particularly disappointing since the conflict between multilateralism and regionalism is at the very heart of the economic and social transformation in contemporary industrial society. The outcome of the competition between these two forms of economic and political relations between states will have profound and far reaching social consequences. As this chapter has attempted to demonstrate, important questions remain unanswered: What happens to multilateralism in an era of regionalism and continental integration? What choices exist for countries and regions which find themselves outside the principal trading blocs? What types of influence do continental arrangements have on multinational enterprises? Is a 'trade war' between the EC, NAFTA and the Asian bloc inevitable? How will existing industrial relations, productivity, competitive advantage and organisation of enterprises be remodelled by the increasing fragmentation of the world economy?

Instrumental to the transformation from a liberal trading system embodied in the GATT to the world economy fragmented into regional blocs is the reassertion of the role of the nation-state in regulating international economic relations. Contrary to some commentators who insist that nation-states are becoming increasingly irrelevant, we suggest that the political and economic interests of nation-states allied into regional blocs are the principal factors in reconstructing the international economic environment. The state has a strong and historical role in structuring trade activity. The formation of economic alliances between nation-states and more militant forms of 'unilateral regionalism' are the most vivid examples of this powerful leverage. As this chapter makes clear, the emergence of trading blocs cannot be understood in liberal economic terms alone. States with distinct histories, cultures and political heritages, or what liberal economists would describe as 'external economies', pursue surprisingly similar protectionist trade policies. An explanation of these striking similarities in forms of regionalism demands reference to more general theory of the social and economic transformation of the industrial society.

Perhaps the most important conclusion of this chapter is that multilateralism and regionalism are associated with different modes of regulation. A powerful combination of multilateralism regulated through the GATT with Fordism forged a solid foundation on which the long post-war economic boom was built. Fundamental to this dual regulation was full employment achieved through a series of social compromises between labour, capital and the state. Multilateralism and Fordism are therefore intrinsically linked and the crisis of Fordism created conditions for the crisis of the GATT and multilateralism. Moreover, the principal economic players of the post-war world economy, the United States and western Europe, became increasingly vulnerable to competition from Japan and the newly industrialised countries of South-East Asia. The collapse of the socialist experiment in eastern Europe and the Soviet Union removed the rationale for the close economic cooperation between those powers whose industrial and trade policies until then were subordinated to broader political and military considerations.

The change in the mode of regulation from Fordism to flexible specialisation created a new international climate whereby increased calls for the protection of domestic enterprises evolved into new policies introducing managed trade. In other words, the new mode of economic regulation coincides with the new mode of international economic regulation, i.e. continental integration into trading blocs. Thus trading blocs are the spatial expression of a new mode of international regulation between the principal economic players in the world economy. Of course, the international economy continues to grow, however it is also increasingly moulded by the regulationist attempts of major economic powers eager to strengthen, or at least to maintain, their privileged position. The resulting conflicts and tensions over the costs and benefits of international economic relations will shift the momentum of integration from a global to a continental or even national scale.

Regionalisation as a mode of international regulation is not identical with post-Fordism or flexible accumulation. The central rationale for continental integration is regulation and coordination of two conflicting tendencies in the industrial system of production and consumption. The first tendency is the pursuit of greater flexibility in the deployment of labour, machinery and capital. The second tendency is protection of domestic markets and enterprises from outside competition. Continental integration is an attempt at institutional regulation on the international scale in response to these contradictory requirements of capital. Of course, a search for a greater flexibility in deployment of the factors of production or protectionist sentiments alone cannot explain the strength of the recent surge of regionalism. These economic trends coincided, however, with the end of the Cold War and the consequent decline of political and economic relations between the United States and the European Community, as well as the spectacular growth of Asian economies. These historical circumstances, combined with the change in the mode of regulation provided the foundation for continental integration. Thus regionalism is not the result of some 'universal' laws of capitalist development or an 'inevitable' historical transformation, but rather an outcome of a largely unpredictable coincidence of historical, economic and social circumstances.

The emphasis on continental integration in this chapter, and competition between the 'Triad' of NAFTA, the EC and the Asian bloc, should not be interpreted as the only consequence of the decline of multilateralism. None of the trading blocs is a monolithic entity. In strictly economic terms, none of the 'free trade areas' examined in the following chapters is truly free and without any internal barriers and impediments to trade. On the contrary, examples of acrimonious EC and NAFTA negotiations demonstrate that neither bloc is free from internal tensions and contradictions. In fact, competition within each bloc will most certainly intensify as the social and political effects of these agreements produce new tensions and conflicts (Wise and Gibb, 1993). The existing configuration of spatial alliances between individual countries within each bloc should not be taken for granted. Continental integration and regionalisation are processes, not a static state of affairs. The increasing pressure on countries left outside these regional clubs will intensify competition for early and full membership.

Regionalism in the world economy is a powerful process fundamentally transforming the relationship between the world's principal economic players. Although trading blocs are not new, the 1980s and 1990s witnessed a gradual fragmentation of the multilateral world economy into increasingly protectionist economic blocs. Consequently the world economy is not becoming more global in the sense outlined earlier in this chapter. Instead, it is becoming more regional, or rather interregional, as regional trading blocs continue to gain ground. Multinational enterprises, far from becoming footloose economic giants answerable to nobody but themselves, become allied with governments eager to protect their political power projected through a network of regional subsidiaries.

It is unlikely that an open and all-out trade war will develop between trading blocs, although such a possibility cannot be ruled out entirely considering the history of the twentieth century. A far more plausible scenario is the slow but persistent erosion of confidence in the practice of multilateralism translated into 'fair' trade policies regulated and managed by the state. The test of 'fairness' will no longer be the aggregate increase in the collective economic well being or volume of world trade as defined by liberal ideologues, but instead a much more narrowly defined 'trade balance' or 'market share'. Thus the emerging new mode of international regulation will determine the nature of international economic relations for years to come, and well into the next century. In the following chapters of this book, the practice, responses and policies of several trading blocs will be examined in much greater depth than was possible in this introductory section.

References

Aglietta, M. (1979) *A Theory of Capitalist Regulation*, New Left Books, London.

Amin, A. (1989) Flexible specialisation and small firms in Italy: myths and realities, *Antipode*, 21, 13–34.

Amin, A. and Robins, K. (1990) The re-emergence of regional economies? The mythical geography of flexible accumulation, *Environment and Planning D: Society and Space*, 8, 7–34.

Anderson, K. (1991) Europe 1992 and the western Pacific economies, *The Economic Journal*, 101, 1538–1552.

Balassa, B. (1991) *Economic Integration in Eastern Europe*. Office of the Vice-President, The World Bank, Washington, D.C.
Baldwin, R.E. (1992) Assessing the fair trade and safeguards laws in terms of modern trade and political economy analysis, *The World Economy*, 15, 185–202.
Bhagwati, J.N. (1988) *Protectionism, The 1987 Ohlin Lectures*, MIT Press, Cambridge, Mass.
Bhagwati, J.N. (1990a) *Multilateralism at Risk: The Seventh Harry G. Johnson Lecture*, Princeton University Press, Princeton, NJ.
Bhagwati, J.N. (1990b) Departures from multilateralism: regionalism and aggressive multilateralism, *The Economic Journal*, 100, 1304–1317.
Bhagwati, J.N. (1991) *The World Trading System at Risk*, Harvester-Wheatsheaf, Hemel Hempstead.
Bhagwati, J.N. (1992b), The threats to the world trading system, *The World Economy*, 15, 443–456.
Bhagwati, J.N. (1992c) Regionalism versus multilateralism, *The World Economy*, 15, 535–555.
Bhagwati, J.N. and Patrick, H.T. (eds.) (1990) *Aggressive Unilateralism: America's 301 Trade Policy and the World Trading System*, Harvester-Wheatsheaf, Worcester.
Boyer, R. (1990a) *The Theory of Regulation: A Critical Analysis*, Columbia University Press, New York.
Boyer, R. (1990b) *The Regulation School: A Critical Introduction*, Columbia University Press, New York.
Brusco, S. (1982) The Emilian model: productive decentralization and social integration, *Cambridge Journal of Economics*, 6, 167–184.
Bulgakov, S. (1982) *Filosofiia Khoziaistva* (first published 1912), Chalidze Publications, New York.
Coriat, B. (1990) *L'atelier et le robot*, Bourgois, Paris.
Coriat, B. (1991) *Penser à l'envers*, Bourgois, Paris.
de Melo, J. and Panagariya, A. (1992) *The New Regionalism in Trade Policy*, The World Bank, Washington, D.C.
Dicken, P. (1992a) *Global Shift: The Internationalization of Economic Activity* (2nd Edition), Chapman, London.
Dicken, P. (1992b) International production in a volatile regulatory environment: the influence of national regulatory policies on the spatial strategies of transnational corporations, *Geoforum*, 23, 303–316.
Dixit, A. and Norman, V. (1980) *Theory of International Trade*, Cambridge University Press, Cambridge.
Drache, D. and Gertler, M.S. (eds.) (1991) *The New Era of Global Competition: State Policy and Market Power*, McGill-Queen's University Press, Montreal.
Dunning, J.H. (1993) *Multinational Enterprises and the Global Economy*, Addison-Wesley, Wokingham.
The Economist (1993a) Multinationals: back in fashion, *The Economist*, March 27, 1–28.
Ethier, W. (1982) National and international returns to scale in the modern theory of international scale, *American Economic Review*, 72, 389–405.
Gaile, G.L. and Grant, R. (1989) Trade, power, and location: the spatial dynamics of the relationship between exchange and political economic strength, *Economic Geography*, 65, 329–337.
Gertler, M.S. (1988) The limits to flexibility: comments on the post-fordist vision of production and its geography, *Transactions of the Institute of British Geographers*, 13, 419–432.
Gertler, M.S. (1989) Resurrecting flexibility? A reply to Schoenberger, *Transactions of*

the Institute of British Geographers, 14, 109–112.
Gertler, M.S. (1992) Flexibility revisited: districts, nation-states, and the forces of production. *Transactions of the Institute of British Geographers*, 17, 259–278.
Gertler, M.S. and Schoenberger, E. (1992) Commentary. Industrial Restructuring and continental trading blocs: the European Community and North America, *Environment and Planning A*, 24, 2–10.
Glasmeier, A., Thompson, J.W. and Kays, A.J. (1993) The geography of trade policy: trade regimes and location decisions in the textile and apparel complex, *Transactions of the Institute of British Geographers*, 18, 19–35.
Globerman, S. (eds) (1991) *Continental Accord: North American Economic Intergration*, Fraser Institute, Vancouver.
Grant, R. (1993) Trading blocs or trading blows? The macroeconomic geography of US and Japanese trade policies, *Environment and Planning A*, 25, 273–291.
Greenaway, D., Hyclak, T. and Thornton, R.J. (eds.) (1989) *Economic Aspects of Regional Trading Arrangements*, New York University Press, New York.
Hanink, D.M., (1989) Introduction: trade theories, scale, and structure, *Economic Geography*, 65, 267–270.
Harvey, D. (1989) *The Condition of Postmodernity: An Enquiry into the Origins of Cultural Change*, Blackwell, Oxford.
Harvey, D. and Scott, A. (1988) The practice of human geography: theory and empirical specificity in the transition from Fordism to flexible accumulation, in: W.D. MacMillan (ed.) *Remodelling Geography*, Blackwell, Oxford, 217–229.
Hirst, P. and Thompson, G. (1992) The problem of 'globalization': international economic relations, national economic management and the formation of trading blocs, *Economy and Society*, 21, 357–396.
Hirst, P. and Zeitlin, J. (1991) Flexible specialization versus post-Fordism: theory, evidence and policy implications, *Economy and Society*, 20, 1–56.
Jacques, M. and Hall, S. (eds.) (1989) *New Times*, Lawrence & Wishart, London.
Jessop, R. (1990) Regulation theories in retrospect and prospect, *Economy and Society*, 19, 153–216.
Jessop, R. Kastendick, H. Nielsen, K. and Pedersen, O.K. (eds.) (1991) *The Politics of Flexibility: Restructuring State and Industry in Britain, Germany and Scandinavia*, Elgar, Aldershot.
Johnston, R.J. (1989) Extending the research agenda, *Economic Geography*, 65, 338–347.
Kalecki, M. (1990) *Collected Works. Vol. 1 – Capitalism: Business Cycle and Full Employment*, Clarendon, Oxford.
Kalecki, M. (1991) *Collected Works. Vol. 2 – Capitalism: Economic Dynamics*, Clarendon, Oxford.
Keenwood, A.G. and Loughead, A.L. (1992) *The Growth of the International Economy 1820–1990*, Routledge, London.
Kowalik, T. (1971) *Róża Luxemburg, Teoria akumulacji i imperializmu* (Rosa Luxemburg, Theory of Accumulation and Imperialism), Państwowy Instytut Wydawniczy, Warsaw.
Krueger, A.O. (1992) Global trade prospects for the developing countries, *The World Economy*, 15, 457–474.
Krugman, P. (1980) Scale economies, product differentiation and the pattern of trade, *The American Economic Review*, 70, 950–959.
Krugman, P. (1983) New theories of trade among industrial economies, *The American Economic Review*, 73, 343–347.
Krugman, P. (1986) *Strategic Policy and the New International Economics*, Cambridge

University Press, Boston.
Krugman, P. (1990a) *Increasing Returns and Economic Geography*, National Bureau of Economic Research, Working Paper series No. 3245, Cambridge, Mass.
Krugman, P. (1990b) *Rethinking International Trade*, MIT Press, Cambridge, Mass.
Krugman, P. (1991) *Geography and Trade*, MIT Press, Cambridge, Mass.
Krugman, P. (1992) Does the new trade theory require a new trade policy?, *The World Economy*, 15, 423–441.
Lancaster, K. (1980) Intra-industry trade under perfect monopolistic competition, *Journal of International Economics*, 10, 151–175.
Lash, S. and Urry, J. (1987) *The End of Organized Capitalism*, Polity, Cambridge.
Leborgne, D. and Lipietz, A. (1988) New technologies, new modes of regulation: some spatial implications, *Environment and Planning D: Society and Space*, 6, 263–180.
Leborgne, D. and Lipietz, A. (1989) *Pour éviter L'Europe à deux vitesse*, CEPREMAP, Paris.
Leyshon, A. (1992) The transformation of regulatory order: regulating the global economy and environment, *Geoforum*, 23, 249–267.
Lipietz, A. (1977) *Le capital et son espace*, Maspero, Paris.
Lipietz, A. (1986) New tendencies in the international division of labour: regimes of accumulation and modes of regulation, in: A.J. Scott and M. Storper (eds.) *Production, Work, Territory: The Geographical Anatomy of Industrial Capitalism*, Unwin & Hyman, Boston, 16–40.
Lipietz, A. (1987) *Mirages and Miracles: The Crisis of Global Fordism*, Verso, London.
Lipietz, A. (1992) *Towards a New Economic Order: Postfordism, Ecology and Democracy*, Oxford University Press, Oxford.
Lipietz, A. (1993) The local and the global: regional individuality or regionalism?, *Transactions of the Institute of British Geographers*, 18, 8–18.
Luxemburg, R. (1951) *The Accumulation of Capital* (first published 1913), Routledge & Kegan, London.
Machlup, F. (1977) *A History of Thought on Economic Integration*, Columbia University Press, New York.
Marx, K. (1967) *Capital: A Critique of Political Economy. Vol. 2: The Process of Circulation of Capital* (first published 1893), Progress Publishers, Moscow.
Michalak, W.Z. (1993) Foreign direct investment and joint ventures in East-Central Europe: a geographical perspective, *Environment and Planning A*, 25, 1573–1591.
Michalak, W.Z. and Gibb, R.A. (1992a) Political geography and eastern Europe, *Area*, 24, 341–349.
Michalak, W.Z. and Gibb, R.A. (1993) Eastern Europe and the World System: an anti-systemic reply to Taylor and Johnston, *Area*, 25, 305–309.
Milner, H. (1988) *Resisting Protectionism: Global Industries and the Politics of International Trade*, Princeton University Press, Princeton, NJ.
Moulaert, F. and Swyngedouw, E.A. (1989) Survey 15: A regulation approach to the geography of flexible production systems, *Environment and Planning D: Society and Space*, 7, 327–345.
Murphy, A.B. (1992) Western investment in East-Central Europe: emerging patterns and implications for state stability, *Professional Geographer*, 44, 249–259.
Offe, K. (1985) *Disorganized Capitalism*, MIT Press, Cambridge, Mass.
Ostry, S. (1992) The domestic domain: the new international policy area, *Transnational Corporations*, 1, 7–26.
Pearson, C. and Riedal, J. (eds.) (1990) *The Direction of Trade Policy*, Blackwell, Oxford.
Perlmutter, H. (1969) The tortuous evolution of the multinational enterprise, *Columbia*

Journal of World Business, 4, 9–18.
Petit, P. (1984) *Slow Growth and the Service Economy*, Pinter, London.
Piore, M.J. and Sabel, C. (1984) *The Second Industrial Divide*, Basic Books, New York.
Pipes, R. (1970) *Struve, Liberal on the Left, 1870–1905*, Harvard University Press, Cambridge, Mass.
Pipes, R. (1980) *Struve, Liberal on the Right, 1905–1944*, Harvard University Press, Cambridge, Mass.
Pollert, A. (1988) Dismantling flexibility, *Capital and Class*, 34, 42–75.
Pollert, A. (ed.) (1991) *Farewell to Flexibility?*, Blackwell, Oxford.
Prestowitz, C.V., Tonelson, A. and Jerome, R. (1991) The last gasp of GATTism, *Harvard Business Review*, March–April, 130–138.
Richardson, J.D. (1990) The political economy of strategic trade policy, *International Organisation*, 44, 107–135.
Rugman, A.M. and Verbeke, A. (1990) *Global Corporate Strategy and Trade Policy*, Routledge, London.
Sayer, A. (1989) Postfordism in question, *International Journal of Urban and Regional Research*, 13, 666–695.
Sayer, A. and Walker, R. (1992) *The New Social Economy: Reworking the Division of Labour*, Blackwell, Oxford.
Schoenberger, E. (1988) Multinational corporations and the new international division of labour: a critical appraisal, *International Regional Science Review*, 11, 105–119.
Schott, J.J. (ed.) (1990) Completing the Uruguay Round: A Results-Orientated Approach to the GATT Trade Negotiations, *Institute for International Economics*, Washington, D.C.
Schott, J.J. (1991) Trading blocs and the world trading system, *The World Economy*, 14, 1–17.
Scott, A.J. (1988) *New Industrial Spaces*, Pion, London.
Scott, A.J. and Storper, M. (eds.) (1986) *Production, Work, Territory: The Geographical Anatomy of Industrial Capitalism*, Unwin & Hyman, Boston.
Storper, M. (1992) The limits to globalization: technology districts and international trade, *Economic Geography*, 68, 60–93.
Storper, M. and Christopherson, S. (1987) Flexible specialization and regional industrial agglomeration: the case of the US motion picture industry, *Annals of the Association of American Geographers*, 77, 104–117.
Storper, M. and Scott, A.J. (1989) The geographical foundations and social regulation of flexible production complexes, in: J. Wolch and M. Dear (eds.) *The Power of Geography: How Territory Shapes Social Life*, Unwin & Hyman, Boston, 21–40.
Storper, M. and Walker, R. (1989) *The Capitalist Imperative: Territory, Technology, and Industrial Growth*, Blackwell, Oxford.
Taylor, F.W. (1947) *The Principles of Scientific Management* (first published 1911), Greenwood Press, Westport, Conn.
Taylor, M.J. and Thrift, N.J. (eds.) (1986) *Multinationals and the Restructuring of the World Economy*, Croom Helm, Beckenham.
Thrift, N.J. (1990) The perils of the international financial system, *Environment and Planning A*, 22, 1135–1136.
Thrift, N.J. and Leyshon, A. (1988) 'The gambling propensity': banks, developing country debt exposures and the new international financial system, *Geoforum*, 19, 55–69.
Tickell, A. and Peck, J.A. (1992) Accumulation, regulation and the geographies of post-Fordism: missing links in research, *Progress in Human Geography*, 16, 190–218.
Tomlinson, J. (1988) Can governments manage the economy?, *Fabian Tracts*, January, 524.

Tugan-Baranovsky, M.I. (1913a) *Les crises industrielles en Angleterre*, Giard et Briere, Paris.

Tugan-Baranovsky, M.I. (1913b) *Sociale Theorie der Verteilung*, Scholz, Berlin.

Tyson, L. (1987) *Creating Advantage: Strategic Policy for National Competitiveness*, BRIE Working Paper, Berkeley.

Tyson, L. (1992) *Who's Bashing Whom? Trade Conflicts in High-technology Industries*, Institute for International Economics, Washington, D.C.

Tyson, L., Dickens, W.T. and Zysman, J. (1988) *The Dynamics of Trade and Employment*, Ballinger, Cambridge, Mass.

United Nations (1990a) *Regional Trading Blocs: A Threat to the Multilateral Trading System: Views and Recommendations of the Committee for Development Planning*, Department of International Economic and Social Affairs, United Nations, New York.

Vernon, R. (1971) *Sovereignty at Bay: The Multinational Spread of US Enterprises*, Basic Books, New York.

Vernon, R. (1977) *Storm Over the Multinationals: The Real Issues*, Harvard University Press, Cambridge, Mass.

Vernon, R. (1985) *Exploring the Global Economy: Emerging Issues in Trade and Investment*, Center for International Affairs, Harvard University.

Webber, M.J. (1991) The contemporary transition, *Environment and Planning D: Society and Space*, 9, 165–182.

Winters, L.A. (1990) The road to Uruguay, *The Economic Journal*, 100, 1288–1393.

Winters, L.A. (1992) Goals and own goals in European trade policy, *The World Economy*, 15, 557–574.

Woolcock, S. (1991) *Market Access Issues in EC-US Relations: Trading Partners or Trading Blows?*, Royal Institute of International Affairs, Pinter, London.

Yamazawa, I. (1992) On Pacific economic integration, *The Economic Journal*, 102, 1519–1529.

Yoffie, D.B. (1983) *Power and Protectionism: Strategies of Newly Industrialized Countries*, Columbia University Press, New York.

Yoffie, D.B. (1990) *International Trade and Competition: Cases and Notes in Strategy and Management*, McGraw-Hill, New York.

Yoffie, D.B. and Milner, H.V. (1989) An alternative to free trade or protectionism: why corporations seek strategic trade policy, *California Management Review*, 31, 111–131.

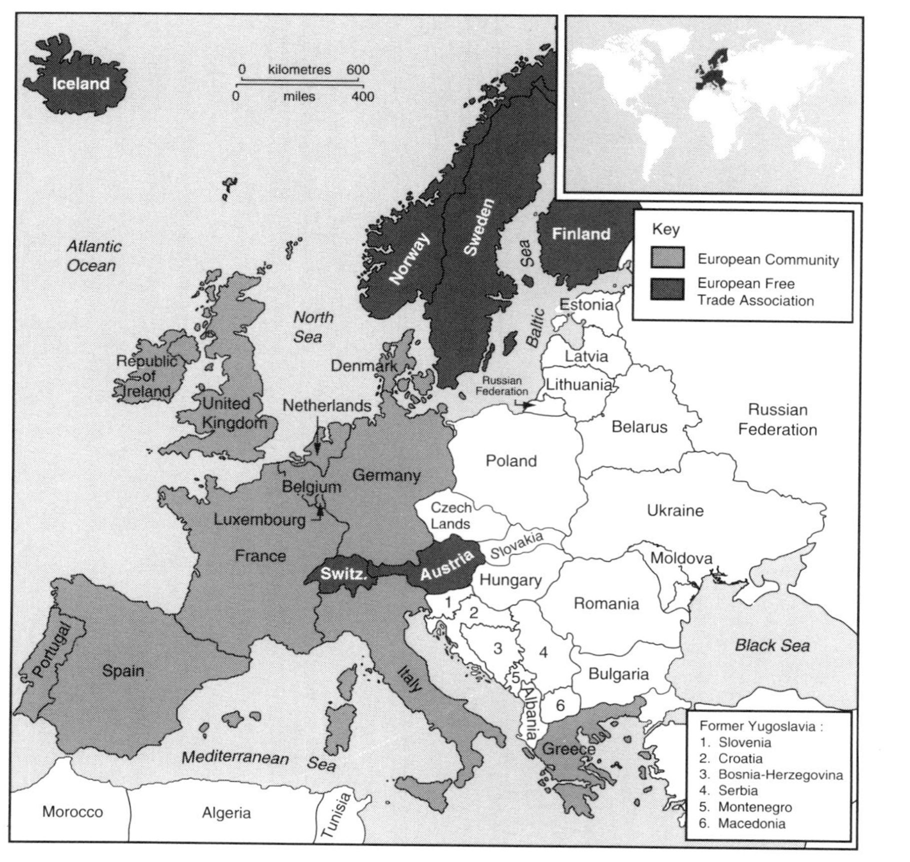

Figure 3.1 Member states of the European Community and the European Free Trade Association, 1993

3 The European Community

Mark Wise
University of Plymouth

INTRODUCTION

The European Community (Figure 3.1) is sometimes seen as a role model for other regional international communities. However, superficial descriptions of the Community as a 'common market' have disguised the fact that it has always harboured political ambitions extending far beyond the free trading arrangements sought by other multistate regional economic organisations. A central aim of this chapter is to describe these unique economic *and* political characteristics of the European Community (EC) in order to prevent excessively simplistic comparisons with international regionalism elsewhere. The chapter also seeks to show that, as a product of diverse and often contradictory motivations, the movement to unity in western Europe has been moulded in a persistent tension between those who want to construct a distinct political–economic regional entity along essentially federal lines and others who desire little more than a loose free trade association.

This chapter also aims to clarify the notion that geopolitical considerations – such as the maintenance of peace in Europe, the containment of Germany, the ability to counterbalance superpowers like the USA, the former USSR and, more recently, Japan – have made *political* rather than *economic* unity the overriding goal for many western European leaders. Economic integration, although important in itself, has often been seen as a means to more fundamental political ends. However, although an important objective of this chapter is to demonstrate the uniqueness of the EC, there is also an attempt to identify issues of European integration which have more general implications for other groups of countries trying to come together into some form of regional economic community. In particular, the difficulty of integrating at a simple 'free trade' level without taking *political* action in wider economic, social and environmental fields is revealed. In order to achieve these aims, clarify the debate and make sense of a complex, multi faceted Community, some of the major conceptual strands running through the ongoing debate about 'European integration' are first identified.

Continental Trading Blocs: The Growth of Regionalism in the World Economy
Edited by R. Gibb and W. Michalak

FEDERALISTS, FUNCTIONALISTS, NATIONALISTS AND 'SPILLOVER THEORY'

Federalism is a major concept shaping the process of European integration in that it seeks to create cooperative unity amongst diverse human groups by building common political institutions within which they can make common policies and laws. It does not seek to eliminate this diversity nor all the distinctive political institutions associated with it; most federalists would view such homogenising aims as both unrealistic and undesirable. Confederalists set similar aims, but stress the desirability of paramount powers resting with the constituent member states, while federalists accept more readily that sovereignty in certain key policy areas such as defence, foreign and monetary control should reside with the central institutions. In a 'real world', careless about precise semantic definitions, European 'federalists' incorporate those who want substantial powers for central institutions and those who favour very loose decentralised confederal arrangements which differ little from the view of those who can be described as 'functionalists'.

Functionalists also seek to build an international system fostering cooperation rather conflict (Mitrany, 1933; 1966; 1975). But they mistrust the idea of large multistate federations, fearing a replication of national rivalries on a larger 'regional' scale and a widening of the gap between government and governed in centralised mass societies. Therefore they advocate the creation of numerous international agencies each 'functioning' to deal with a specific problem; for example, the international management of the Rhine's polluted and much-used waters or the regulation of international trade in some region of the world. Thus, instead of highly visible international communities based on federal constitutions applied across a clearly defined geographical territory, a piecemeal patchwork of overlapping agencies of varying spatial extent would emerge adjusted to the needs of the particular problem at hand. States would thus develop habits of constructive international cooperation without feeling that their national sovereignty was being undermined by some federal 'superstate'.

Neofunctionalists believe that functionalist agencies will tend to take on an increasing number of roles through a process of 'spillover' (Haas, 1968; Lindberg, 1963; Lindberg and Scheingold, 1970). Thus functional bodies set up to promote free trade would create pressures to integrate other sectors such as industrial policy (should not national subsidy levels be harmonised in order to avoid a distortion of free trade?); regional policy (should not poorer regions be helped to compete on equal terms with richer areas in an open international market?); social policy (should there not be coordination to prevent companies in countries with low welfare costs unfairly undercutting those in states with generous and costly social provisions?); environmental policy (should there not be common measures to ensure that industries paying the costs of pollution control in certain countries are not undermined by competition from those free of such constraints?); and monetary policy (can 'free' trade be 'fair' trade if national currencies can be devalued at will to produce a competitive advantage?). According to neofunctionalists, such questions inevitably lead to demands for additional common policies to complement the original single-sector initiative. Thus action in one policy domain will 'spillover' into others in a cumulative fashion, drawing states together almost imperceptibly into an increasingly integrated regional entity with economic fusion leading to more overtly political union as well.

Nationalism also remains a formidable force which no discussion of European regional integration can ignore. Some nationalists remain inward looking, ever ready to reinforce the barriers around their nation-state. However, others manifest a nationalism which often merges into federal or functional concepts when the pursuit of European integration is justified as a defence of national interest. This paradox of the 'nationalist-European' (Wise and Gibb, 1993, 36) stems from the belief that, within the larger protective framework of European unity, the relatively small nation-states of the continent can survive and influence world events; outside they would be economically, politically and culturally overwhelmed by the world's superpowers.

FREE TRADE AND COMMON MARKET PROTECTIONISM

The overriding *economic* debate running through efforts to integrate western Europe has revolved around the question of whether the EC should be an open, free trading organisation, striving to break down barriers worldwide, or a more protectionist regional bloc based on a common internal market ready to restrict imports from the rest of the world. Free-traders, particularly strong in Britain, believe that the greatest welfare benefits are generated by eliminating barriers to trade over the widest possible geographical area (see Chapter 1). However, most free-traders, while rejecting *national economic protectionism*, defend the idea of *national political sovereignty*. Reluctant to concede that international free trade involves some surrender of sovereign state rights, they portray themselves as *economic pragmatists* in opposition to *political ideologues* given to federalist fantasies. Free-traders may be content to pursue their goals within a limited regional context of a continent, but often favour worldwide liberalisation of trade. In contrast, supporters of a common European market are more ready to accept a degree of continental regional protectionism. Whilst admitting that the small domestic markets of Europe's nation-states are inadequate for modern economies of scale, they fear a global 'free-for-all' in trade where cheap imports from low-wage countries might undermine key economic sectors, increase unemployment and ferment social unrest.

SUPRANATIONAL, INTERNATIONAL AND INTERGOVERNMENTAL INSTITUTIONS

Erroneous comparisons of the EC with other multistate regions often stem from a failure to define terms. Most forms of regional integration can best be described as 'international' in character, but the distinctive concept of 'supranationality' is essential to understanding the EC. A multitude of agreements between states can be described as 'international', not least multilateral trade agreements like the GATT. Within them, signatory states retain their political sovereignty and recognise no higher authority. In contrast, supranationalism requires the creation of additional institutions which are literally 'above' those of individual member states. Members effectively pool part of their political sovereignty in a distinctive political body which has the power, in specified domains, to make common policies and laws for all member states.

The permanent institutions of the EC are intentionally supranational in character. At their heart lie the European Commission and the Council of Ministers. The Commission, led by powerful political figures appointed by EC Member States, has the task of proposing European policies and implementing them once they have been decided. However, the final power of decision still resides in a Council of Ministers made up of ministers from Member State governments. Decisions are made by qualified majority vote or unanimity, depending on the question under consideration. These decisions are translated into Community policy and law which become applicable in one form or another throughout the entire territory of the EC. Disputes over EC law are ultimately dealt with in the Court of Justice whose decisions take precedence over national jurisdiction. The European Parliament is not a legislative body in the manner of Member State parliaments, but its powers to influence the making of EC policy have been steadily increasing (Nugent, 1989; Lintner and Mazey, 1991). However, EC supranationalism remains essentially confederal in that it maintains an extremely strong role for the individual Member States which still wield a virtual 'veto' on matters considered to be of essential national importance. This continuing strength of nation-state power has led some to prefer the term 'intergovernmental' when describing the EC's decision-making processes (Wallace et al., 1983).

With the aid of these clarifying concepts, this chapter now outlines how they have intertwined through time to produce contemporary patterns of western European international regionalism. Then the more recent developments highlighted by the drive towards a Single European Market (SEM) by the end of 1992 and the Treaty on European Union (European Commission, 1992) agreed at Maastricht are analysed. This will reveal the continuing tensions existing between different visions of the form European unity should take, as well as highlighting the problems of promoting international economic cooperation (for example, free trade areas) without taking account of wider political and social dimensions (the 'spillover' factor). Finally, the concluding section will summarise the fundamental problems still faced by those seeking to unite 'Europe' and the relevance of the EC experience for other would-be international regional organisations.

CONFLICTING IDEAS

Calls to unite the states of Europe recur across the centuries, usually in response to the devastating wars which have scarred the continent (Wise, 1991). Thus in 1923 the Pan-European Union was founded following the First World War, winning the support of various European leaders, most notably in a France preoccupied with European security (Urwin, 1991). Indeed the French Foreign Minister, Aristide Briand, proposed that a 'confederal' bond be forged among the states of Europe based upon the economic foundations of a 'common market'. But his initiative, made in 1929, was swept away in the wake of the Great Depression and the rise of aggressive nationalism. National economic protectionism flourished, international free trade withered and Europe moved once again towards becoming the epicentre of a second World War. Although Europe emerged from this conflict divided into two ideological camps, dominated respectively by the United States to the west and the Soviet Union to the east, the 'European idea' again survived to fuel a drive towards the integration of its western part.

However, this post-war movement towards greater unity in western Europe has been driven by a complex mixture of often incompatible motives. The European Union of Federalists continued to promote an ideal designed to surpass nationalism, while more nationalistic politicians advocated looser forms of European cooperation as a means of renovating the nation-states to which they remained primarily attached. A geopolitical concern to contain Germany also provided a stimulus for integration, not only in western Europe but also in the USA. Although these differing interest groups came together at the 1948 Congress of Europe, nothing of importance emerged. Their deliberations led only to the Council of Europe, a weak body with negligible powers. In effect, it was a compromise between those with differing visions of what form European unity should ultimately take. As such it reflected a tension between competing goals which continues up to the present.

The ECSC, EEC and EURATOM

Realising that a federal Europe could not be created in one mighty political action, people like Jean Monnet pursued 'federalist' ends by more gradual 'functional' means. They identified specific sectors where states could perceive a national interest in working together. Thus European regional integration would progress step by incremental step fuelled by both federalist ideals and the pursuit of national interest. In May 1950 the French Foreign Minister, Robert

Schuman, put this approach into practice by proposing a European Coal and Steel Community (ECSC) where national output of these products would be placed under the control of common European institutions (European Commission, 1987). Behind its apparently limited economic objectives, this 'Schuman Plan' was highly political in character. The supranational institutions designed to run it were protofederal in character with Schuman clearly stating that the initiative was 'the first concrete foundation of a European federation which is indispensable to the preservation of peace' (Schuman, 1950). These federalist aspirations were ingeniously linked to the national interests underpinning the ECSC. The French wanted to ensure that a vital element of Germany's economic recovery (now desired by an American-dominated 'West' concerned to contain communism) took place within the controlling context of a common European body. For the Germans, the ECSC offered a way back into the international community as an equal partner. Similar mixtures of national interest, federal idealism and geopolitical realism led Italy and the Benelux countries to join the original 'Six' ECSC countries.

The British, however, refused to join, still seeing themselves as an undefeated world power at the head of a politico–economic system based on Empire, Commonwealth and a 'special relationship' with the USA. Such factors also explain Britain's refusal to participate in subsequent steps towards greater European integration, namely the European Atomic Energy Community (EURATOM) and the European Economic Community (EEC) established by the 1957 Rome Treaties (European Commission, 1987). Both bodies reflected the familiar mix of (con)federalism, functionalism and national interest, with a constant geopolitical backdrop of the strategic aims of building a western European bulwark against the Soviet Union, counterbalancing the USA and binding West Germany into European institutions. However, in requiring a common market, the EEC went beyond a simple sectoral approach. Here the process of 'spillover' was being primed, as the free movement of goods, people and capital was bound to trigger further common policy demands.

Britain's attempt to promote the 'free trade' alternative

Having refused to join these new European Communities, Britain promoted an alternative model of a western European regionalism loosely linked by free trading arrangements for industrial products. This would suit British interests by opening up a larger European market for its industrial goods while, at the same time, permitting the continuation of preferential trading arrangements with the Commonwealth, especially with regard to the importation of cheap food. But the UK attracted only the smaller, more peripheral European states into the European Free Trade Association (EFTA) which began operation in May 1960 (Figure 3.2). These states shared the British dislike of the EC's supranational institutions and wider political ambitions. They also did not want to be confined

Figure 3.2 Member states of the European Free Trade Association, 1960–1973

in a common market structure with its common external tariff, common foreign trade policy and other common policies decided within common European institutions.

But soon after EFTA came into force, Britain applied for EC membership. A number of factors explain this volte-face, including Britain's loss of imperial great power status, the need to find a new means of influencing global affairs and the hope that greater competition within the Common Market would arrest the UK's relative economic decline. When France blocked UK entry into the EEC the reality of Britain's diminishment was made even starker. Furthermore, the French President, de Gaulle, justified his stance in terms which stressed the differing visions of European regional integration which still hold sway on opposite sides of the Channel. After describing the global free trading outlook of the British and their desire to maintain preferential links with the 'anglo-saxon' world, he argued that UK membership would lead to a much enlarged and diverse organisation which would ultimately merge into 'a colossal Atlantic Community under American domination'. In contrast, France wanted to build a more 'strictly European construction' (Urwin, 1991)

Nevertheless, Britain eventually entered the EC in 1973 along with Denmark and Ireland, after negotiations revealing that the battle between those favouring an open free trade association and those wanting a more comprehensive form of regional integration had not been resolved. For example, Britain fought hard to continue importing agricultural foodstuffs from places like New Zealand and the West Indies, while the French insisted that Community farm production should be protected from third-country imports. Such different concepts of integration have continued to cause tensions. 'Maximalists' press for the strengthening of EC institutions along with an extension of EC competence into all areas of policy. In general, this group is also more ready to adopt measures to protect European industry for a variety of strategic and social reasons. In contrast, 'minimalists' cling to national sovereignty, defend intergovernmentalism and fight to restrict common action to the minimum necessary to facilitate free trade in a 'common market'.

THE EC AND WORLD TRADE

The world's largest trading bloc

A 'maximalist' view of the EC has prevailed in the way it organises trade with non-Community countries. Here, the EC acts as a single body operating a Common Commercial Policy (CCP). It sets the common tariff rates around the Community as well as negotiating trade agreements in bodies like the GATT. Individual Member States cannot take unilateral action on foreign trade which overrides EC powers. A crucial element of the CCP is the Common Customs Tariff (CCT) required by the Rome Treaty (Article 3). As a key element in the common market structure (see Chapter 1) upon which the EC is based, Member States must agree to a CCT and, unlike those within simpler free trade associations, cannot fix separate national tariffs on trade with third countries. The French insisted on this supranational system of foreign trade policy (Pinder, 1991). With their long tradition of national protectionism, they were worried that their industries might be undercut by cheap imports coming through Member States with lower tariffs. France was also determined to establish the concept of 'Community preference', whereby internal EC production, particularly agricultural, would, if social and economic reasons required, be protected against third-country imports. The familiar arguments about ensuring 'fair' competition within a common market also played a persuasive role. Why should some countries enjoy a productive advantage because of their access to cheaper imports of food, fuel or whatever? Another attraction of the CCP with its CCT was that it would give the EC and its Member States more power to bargain effectively in international trade negotiations, often thought to be dominated by the United States. Finally, in that it obliged Member States to act together to produce a truly common policy, the CCP/CCT was intended to enhance the political cohesion of the EC.

As we have seen, Britain originally opposed this 'common market' concept of European regional integration, promoting instead its free trade vision within EFTA. Thus a more protectionist EC approach to world trade (inspired largely by the French) is often contrasted with a more liberal view (promoted primarily by the British). However, EC trade policy can promote either protectionist or free trade ends (or, most likely, a mixture of both). Parts of the EEC Treaty, notably those dealing with agriculture, have indeed permitted the establishment of protective measures according to the principle of Community preference. However, Article 110 of the Rome Treaty sets the free trade aim of progressively lowering tariffs and abolishing restrictions on international trade. In fact, the general trend has been for the EC to accept the general lowering of tariffs in successive GATT negotiations since the 1960s, although agriculture has been a notable exception. The achievement of successive GATT deals in lowering tariffs also demonstrates the ability of EC Member States to master internal conflicts over international trade policy in order to come to a common position and give the European Commission a mandate to negotiate on their collective behalf.

In addition to forging deals with the USA, Japan and others within the GATT, the EC has developed preferential trading links with large parts of the poorer world. Closest to Europe, privileged arrangements have been extended to most of the countries around the Mediterranean Sea (Featherstone, 1989). Greece was brought into a customs union with the EC in 1962, as a prelude to its full membership in 1981. Other Mediterranean countries are likely to remain EC associates. For example, although Turkey has obtained an Association Agreement, the EC remains reluctant to grant the full membership sought by the Turks since 1987; there is no great desire to incorporate a 'Moslem' country regarded as beyond the European cultural realm. Most North African states enjoy free access to the EC market for industrial products along with tariff preferences for agricultural goods. By the early 1970s Morocco, Algeria, Tunisia, Egypt, Jordan, Lebanon and Syria had been drawn into this EC-centred Mediterranean trade network along with Israel. The 'European' mini-states of Malta and Cyprus obtained full association status on the Turkish model, while a cooperation agreement was granted to the former Yugoslavia as a reward for distancing itself from the former Soviet bloc. With Spain and Portugal becoming full members of the EC in 1986, only Libya and Albania, through choice, remain outside this system of Mediterranean trade preferences, which also include EC aid packages and cooperation accords touching on such issues as migrant labour, the environment, education, science and technology. Although their spatial spread around the Mediterranean Basin is almost complete, these trading networks are not uniform in character. In general, they all permit a high degree of tariff-free entry into the EC for industrial goods and reduced tariffs for many agricultural products. In most cases, there are no reciprocal reverse preferences for EC goods going to these southern partners, although this is not the case with regard to Turkey, Cyprus and Malta (which are moving towards a customs union with the EC) and Israel (which accepts free trade with the EC).

Beyond the Mediterranean, the EC has preferential agreements spreading across the developing world. These grew out of a French insistence that the preferences accorded to its former colonies in Africa and elsewhere be extended to all the EC states. The original Yaoundé Convention, enshrining this arrangement, was replaced in 1975 by the Lomé Convention which pulled in poorer British Commonwealth and other countries from Africa, the Caribbean and the Pacific (the ACP states); there are now over 60 of them linked in a system of preferential trade and aid focused on the Community. The Lomé agreement also removed the system of reverse preferences accorded to EC exporters (seen as a relic of the old imperial order).

Other third world producers wanting access to the EC's gigantic market suspected that EC countries would only enter into preferential agreements with former colonies which provided them with primary products but posed little challenge to Europe's manufacturing industries. Such arguments helped lead to the Generalised System of Preferences (GSP) introduced in 1971. This liberalised trade between the EC and the third world still further to include the remaining countries in Africa, as well as those in Asia and Latin America. The GSP permits tariff-free entry into the EC for a range of manufactured and semi-manufactured goods from less-developed countries. However, important quota restrictions are placed on a substantial number of 'sensitive' imports produced by cheap labour which threaten jobs in the EC. This is especially true of textiles which are largely excluded from the GSP under a series of Multi-Fibre Agreements (Penketh, 1992). Although some agricultural products are included in the GSP, tariffs are imposed at reduced rates rather than eliminated so as to protect both EC and ACP farmers.

Thus, although protectionist conflicts monopolise political attention, the post-war period has seen a progressive liberalisation of global trade with the EC playing a leading role. This is not surprising given the continuing importance of trade beyond the EC for most of its Member States (Table 3.1). Although there is an assumption that the creation of the Common Market has had a trade diversionary effect, increasing the relative importance of intra-Community trade, it is extremely difficult to assess the precise degree to which this has occurred (Newman, 1991). It is impossible, for example, to know what would have happened if the Common Market had not been created. In addition, the spatial growth of the Community from 6 to 12 members makes comparative analyses

Table 3.1 EC imports and exports by partner countries in 1990

EC Member State	*Total % imports (exports)*	*EC 12*	*USA*	*Japan*	*Rest of world*
U.K.	100.0	51.0	12.7	5.4	30.9
	(100.0)	(52.6)	(12.6)	(2.6)	(32.2)
Denmark	100.0	53.7	5.5	3.4	37.4
	(100.0)	(52.1)	(5.2)	(3.3)	(39.4)
Germany	100.0	54.1	6.1	5.4	34.4
	(100.0)	(53.6)	(7.2)	(2.7)	(36.5)
Italy	100.0	57.4	5.1	2.3	35.1
	(100.0)	(58.2)	(7.6)	(2.3)	(31.8)
Spain	100.0	59.1	8.3	4.1	28.4
	(100.0)	(64.9)	(5.5)	(0.9)	(28.7)
Netherlands	100.0	59.9	8.1	4.0	28.0
	(100.0)	(76.6)	(3.9)	(0.8)	(18.7)
Greece	100.0	64.1	3.7	5.9	26.2
	(100.0)	(64.0)	(5.6)	(1.0)	(29.5)
France	100.0	64.8	7.3	2.8	25.0
	(100.0)	(62.7)	(6.1)	(1.9)	(29.2)
Portugal	100.0	69.1	3.9	2.6	24.4
	(100.0)	(73.5)	(4.8)	(1.0)	(20.7)
Belgium/	100.0	70.7	5.8	3.5	20.0
Luxembourg	(100.0)	(75.1)	(4.3)	(1.3)	(19.2)
Ireland	100.0	70.8	14.2	4.4	10.7
	(100.0)	(74.8)	(8.2)	(1.8)	(15.2)
EC applicant countries					
Switzerland	100.0	71.7	6.1	4.4	17.8
	(100.0)	(58.1)	(8.0)	(4.8)	(29.2)
Austria	100.0	68.8	3.6	4.5	23.2
	(100.0)	(65.2)	(3.2)	(1.6)	(30.0)
Sweden	100.0	55.3	8.7	5.1	30.9
	(100.0)	(54.3)	(8.6)	(2.1)	(34.9)
Finland	100.0	46.3	6.8	6.4	40.5
	(100.0)	(46.9)	(5.8)	(1.4)	(45.9)
Norway	100.0	45.8	8.1	4.4	41.7
	(100.0)	(64.9)	(6.5)	(1.7)	(26.9)
Turkey	100.0	41.9	10.2	5.0	42.8
	(100.0)	(53.3)	(7.5)	(1.8)	(37.4)
Economic superpowers					
USA	100.0	18.6	–	18.1	63.3
	(100.0)	(25.0)	–	(12.4)	(62.7)
Japan	100.0	15.0	22.5	–	62.5
	(100.0)	(18.8)	(31.7)	–	(49.5)
Other					
ex-USSR	100.0	25.3	3.1	3.0	68.7
	(100.0)	(27.2)	(0.9)	(2.4)	(69.5)

Source: Eurostat, 1992.

fraught with complications. However, whatever these uncertainties, it is beyond doubt that non-EC trade remains extremely important for Community states. By 1990, around 60 per cent of visible EC trade took place between its 12 Member States (Eurostat, 1992). This represents an increase from the equivalent 35 per cent between the original 'Six' when the EEC began in the late 1950s, but still leaves some 40 per cent taking place beyond the Single Market.

For some states, notably the bigger ones, the importance of global trade is even higher. The proportion of visible British trade with the Community has certainly grown from around 33 per cent when it joined in 1973 to over 50 per cent in 1990, but nearly half still involves the rest of the world. This helps to explain why the British continue to lead the global free trade faction within the Community. But the trading geography of other large Member States is not so dissimilar. The Germans and Danes are only marginally more dependent on EC trade (with about 54 per cent of their imports and exports linked to other Community countries) followed closely by the Italians (some 58 per cent). Even the French, who lead the more protectionist lobbies within the Community, carry out over a third of their visible trade with the world beyond the EC. The smaller countries typically carry on between 25 per cent and 30 per cent of their trade with non-Community countries. The importance of world trade to the EC can also be grasped by noting that, during the mid-1980s, exports from the EC accounted for some 20 per cent of the world total and represented around 14 per cent of the Community's GDP; the equivalent figures for the USA were 13 per cent and 6 per cent (Penketh, 1992). Nevertheless, despite the global spread of much EC trade, the fear of a 'Fortress Europe' retreating behind reinforced protective bastions has developed since the mid-1980s.

A 'Fortress Europe'?

There have always been elements of protectionism in the EC's commercial policy, just as in its major trading partners like the USA and Japan. Some barriers to trade are covert, but others are openly defended as legitimate means to protect important social, economic and strategic interests or to shield EC producers against unfair trade. Article 115 of the EEC Treaty and other arrangements permit the Community to take a range of legitimate protective actions including: Voluntary Export Restraints (VERs), Orderly Marketing Agreements (OMAs), safeguard measures and the imposition of anti-dumping duties. All trading nations carry out similar practices, often in conformity with the GATT regulations. The Multi-Fibre Agreements have already been cited as an example of how trade liberalisation is often a question of a controlled lessening of barriers rather than a *laissez-faire* 'free-for-all'. Another prominent example of 'managed' trade is provided by the automobile industry. A series of bilateral accords between individual EC countries and Japan, sanctioned by the Commission, meant that Japanese cars flowed into a supposedly 'common' market at differential rates during the 1980s; thus they could take up 0.7 per cent of the Italian car market, 2.9 per cent of the French, 10 per cent of the UK and 15 per cent of the German. Other EC states without a large domestic car industry (considered a vital national asset by the French, Italians and Germans) did not impose such restrictions. A typically ambiguous EC–Japan pact signed in 1991 commits the large EC car-producing nations to reaching the same liberal goal by the end of the century. Nevertheless, serious disquiet remains, especially in France and Italy where people like Jacques Calvet, the managing director of Peugeot, accuse Community leaders of a naive 'free-tradism' in dealings with Japan which will undermine vehicle manufacture in the EC (Abescat, 1993).

Yet, although protectionist measures are not new in the EC or elsewhere, expressions of fear about a would-be 'Fortress Europe' have flourished in recent years, especially in the USA and Japan, the other two major poles of world trade (Penketh, 1992). Voices within the EC itself echoed these anxieties, especially in Britain; in her capacity as UK Prime Minister, Mrs Thatcher, a fervent believer in 'free markets', insisted that:

> It would be a betrayal if, while breaking down constraints on trade within the [EC] Single Market, the Community were to erect greater external protection. We must make sure that our approach to world trade is consistent with the liberalisation we preach at home. (Thatcher, 1988)

The majority of EC leaders, however, described the fears as exaggerated, while one observer dismissed the 'Fortress Europe' slogan as simply silly (Montagnon, 1989). Sensible measures to manage freer trade did not merit such lurid language!

However simplistic the slogan may be, what perceptions and realities lay behind it? The rapid growth of unemployment in the EC – increasingly involving the young and long-term jobless – was one factor fuelling fears of a 'Fortress Europe'; the total without work had surged to around 15 million in the mid-1980s, nearly 11 per cent of the total EC labour force (Eurostat, 1992). Inevitably, political pressures grew to protect domestic EC jobs threatened by cheaper imports. As usual, French voices led the argument that social considerations should not be eliminated from the formulation of international trade policy. From this perspective it was not enough simply 'to leave everything to the market'; governments had a responsibility to their electorate. If Community workers lost jobs to countries where wages were low, social provisions limited or non-existent, and environmental issues ignored, there was a moral argument for some degree of protection in sensitive areas. Without necessarily advocating widespread protectionism, these voices rejected the 'GATT dogma' inspired by 'ultra-liberal countries' such as 'the United Kingdom, Germany, the Netherlands or Denmark [which] remain fanatical free-traders' (Delavennat, 1993). They were also highly sceptical of the estimates bandied around by some politicians and economists about the gains to be won from greater free trade:

> The GATT is certainly not a panacea. Those in London and Washington who have asserted that a [GATT] agreement in Geneva is going to stimulate world trade by some $200 000 million per year cannot be taken seriously. The OECD study to which they refer anticipates this effect in ten years time, if all goes well. (Dauvergne, 1992)

A concern to maintain the high standards of social welfare and working conditions in parts of the EC also provoked demands for import controls. Labour costs in the richer EC states are very high, not least because of the large contributions made by industry to support an array of social provisions more extensive than elsewhere in the world. Increasingly questions are asked about whether Community countries can maintain these welfare systems as unemployment grows, populations age and competition from countries with alleged 'sweatshop labour' intensifies. European politicians find it hard to argue for reduced social provision in the face of their electorates. Therefore, the pressures to succumb to protectionist pressures in an effort to maintain European wage and social contribution levels are great. In the revealing emotive words of one French government official:

> Whether you like it or not, given the state of the economy, importing T-shirts made by children in Thailand or political prisoners in China won't do any more (Economist, 1993c, 8)

Protectionist demands within the EC also stem from the high-technology end of the economic spectrum. During the 1980s, worries about the penetration of the EC market by high-technology goods from the USA and Japan intensified as the balance of trade in this sector worsened to the Community's disadvantage. Again, it was the French who were most insistent that the EC could only avoid a dangerous subservience in this vital domain by checking imports in order to give European companies time to develop a comparable competitive capacity. However, they enjoyed well-publicised support from groups across the Community, including the Commission. Not surprisingly, high-technology exporters in America and Japan raised the spectre of a 'Fortress Europe' in response to such arguments. Finally, and perhaps most importantly, the enormous publicity associated with the '1992' Single European Market (SEM) project raised consciousness about a developing EC around the world, not least in its major trading competitors, the USA and Japan. Although the SEM programme was only a renewed effort to create the genuine common market required by the 1957 Rome Treaty, there was a fear that the removal of remaining national barriers to trade and other economic activity within the EC would have a trade-diversionary effect operating to the disadvantage of non-EC states.

However, despite the real pressures and actions to protect certain sectors for specific social, economic and strategic ends (think, for example, of food, computers, vehicles and aircraft), the image of a 'Fortress Europe' appears rather exaggerated. The EC's major trading rivals, the USA and Japan, also provide examples of protectionism. Furthermore, there are indications that the SEM, far from closing the Community to imports, is likely to make it more open. For example, as internal barriers to trade within the EC are eliminated, the ability of individual Member States to impose their national controls on imports has been seriously weakened. Once in the Single Market, a product can be moved from one country to another with increasing ease. The fears of European car manufacturers in this respect have already been touched on above, but leaders of other major industries, such as Alain Gomez of Thomson (France's major electronics company) complain of a '*Europe passoire*' with the protectionist capacity of a sieve (Delavennat, 1993, 61). They point anxiously to the EC's large trade deficit with the rest of the world during most of the 1980s, which rose to some ECU 42 906 million in 1990. Moreover, the EC ran substantial trade deficits, not only with Japan but with the USA as the 1990s began (Eurostat, 1992). Furthermore, those who insist that the EC is anything but an enclosed trading citadel maintain that the USA and Japan have more political capacity to be protectionist than the EC. Lacking a strong central authority, the Community has to contend with 12 powerful national governments with diverse political leanings often in pursuit of conflicting interests; for example, the UK government's enthusiasm for Japanese car-manufacturing investment in Britain is definitely not shared by the automobile industry elsewhere in the EC! (Delavennat, 1993).

In effect, it appears that demands for a measure of protection from prominent industries like agriculture and automobile manufacture, notably in France, have generated slogans which magnify the amount of actual trade-wall building around the EC. These protectionist demands are themselves a reaction to a reality where the overall movement is towards freer trade. It is noteworthy that recent French governments which, whatever their political persuasion, lead the 'managed trade' lobby, do not deny the basic desirability of greater trade liberalisation. After all, large and dynamic sectors of French industry depend heavily on exports. In effect, the French are usually seeking a managed 'middle way' between complete free trade and levels of protection tailored to the needs of the economic sector under consideration; in this context they ask:

> Are there not some [free trade] thresholds that can only be crossed progressively, and, first of all, within limited regional areas? (Boissonat, 1992, 13)

It is also possible that protectionist voices in the EC have been amplified by Americans and Japanese who fear that their powerful position in world trade is threatened by growing EC strength. At the end of the Kennedy Round of the

GATT negotiations in 1967, a US observer had already noted:

> The dominant position of the United States in GATT evaporated with the implementation of the Rome Treaty ... the Common Market is now the most important member of GATT, and can determine in large measure the success or failure of any attempt to liberalise trade. When Europeans instruct Americans in the realities of the new international economic situation they are demonstrating the change in relative power that has taken place. (Krause, 1968, 224–225)

The example of agricultural trade

The bitter struggle over agricultural trade within the GATT Uruguay Round illustrates the underlying power struggle between the USA and the EC, as well as the way in which the 'Fortress Europe' slogan hides the serious social and economic issues at stake. First, the agricultural conflict emphasises that the EC is not the monolithic citadel some perceive; the French often find themselves at odds with their EC partners as they defend a vital national interest. The roots of this USA–EC/France clash are revealed by a few basic statistics. The USA is the world's primary exporter of cereals with exports totalling around 82 000 tonnes per annum in the early 1990s, about one-third of its production (L'Expansion, 1992a). During the same period France, the world's second largest exporter of cereals, sold abroad around 26 000 million tonnes per annum, about half its production. It is clearly the EC's largest cereal grower, producing rather more than the Germans and the British together (Eurostat, 1992). Over recent decades, the US share of the world's cereal trade has been declining. Between 1982 and 1992, it fell from 50 per cent to 43 per cent of the total, while the EC share rose from 10 per cent to 14 per cent (L'Expansion, 1992b). These trends have inevitably brought the American and European/French agricultural industries, both electorally important, into conflict as they fight for global export outlets for their surpluses.

American discontent is aggravated by the fact that EC agriculture is more heavily subsidised than that of the USA. An OECD report in 1991 showed that EC farm subsidies totalled $83 600 million as opposed to a USA equivalent figure of $34 700 million (L'Expansion, 1992b). However, in per capita terms, the balance tips in Europe's favour; these subsidies amounted to some $9 630 per person employed in EC farming and some $10 342 for the US counterpart. Nevertheless, on world export markets the proportion of subsidy in the price obtained by EC farmers has often been markedly higher than that enjoyed by American producers. For example, in the early 1990s it made up 61 per cent of the price for EC wheat, but just 50 per cent of its USA equivalent. In defence of subsidised farm exports, EC representatives present an array of arguments about maintaining adequate levels of Community self-sufficiency in foodstuffs and responding to the social needs of farming communities (Wise, 1989). This social dimension of agricultural policy reflects attitudes profoundly embedded in continental Europe, although not in Britain. The EC's CAP specifically requires that account is taken of 'the particular nature of agricultural activity, which results from the social structure of agriculture' (European Commission, 1987, Article 39.2.a). In practice this means that European farmers are not just seen as producers of food in some narrow industrial sense. True, electoral considerations do much to explain a European readiness to subsidise farming, but a desire to preserve attractive landscapes as well as the family farms and rural communities which conserve them has also played a role. Besides, from this 'social' perspective, what gain is there in accepting a global free trade policy which would entail the economic costs of coping with a rapid collapse of an ageing, relatively poor farming population with no alternative form of employment in prospect?

Consideration of such social factors has given US-EC/French disputes about farm exports an ideological complexity which cries of 'Fortress Europe' do little to illuminate. It also explains why agriculture was excluded from the GATT deals which preceded the Uruguay Round. However, once farm products were included, an inevitable battle ensued between the EC and the USA. The latter were supported in the so-called 'Cairns group' by other agricultural exporters, like Australia and Canada, with large land areas, small populations and a more purely 'economic' approach to farming. American demands for massive cuts in farm export subsidies were met by determined French resistance, with other EC countries arrayed behind them in varying postures of support and opposition. Tortuous GATT negotiations led to the 'Blair House' compromise which, while envisaging a diminution of some 21 per cent or more in subsidised farm exports from the Community by 1999, still maintained a degree of protection; the 100 per cent reduction called for by the USA at the start of the Uruguay round was not achieved. But France became somewhat isolated from the majority of its EC partners in the GATT farm negotiations and significant breaches in the walls of the EC's agricultural 'fortress' were eventually made by the final agreement reached in December, 1993. Electorally important as they are, French farmers were not the only group putting pressure on their government; others wanted the overall GATT negotiation to succeed; for example the threat of imposing a 200 per cent tax on imports of EC wine (France is the world's largest exporter of wine) revealed that the USA wielded substantial power in the GATT game and that by no means all Europeans (and French) who live off the land want to retreat behind a protective barrier!

TOWARDS '1992' AND DEEPER ECONOMIC, SOCIAL AND POLITICAL INTEGRATION

The Single European Act and the Single European Market

It has been explained how the successful promotion of the SEM project (European Commission, 1985) reinforced perceptions that European integration was deepening and, perhaps, turning in on itself. Certainly, in the period leading up to the much-trumpeted year of '1992', those with 'maximalist' visions of European unity appeared to be gaining the upper hand over the 'minimalists'. The 'spillover' process appeared to be pushing inexorably towards further regional integration on all fronts. The 1986 Single European Act (SEA) (European Commission, 1987) was not just about creating a Single Market, but also reflected the wider political ambitions of those who, for example in the early 1980s, won an overwhelming majority in the European Parliament for a plan to create a European Union of undoubted confederal character. Although the British government, at the head of the 'minimalist' tendency, dismissed such proposals as 'airy-fairy' most Member States were disposed to make some moves in this direction (Lintner and Mazey, 1991). The resulting SEA not only required the creation of an SEM, but also reinforced the Community's supranational institutions and set new objectives for common policies across a wide range of fields, including monetary union, social and regional policy, foreign policy, security, environmental policy, research and technological development (Wise and Gibb, 1993). Moreover, it extended the use of qualified majority voting in the Council of Ministers and strengthened the role of the European Parliament in the making of Community legislation.

Much of this was not to the taste of a British government wedded to a minimalist view of European integration. However, what attracted the free-trading British Conservatives was the aim of creating a genuine common market – now called a 'single market' – by the end of 1992. All the non-tariff barriers (NTBs) which continued to hinder intra-EC trade (including the provision of services across national borders) were to be removed along with the remaining impediments to the free movement of people (labour) and capital. The UK government also saw the Single Market not so much as a self-contained entity, but as another step towards a global liberalisation of trade, with Britain leading the free-traders in the formulation of EC policy towards the GATT 'Uruguay Round'. This enthusiasm for open international markets also explains why the UK government accepted the view that decisions about eliminating NTBs in the EC should be made by the essentially federal practice of qualified majority voting. It was a means of overcoming the potential vetoes of more protectionist governments less keen on the 'logic of the market'.

The overall economic performance of the EC was lagging behind that of the USA and Japan, those economic superpowers whose dominating influence was so much disliked in countries with protectionist-nationalist traditions like France. Moreover, competition from newly industrialising countries (NICs) was posing a growing threat to sluggish European industries. The term 'Euro-sclerosis' described increasing European dependence on high-technology imports (e.g. computers) along with low rates of growth and innovation. Soaring unemployment, as old industries collapsed faster than new ones could be created, added to this sense of economic stagnation. A broad consensus thus developed that, within an SEM, competitive forces would reinvigorate European industry and provide the benefits of a large 'domestic' market of the sort enjoyed by the USA and Japan. The costs of 'non-Europe' would be eliminated as costly border delays disappeared, competitive tendering could take place across national frontiers (no more national monopolies!) and European companies joined forces to produce more efficiently for a 'home market' of some 320 million people. Similarly, the huge research and development costs of modern industry would be more effectively met as capital flowed across borders along with people and ideas to stimulate new 'synergies'. The European Commission calculated that, in the medium term, the SEM would add 4.5 per cent to the EC's GDP, deflate prices by 6.1 per cent and create 1.8 million new jobs (Cecchini, 1988). Both 'federalist' and 'free trade' traditions united around these optimistic economic scenarios of a 'Europe without frontiers'.

A 'Social Europe' to balance an 'Economic Europe'?

However, some challenged the euphoric optimism of the Single Market message (Wise and Gibb, 1993). Not all industries would thrive as protective national barriers were removed. Competition can also lead to unemployment and poorer working conditions. Consequently, the more 'maximalist' elements in the Community insisted that the SEA revive the EC's dormant 'social dimension' to counterbalance the resurrected common market ideal (Wise and Gibb, 1993). Thus EC legislation designed to improve the 'working environment of workers' should be adopted by a qualified majority in the EC Council of Ministers (European Commission, 1987; Single European Act, 1986, Article 21). Furthermore, the SEA required that the resources devoted to the EC's Structural Funds (designed to stimulate employment, reduce regional disparities in welfare and modernise agriculture) be increased substantially; consequently they were doubled over the period from 1988 to 1993 in parallel with the elimination of economic barriers (Wise and Gibb, 1993). This determination to build a 'Social Europe' reflected fears that much of the general public would become politically disaffected by a 'Businessman's Europe' based primarily on free market economic principles. Thus the European Commission insisted that:

> A social consensus is an essential factor in maintaining sustained economic growth. Europe cannot be built against the opinions of the employers or of the

> workers or of the general public, and efforts must be made ... to prevent distortions of competition from leading to forms of social dumping. (European Commission, 1989, 1)

This concept of 'social dumping' encapsulates the fear that free movement of goods, services, capital and labour within a single market will lead to ruthless rivalry between companies, countries and regions, thus undermining the high levels of social protection won by workers in the EC's richer countries and preventing the poorer states from building them up. Unions forecast fiercely competitive situations where multinational companies, free from national controls on capital movements in the SEM, would shift their investment away from workers who do not accept reductions in their wages and working conditions (Boyer, 1988). The enlargement of the EC to include the poorer countries of Greece, Spain and Portugal had made the existing divergences in working conditions among Member States enormous, thus intensifying these misgivings.

For example, Table 3.2 shows that in 1988, overall labour costs per hour in Portugal (the poorest Member State) were less than one-sixth of those in West Germany, the richest Member State (European Commission, 1988). These wage gaps had already encouraged companies like Volkswagen to invest in countries such as Spain where wages, although high by Spanish standards, were some 50

Table 3.2 Hourly labour costs in industry (manual and non-manual workers) within EC member states, 1988

Member State	*Costs in PPS*	*Costs in ECU*
Netherlands	20.26	16.31
West Germany	19.57	18.11
Belgium	19.18	15.43
Italy	18.41	13.70
France	17.80	14.95
Luxembourg	17.47	13.49
Denmark	14.97	15.45
United Kingdom	14.83	10.72
Spain	14.36	8.95
Ireland	13.80	10.33
Greece	9.92	5.12
Portugal	6.79	2.87

Note:
ECU = European Currency Unit
PPS = Purchasing Power Standard
PPS cancels out differences in price levels existing between different countries in order to make more realistic comparisons. Simple monetary parities do not reflect the domestic purchasing power of national currencies or the ECU. For example, the 'low' income of a Portuguese worker will sometimes buy more in Portugal than a nominally 'higher' German salary can buy in high-cost Germany.

Source: Eurostat, 1992.

per cent of those enjoyed by German workers and employee rights were less extensive and expensive. Spanish readiness to accept less than their German counterparts is easy to understand; unemployment in Spain through the 1980s hovered around 20 per cent (compared to some 5 per cent in West Germany). Fears of social dumping were also intensified by competitive pressures from third countries such as Japan; for example, Volkswagen's investment in Spain had been impelled by a wish to remain competitive with the Japanese in the small car sector. With the GATT committed to the continuing liberalisation of global trade, the tension between the need to remain competitive and the desire to maintain the high standards of the 'social market model' developed in the richer EC countries is bound to continue.

All this led to calls for a minimum (not common) level of social protection throughout the Community in order to prevent competitive pressures in the SEM tempting governments and employers to depress working conditions in order to win markets and investment. In other words 'a so-called level playing field' of broadly similar social provisions should be laid down on which companies could compete fairly across the different national zones of the new EC 'economic space'. However, opponents of this approach, headed by the British government, saw 'excessive' social provision for employees as a factor in the EC's relative economic stagnation *vis-à-vise* the USA and Japan. How could companies keep their competitive edge when burdened with the high costs of maintaining minimum wages, short working weeks, long annual holidays, and contracts making it difficult to dismiss employees? Given Europe's socio-economic and cultural diversity, it was folly, from this standpoint, to impose uniform EC-wide legislation. If workers in the poorer, peripheral countries (Spain, Portugal, Greece, Ireland and Britain) were prepared to work longer hours for less pay than those in the richer core countries with higher social standards (Germany, France and Benelux), then so be it.

However, the majority of EC political leaders remained faithful to the European social market model. Thus in 1989, all Member States but Britain adopted the 'Community Charter of the Fundamental Social Rights of Workers', the so-called 'Social Charter' (European Commission, 1990). It called for common action across a very broad swathe of social policy. But the conversion of this ambitious statement of social intent into concrete action has lagged far behind the measures adopted to construct the Single Market (Wise and Gibb, 1993; Teague, 1989; Bailey, 1992). British opposition to EC social proposals has been resolute and others across Europe (notably employers), increasingly worried by the problems of competing with countries unburdened by such welfare costs, have been happy to shelter behind the UK Conservative government.

The SEA contained other ambitions to build a regional organisation with characteristics extending far beyond a free-trading bloc. However these goals, and others, were reiterated even more emphatically in the Treaty on European

Union (European Commission, 1992) signed by Member State governments at the end of 1991 in the Dutch city of Maastricht.

The Treaty on European Union: towards an ever closer union?

Characteristically, this additional EC Treaty resulted from mixed motivations. The persistent efforts of those with federalist ambitions were helped by the unification of Germany. This provided a geopolitical catalyst which pushed the Member State governments from declarations of intent to deeds. Germany wanted to reassure its neighbours by encapsulating its new-found national oneness in a larger European unity. France, ever concerned to anchor Germany to containing European institutions, was similarly ready to respond to confederal pressures in the wake of the collapsing East-West divide across the continent. Others followed the lead set by the Franco-German axis with varying degrees of enthusiasm to produce a Treaty which *facilitates*, but does not *compel*, moves towards a comprehensive degree of political–economic integration of a confederal kind.

For a start, 11 Member States wanted to strengthen the EC's 'social dimension' by incorporating a 'Social Chapter' in the Treaty. However, in the face of absolute British governmental opposition, they were forced to withdraw this section from the Treaty proper and establish it as a separate Protocol to which the UK was not signatory. Nevertheless, there was agreement to enlarge the existing EC Structural Funds designed to stimulate employment, help the EC's poorer regions and modernise agriculture. In addition, a new Cohesion Fund was created to transfer yet more investment from the richer to the poorer countries of the Community. The EC's 'social' face was also reinforced by extending the already substantial common environmental policy along with a large raft of other provisions aimed a building a 'People's Europe'. Thus the competence of the Community to act in areas such as public health, consumer protection, education and culture was formalised and widened (Articles 126 to 130). Moreover, the Treaty provided that 'every person holding the nationality of a Member State shall be the citizen of the Union' (Article 8:1). Thus, once the Treaty is ratified, all Member State citizens will become Community citizens as well, enjoying additional rights including the ability to vote in local and European elections when living outside their national country. Similarly, aspects of justice and home affairs were also brought under the 'Maastricht' umbrella; the opening of internal borders for trade within the Single Market posed common problems of international crime control, cross-border terrorism and illegal immigration (Title VI, Articles K1-K9). The Treaty's proclamation that 'a common foreign and defence policy is hereby established' (Title V, Article J) also reveals confederalist ambitions. However, the tensions between Member States in this highly sensitive area are very apparent in the machinery set up for this purpose (Articles J.1 to J.5). Although the system seeks to pull Member States towards 'joint action' on foreign affairs and defence, the national right to 'opt out' remains. Fundamental differences in national policies will make common agreement in this area very difficult, as the EC's incapacity to control the nationalist conflicts in former Yugoslavia has made clear.

A Single Currency for a Single Market?

A similar sense that national attitudes may overwhelm confederalist ambitions flows from analysis of the aim to create a single European currency by January 1999. The logic that a single market cannot truly exist without a single currency forms the core of the Treaty. Compared to the USA with its single dollar, the costs and complexities of commercial dealings within the EC are far greater as 11 national currencies shift in value one against another. Furthermore, the ability of individual Member States to stimulate exports and reduce imports by unilateral devaluations of their currencies makes direct comparisons with the domestic markets of the USA and Japan invalid. National currency depreciations within the EC inevitably produce the cry that 'free' trade within the SEM must also be 'fair'. Why should one Member State open its national market to a partner which uses competitive devaluations in order to gain an 'unfair' export advantage?

Such considerations originally led EC leaders to set the objective of attaining European Monetary Union (EMU) by 1980! However, the shift from principle to practice on this point has proved immensely difficult (Gros and Thygesen, 1992). The enormous divergences among the national economies of the EC members have proved far too wide for one currency alone to bridge. In 1979 the core members of the Community, led by France and Germany, established the European Monetary System (EMS) with an Exchange Rate Mechanism (ERM) at its heart. This set up a system of semi-fixed exchange rates whereby national currencies move against each other within bands 2.25 per cent or 6 per cent either side of an agreed rate. The European Currency Unit (ECU) was also developed as a kind of common currency composed of a 'basket' of national monies. Britain, ever suspicious of being trapped in a European straightjacket, originally refused to join the ERM, but eventually did so in late 1989 as the EC moved confidently towards the completion of the Single Market by the end of 1992; other countries with weaker economies like Greece, Spain and Portugal did likewise.

It was against this optimistic background of apparent monetary convergence that the Maastricht Treaty mapped out a series of stages to achieve a single currency by the end of the century. But the Treaty does not guarantee that this momentous goal will be attained. The very diverse national economies of the EC must converge sufficiently for EMU to occur around a model of strict financial discipline based on the example set by the German Bundesbank over the period leading up to Germany's unification. A European Central Bank is envisaged as a guarantor of this new financial order. Once again, however, Britain insisted on its sovereign right to opt out of these plans if it so chooses; Denmark has done likewise. British 'Euro-sceptics' often revealed the deep national passions underlying the economic technicalities of the debate when describing EMU as another means by which Germany would extend its financial, economic and political dominance across the continent; for example, the Chairman of the British Anti-Federalist league, claiming that clauses in the European Union Treaty 'could have come straight out of Hitler's Enabling Act', insisted that it created 'the new superstate of ... [German] dreams':

> Far from containing the new Germany, it will permanently institutionalise German predominance. Its economic provisions will lead to the de-industrialisation of large parts of the continent (including Great Britain), while democracy will be stifled by centralised, unaccountable, bureaucratic rule. (Sked, 1992)

The fragile foundations of this monetary ambition were laid bare when the pound sterling was forced to leave the ERM in September 1992 by international currency markets responding to divergences among the national economies of Member States; the other weaker EC currencies of Spain, Portugal, Italy and Ireland were also forced to devalue. For the architects of European Union, much worse was to follow at the beginning of August 1993 when international speculation against the French franc seriously disrupted the existing EMS and the crucial Franco-German relationship at its heart. Over the previous decade the French government had been successfully adopting the financial disciplines characteristic of West Germany as a move towards monetary integration. However, it was the national integration of their reunited country that eventually proved more important to German leaders. In order to absorb their poorer eastern compatriots, they uncharacteristically accepted large budgetary deficits and awarded salary increases that bore no relation to the real state of the economy in the former East Germany. As a result, inflation was unleashed in a country which, for deep-rooted historical reasons, has a profound fear of it. Consequently, the Bundesbank insisted on more rigorous monetary policies to stem this inflation, notably by imposing high interest rates.

While this made sense from a German national viewpoint, it contradicted French interests. As recession deepened and unemployment soared within France, domestic demands to lower French interest rates in order to stimulate the national economy became louder; in other words, the long-held tactic of tying the French franc tightly to the German Mark should be relaxed or abandoned. The French government, with its own inflation firmly checked and determined to preserve the vital Franco-German axis at the core of the EC, tried to persuade the Germans to reduce their interest rates in an effort to reinvigorate European economies as a whole. In this, they enjoyed the support of other Member States who also wanted to reduce the cost of borrowing money in an effort to break out of recession. But with national unification and control of domestic inflation as main priorities, the Germans kept their interest rates high, thus strengthening the Mark and increasing international speculation against the French franc. Eventually, this clash of interests effectively broke the existing EMS. Although, the system was maintained in theory, the widening of the exchange-rate fluctuation bands on either side of the fixed rate from 2.25 per cent to 15 per cent signified its quasi-dismantlement.

The long-term implications of this event for European unity remain to be seen. The pessimists in the European unity camp fear that it may prove the point where Germany, the powerful pivotal country at the core of the continent, shifted decisively away from the path towards European Union mapped out with the French and others over the post-war period to pursue more independent national policies (Valance, 1993). They noted, for example, the persistent opinion poll findings which showed that a majority of Germans (59 per cent immediately prior to the franc crisis) were opposed to the fusion of the German Mark into a single European currency, suspecting that a similar resistance was to be found in the Bundesbank itself. The Deutschemark is a potent political symbol of western Germany's post-war success, in addition to its more obvious economic importance. In contrast, the European optimists pointed to the extraordinary circumstances created by German unification and hoped that the *de facto* 'suspension' of the EMS would indeed prove but a temporary disruption on the road to monetary union (Casanova, 1993). The EC had weathered other serious crises in a European unification process where a long-term historical view, which looked beyond the chaos of short-term events, had to be taken. A memory of Europe's brutal, nationalistic past would ensure that German, French and other EC leaders remembered the basic political goal of European Union and would arrest centrifugal forces before they went too far. Others, notably in Britain, who have never shared the vision of an integrated Europe going beyond much more than a loose free trade area saw the disintegration of the EMS as further vindication that their view of a Europe dominated by nation-states in a world moulded by 'free markets' was correct.

TOWARDS GREATER UNION, DISINTEGRATION OR A VARIABLE GEOMETRY EUROPE?

Maastricht: a turning point in which direction?

The forward surge of western European integration in the decade leading up to the 'Maastricht' Treaty suggested victory for the comprehensive view of European Union extending far beyond free trade areas and single markets. However, the manifest failure to deal effectively with nationalist conflict in former Yugoslavia and the severe damage done to the EMS in the summer of 1993 starkly revealed that, beneath these aspirations, the pursuit of perceived national interest remains predominant in an international world often beyond the EC's control. Furthermore, within the Maastricht Treaty itself, there are elements which clearly mark the reluctance of Member States to pool more of their national sovereignty in the EC's proto-federal supranational institutions. Indeed, the UK government presented the Treaty to its public as the moment when the traditional British vision of a looser, less federalist body began to gain the upper hand.

Certainly, parts of the European Union's complicated, asymmetrical architecture support this belief. The Treaty did not create a single supranational body with a single constitution of the sort associated with confederal organisations. Instead, the 'Union' is composed of three distinct 'pillars'. The first is founded on the original EC Treaties and supranational institutions as amended by the Single European Act of 1986 and the Maastricht accord. However, alongside this established supranational body, new structures have been constructed which, while making use of the existing EC institutions, are intergovernmental in nature. Here, the dominant role of national governments is emphasised even more than in the existing EC by reducing the influence of the Commission and the European Parliament. One of these intergovernmental pillars deals with home affairs and justice (immigration, international crime, terrorism, etc.) whilst the other is concerned with foreign and security policy. This latter pillar draws in a separate body, the Western European Union (WEU) 'to elaborate and implement decisions and actions of the [European] Union which have defence implications' (European Commission, 1992, Article J4). The WEU is a mutual western European defence organisation, founded in 1948, which had become dormant under an American-led NATO in the post-war period. EC Member States not in the WEU – Greece, Denmark and Ireland – are invited to join, although the 'Maastricht' agreement does not compel them to do so.

Towards a 'variable geometry' Europe?

In reality, beneath the European Union banner, a 'variable geometry Europe' is developing where countries opt in and out of different arrangements. This 'Europe à la carte' was long resisted by the original EC states which feared that it would herald a move towards the looser association of states long favoured by the British, and to a break-up of the Community. If states can choose to join only those parts of the EC of obvious benefit to them without bearing the costs of cooperation, then the potential for mutual recriminations would grow. Thus, talk of a 'two-or-more-speed Europe', where an inner core of EC states advances more rapidly than countries on the geographical peripheries, has traditionally been discouraged. Yet the Maastricht accord clearly accepts that progress on all fronts involving all countries is impossible. There was no way of forcing the UK to accept the Social Chapter or commit itself unequivocally to the single currency. Later, the rejection of the Union Treaty in the first Danish referendum underlined the ability of national electorates to resist coherent constitutional 'blueprints' for a united Europe, thereby obtaining 'opt-out' provisions on matters as diverse as defence and the right to prevent non-Danes buying holiday homes in Denmark, in flagrant contradiction of the concept of Community citizenship. In truth, there have always been flexible arrangements allowing some countries to integrate faster than others; most notably, the attempt to drive towards EMU has always had a marked multi-speed element to it. The European Union Treaty confirms rather than reverses this reality.

The enlargement of the European Community

The pressure to develop a flexible model of European integration where countries belong to overlapping systems of varying spatial extent is bound to increase as the Community takes in new members (Figure 3.3). By 1993, some eight countries had applied to join while most central European states had indicated a desire to do likewise (see Chapter 4). The prosperous countries of the European Free Trade Association (EFTA) head the queue. Since 1977, they have had agreements with the EC securing free trade in most industrial products. These arrangements came under pressure in the 1980s for a number of reasons. First, the SEM threatened to increase internal EC trade at the expense of the EFTA. Secondly, an enormous body of EC legislation was being adopted to ensure that trade was no longer blocked by a mass of non-tariff barriers, including national differences in product quality norms, technical standards, state purchasing practices, professional qualifications, taxation, and so on. The EFTA countries could not ignore these developments if they wished to maintain unimpeded access to the Single Market. Furthermore, the enormous growth of unemployment within the twelve EC states – from some 2 million in 1970 to around 13 million by 1991 – was promoting protectionist demands which could threaten EFTA exports. There was also a feeling that the price of entry to the EC market had to be a contribution to the increasing costs of a growing range of Community policies (Wise and Gibb, 1993).

Such pressures initially led to the signing of the European Economic Area (EEA) agreement in October 1991, bringing together EC and EFTA governments in a market of some 380 million consumers (Figure 3.4). While broadly accepting the 'four basic freedoms' of the SEM, namely free movement of products, capital, labour and services, the EFTA states stayed outside the EC's agricultural policy (not protective enough in their view) and obtained arrangements for the production of fish, energy, coal and steel falling short of a full Single Market agreement. Nevertheless, in exchange for contributing financially to EC policies designed to maintain the Community's 'economic and social cohesion', EFTA states would have some influence in shaping the SEM. In essence, the EFTA countries were trying to stay *in* the EC market while staying *out* of its

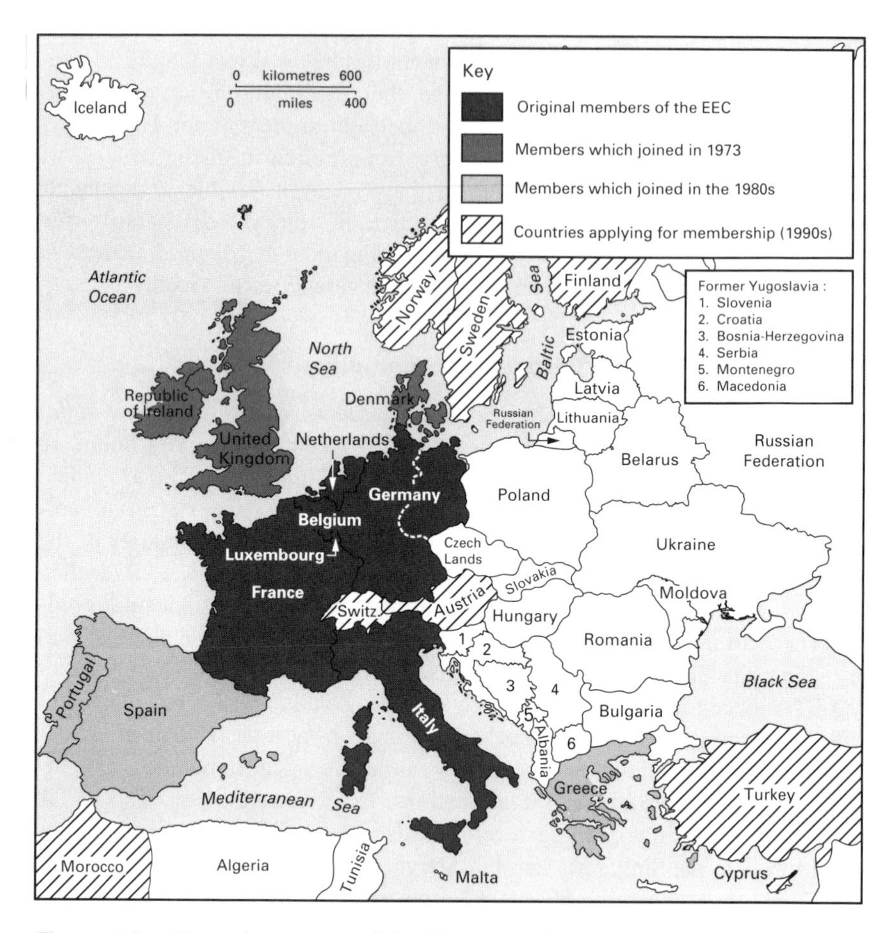

Figure 3.3 The enlargement of the European Community

institutions and wider political ambitions. However, as the EEA negotiations progressed, EFTA leaders increasingly felt that this 'in–out' strategy was putting them into a subservient position. Their economies were increasingly moulded by EC laws they had no direct hand in making while they were being obliged to contribute to the costs of social and regional policies of a Community to which they did not belong. Thus, as soon as the EEA Treaty was signed, the EFTA countries (with the exception of Iceland) announced their intention to apply for full membership of the EC.

Not everyone in the EFTA countries accepts the logic of this 'spillover' process. Apart from the Icelanders (ever fearful that EC principles will erode the national barriers protecting their fishing industry), substantial resistance to full EC membership remains throughout EFTA. The public rejection of the EEA in

Figure 3.4 Signatories of the Europeam Economic Area, 1991

Switzerland's national referendum on the issue in 1993 made this clear. In Norway, the scene is set to repeat the arguments which preceded its 1972 referendum which overthrew the government's decision to enter the Community. For example, the problems of integrating its fishing industry into the EC's Common Fisheries Policy (CFP) remain. This issue has always revealed fundamentally opposed conceptions of European unity (Wise, 1984). The Norwegians, as western Europe's major fish exporter, want free access to the Single European Market. However, they resist the CFP principle that access to fishing grounds under the sovereignty of EC states should be free of national discrimination among Community fishermen. In reality, this so-called 'equal access' provision has never been fully applied as compromises have been struck between Community and national principles. However, a Norwegian fear of not being able to discriminate according to nationality remains. This nationalism also clashes with EC policy over agriculture. Norway wants to keep farmers as well as fishermen in its harsh and vast north-western regions and fears that the CAP will not provide adequate protection for them. If Norwegian farmers were exposed to open competition from farmers further south, many would not survive without enormous subsidies. With CAP reforms going in the opposite direction, Norwegian farmers worry about the EC's relative liberalism. The leader of Norway's Centre Party encapsulated this anti-EC viewpoint thus:

> We object to the economic philosophy of the Treaty of Rome. The EC's 'four freedoms' are good for the multinationals but bad for our farmers, our small companies, our poor, our environment and the third world. (Economist, 1993d, 41)

So even among the EFTA countries which began entry negotiations in February 1993, with a view to joining the EC in 1995, there are doubts about whether Community enlargement will take place. Beyond these relatively rich states, the future is even more uncertain. Deep-rooted doubts about Turkey's 'European' credentials mean that its demand for full membership lie dormant, while the application from the mini-state of Malta raises the question of how to deal with a potentially growing number of small states in the Community. The problem of managing large numbers of nation-states also casts a shadow over the aspirations of central and eastern European states to join the Community. In principle, western EC governments accept the ideal of one day integrating these countries. Nevertheless, with GDP levels per capita there often less than a third of the richer EC states, it is hard to imagine how this old pan-European dream can be realised in the foreseeable future.

Certainly, a merger into the proto-federalist structure evolved by the existing Community would prove extremely difficult, a factor which explains the enthusiasm of many anti-federalists to include the eastern half of the continent. A 'wider' Community would, in their view, inevitably be a 'shallower' one, forced to shed its federalist ambitions while falling back onto more limited free trade objectives. Hungary, Poland, the Czech Republic and Slovakia head the list of likely eastern applicants, having already signed Association Agreements with the EC. But while western investment is able to flow with relative ease across the borders in search of opportunities offered by cheap labour in central and eastern Europe, other barriers are growing in an effort to stop the flow of immigrants. When, if ever, all the 'four basic freedoms' of the Single Market can be extended across the old ideological divide remains a matter for conjecture (see Chapter 4).

CONCLUSIONS

This chapter has shown that the EC has always aspired to be more than a simple regional free-trading bloc. Its origins owe more to the desire to preserve peace in Europe than a belief in market liberalisation. For many European politicians, economic integration has often been more a means to achieve a wider political unity than the primary end in itself. The desire to create a larger political body able to give western European states greater influence in world affairs has also provided a major stimulus for the development of the EC. Consequently, its actions in world economic affairs – for example, in the GATT negotiations – cannot be fully understood outside this wider political context. EC leaders have more to consider than the economic interests directly involved when formulating common positions on global trading problems.

This is not to deny that a belief in the elimination of trade and other economic barriers has been extremely significant in the complex amalgam of motives associated with the quest for European unity. For many, especially in Britain, it has been the most important motivation. The notion, championed by some in France, that the EC should become a coherent political–economic region ready, for example, to take common '*dirigiste*' action both to check imports and promote exports in the interests of 'Europe' (and France!) has generated little enthusiasm in a UK where the global 'free trade' tradition remains strong and the 'special relationship' with the USA is not completely dead (see for example Table 3.1). Other western European countries have navigated between the competing conceptual leads given by France and Britain in the post-war period. Germany's overall economic inclinations pull it towards the global free-tradism of the UK, but the political imperative of coming together with France to promote continental stability proved decisive from 1950 up to the signing of the Treaty of European Union in 1991. Following the relative failure of its EFTA initiative in the early 1960s, Britain also succumbed to the more comprehensive model of integration conceived by Monnet and Schuman. Moreover, with the remaining EFTA countries now negotiating for EC membership and the states of eastern-central Europe trying to edge into the Community's orbit, the 'spillover' processes set in train by the ECSC appear to have triumphed, leaving the continent apparently poised on the threshold of a confederal 'European Union' with political scope far exceeding that of a simple trading bloc.

Yet, in the summer of 1993, the mood had changed to one of deep uncertainty about the EC's future. The Community's ambitions to create a single currency appear to have been undermined by its inability to cooperate effectively in the face of international financial speculation. Its pretensions to formulate common foreign and defence policies seem tragically misplaced when faced with its inability to check nationalistic barbarities in the Balkans; everything that followed the break-up of the Yugoslavian federation – the glorification of the nation-state, the 'ethnic cleansing', the economic fragmentation – directly opposes the deepest Community ideals. Its internal divisions over agricultural trade within the GATT intensified the pessimism of those committed to a confederal vision of EC unity. However, these same events provide succour to those who combat such a comprehensive 'political' vision of Europe, especially in the UK (Le Monde, 1993). For them these developments are seen as vindication of a belief that Europe's deep-rooted national divisions cannot be contained in a closely integrated political entity. Far better to abandon these 'federal' pretensions and return to the vision of a continental free-trade zone with intergovernmental cooperation on other matters according to the 'variable geometry' model where different groups of nation-states work together on matters of mutual interest in a 'functionalist' manner.

The future will tell what version of European unity, if any, will ultimately prevail. However, the EC's past experience already allows some conclusions to be drawn relevant to others seeking some kind of international regional economic integration. First, there will obviously be persistent tensions of a fundamental kind if the countries concerned are not clear about what kind of union they seek. Caught between federalist visions on one hand and more limited free-trade concepts on the other, western Europe was first divided between the EC and EFTA. This ideological divorce was only partly reconciled by the entry of Britain and others into the EC. Within the enlarged Community, incompatible approaches to integration have continued to create strains which help to explain the asymmetrical nature of the European Union Treaty and the uncertain state of the Community as it moves towards the end of the century. The EC experience also suggests the utility of periodically posing the basic question of why have a multistate regional organisation. If political leaders and their electorates become uncertain about the *raison d'être* of the international body they are trying to build, then its ultimate success will inevitably be compromised; public disaffection with the EC reflected in referendums and opinion polls provides salutary lessons here. Indeed, it is remarkable that in a world which in some ways is apparently pushing towards a vision of global free trade in a world economy, this basic question of why have regional groupings at all is not posed more often. In a western European context, a desire to be 'big enough' to match the economic might of the USA and Japan has provided a motivation which doubtless finds echoes elsewhere in the world. But the original architects of European union found their primary answers to this basic question more in the political goals of peace, stability and the ability of Member States to influence world affairs. Despite the prominence given to the elimination of economic barriers as a means to unity, it is doubtful whether a simple European regional trading bloc would last for long without these deeper political motivations. In a wider world economy dominated by multinational companies and international financial markets, whose actions often escape the control of even the bigger nation-states, a regional organisation based only on free-trading objectives could well melt into a larger, somewhat anarchic, global system rather in the way that the EFTA has gradually withered away.

The EC experience also has implications for groups of countries which are apparently clear that they seek only a free trade or common market organisation. 'Spillover' processes of the sort identified in this chapter will certainly create pressures to push beyond the 'purely economic' arrangements initially envisaged. Demands that trade should be 'fair' as well as 'free' will inevitably emerge, leading to pressures for coordinated action in such areas as social, regional, monetary and environmental policy. The naive (or deliberate) division drawn between things 'economic' and 'political' will soon be revealed as a false dichotomy. By no means all countries, regions and groups are necessarily 'winners' as national economic barriers fall; the 'losers' will, as the EC experience shows, exert political pressure to introduce a 'social dimension' to counterbalance economic forces in the larger markets being created. These pressures will, as in the enlarged EC, become all the greater if the socioeconomic disparities between the richer and poorer members of a regional trading organisation are great. Thus some in the wealthier countries will either press to keep out competing products manufactured under less generous working conditions or insist that their partner states introduce social measures comparable to their own. In contrast, the poorer countries will seek political means to resist economic and monetary domination by the richer states with their multinational companies and powerful financial institutions. If social considerations do not mitigate the management of international trade, the growing numbers who see their jobs disappear and working conditions decline in the push for ever greater competitiveness may perceive the cause of their despair, rightly or wrongly, in regional economic integration. Simplistic national solutions could then replace excessively facile free trading expectations as vote winners. The EC in the 1990s already provides signs of such possibilities with the rise, albeit still limited, of nationalist movements. Other would-be 'common market' builders should take note and remember the global ravages wreaked by European nationalism in the past.

References

Abescat, B. (1993) Le désaccord Bruxelles-Tokyo, *L'Express*, 22–28 juillet, 29.

Bailey, J. (ed.) (1992) *Social Europe*, Longman, London.

Boissonnat, J. (1992) Au-delà du GATT, *L'Expansion*, 19 novembre–2 décembre.

Boyer, R. (1988) *In Search of Labour Market Flexibility: European Economies in Transition*, Clarendon Press, Oxford.

Casanova, J.-C. (1993) La spéculation et l'Europe, *L'Express*, 19 août, 16.

Cecchini, P. (1988) *The European Challenge: 1992 – The Benefits of a Single Market*, Wildwood House, Aldershot.

Dauvergne, A. (1992) Le champ des manoeuvres, *Le Point*, 28 novembre–4 décembre, 22–25.

Delavennat, C. (1993) Protectionnisme: la fièvre monte, *Le Point*, 27 février–5 mars, 61.

The Economist (1993d) Norway not hooked, *The Economist*, 22 May, 41.

The Economist (1993c) A survey of the European Community: a rude awakening, *The Economist*, 3 July, 1–24.

European Commission (1985) *Completing the Internal Market*, White Paper from the Commission to the European Council, Official Publications of the EC, Luxembourg.
European Commission (1987) *Treaties Establishing the European Communities and Treaties Amending These Treaties (Single European Act)*, Official Publications of the EC, Luxembourg.
European Commission (1988) The social aspects of the internal market, *Social Europe 1*, Supplement 7/88, D-G for Employment, Social Affairs and Education, Official Publications of the EC, Luxembourg.
European Commission (1989) *Background Report*, ISEC/B25/89, 11 October, Jean Monnet House, London.
European Commission (1990) *Community Charter of the Fundamental Social Rights of Workers*, Official Publications of the EC, Luxembourg.
European Commission (1992) *Treaty on European Union (Maastricht Treaty)*, Official Publications of the EC, Luxembourg.
Eurostat (1992) *Basic Statistics of the Community, 29th Edition*, Official Publications of the EC, Luxembourg.
L'Expansion (1992a) Guerre des tranchées au GATT, *L'Expansion*, 6–19 février.
L'Expansion (1992b) Au GATT, ce sont les pays les plus riches en céréales qui s'affrontent, *L'Expansion*, 5–18 mars.
Featherstone, K. (1989) The Mediterranean challenge: cohesion and external preferences, in: J. Lodge (ed.) *The European Community and the Challenge of the Future*, Pinter, London, 186–201.
Gros, D. and Thygesen, N. (1992) *European Monetary Integration: From the European Monetary System to the European Monetary Union*, Longman, London.
Haas, E.B. (1968) *The Uniting of Europe*, 2nd Edition, Stanford Press, Stanford.
Krause, L.B. (1968) European economic integration and the United States, in: J. Pinder (1991) *European Community: the building of a Union*, Oxford University Press, Oxford, 173–4.
Lindberg, L. (1963) *The Political Dynamics of European Integration*, Oxford University Press, Oxford.
Lindberg, L. and Scheingold, S. (1970) *Europe's Would-Be Polity*, Prentice-Hall, Englewood Cliffs, NJ.
Lintner, V. and Mazey, S. (eds.) (1991) *The European Community: Economic and Political Aspects*, McGraw-Hill, London.
Mitrany, D. (1933) *The Progress of International Government*, Allen & Unwin, London.
Mitrany, D. (1966) *A Working Peace System* (first published 1943), Quadrangle Books, Chicago.
Mitrany, D. (1975) *The Functional Theory of Politics*, Robertson, London.
Le Monde (1993) Deux visions incompatibles, editorial, *Le Monde*, 1–2 août.
Montagnon, P. (1989) EC could not become a trade fortress, *Financial Times*, 24 July.
Newman, M. (1991) Britain and the European Community: the impact of membership, in: V. Lintner and S. Mazey (eds.) *The European Community: Economic and Political Aspects*, McGraw-Hill, London, 146–164.
Nugent, N. (1989) *The Government and Politics of the European Community*, Macmillan, London.
Penketh, K. (1992) External trade policy, in: F. McDonald and S. Dearden (eds.) *European Economic Integration*, Longman, London, 146–158.
Pinder, J. (1991) *European Community: The Building of a Union*, Oxford University Press, Oxford.
Schuman, R. (1950) The Schuman Declaration, in, European Commission, *Jean Monnet: a Grand Design for Europe*, European Documentation 5/1988, Official Publications

of the EC, Luxembourg, 43–45.
Sked, A. (1992) The case against the Treaty, in: The European, *Maastricht Made Simple*, The European, special guide No. 1, 27.
Teague, P. (1989) *The European Community: The Social Dimension*, Longman, London.
Thatcher, M. (1988) *Britain and Europe, Text of the Prime Minister's Speech at Bruges on 20 September, 1988*, Conservative Political Centre, London.
Urwin, D.W. (1991) *The Community of Europe*, Longman, London.
Valance, G. (1993) France-Allemagne: le jour où tout a craqué, *L'Express*, 5 août, 16–21.
Wallace, H., Wallace, W.V. and Webb, C. (eds.) (1983) *Policy-Making in the European Community*, Macmillan, London.
Wise, M. (1984) *The Common Fisheries Policy of the European Community*, Methuen, London.
Wise, M. (1989) France and European Unity, in: R. Aldrich and J. Connel (eds.) *France in World Politics*, Routledge, London, 37–73.
Wise, M. (1991) War, peace and the European Community, in: N. Kliot and S. Waterman (eds.) *The Political Geography of Conflict and Peace*, Belhaven, London, 110–125.
Wise, M. and Gibb, R.A. (1993) *Single Market to Social Europe: the European Community in the 1990s*, Longman, London.

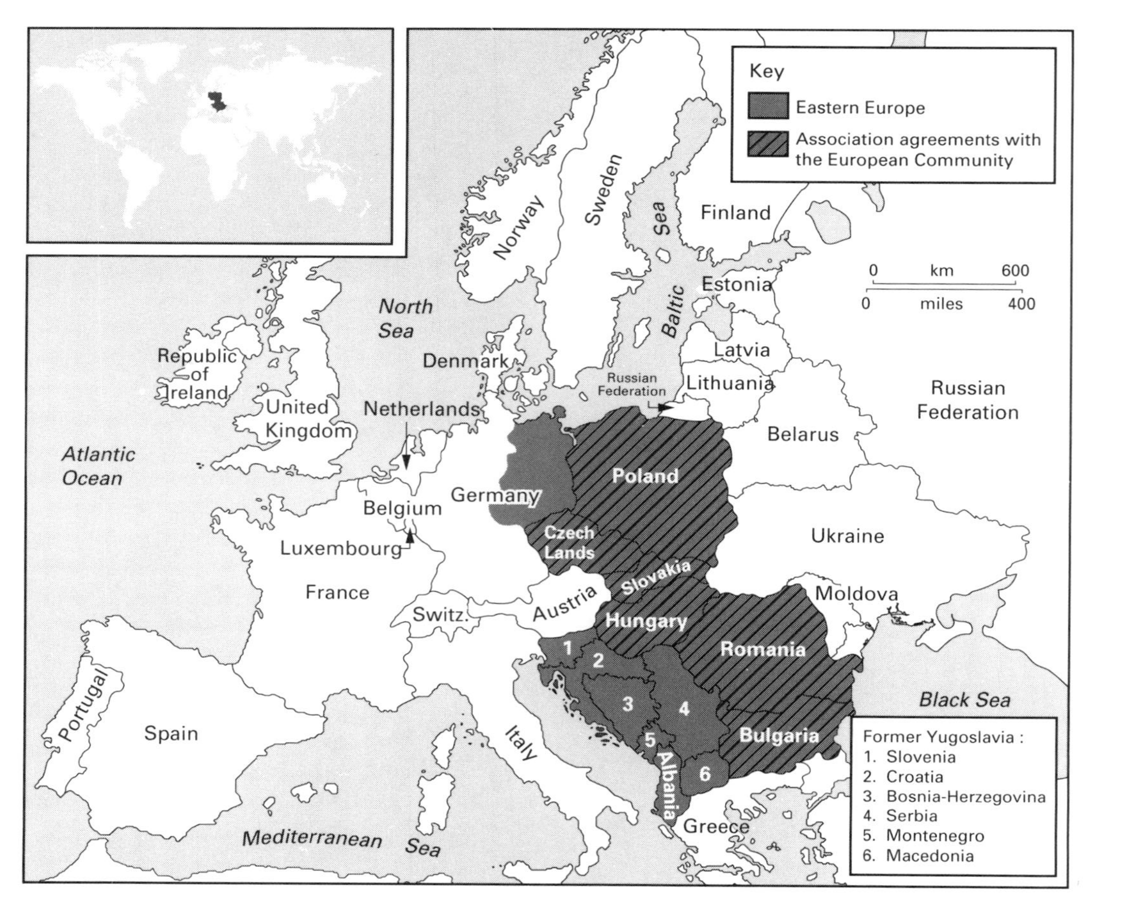
Key
Eastern Europe
Association agreements with the European Community
0 km 600
0 miles 400
Norway
Sweden
Finland
Baltic Sea
Estonia
Latvia
Lithuania
Russian Federation
Russian Federation
Belarus
Ukraine
Moldova
Black Sea
North Sea
Denmark
Republic of Ireland
United Kingdom
Netherlands
Belgium
Luxembourg
Germany
Poland
Czech Lands
Slovakia
Austria
Hungary
Romania
Bulgaria
Switz.
France
Atlantic Ocean
Portugal
Spain
Italy
Albania
Greece
Mediterranean Sea
1
2
3
4
5
6
Former Yugoslavia :
1. Slovenia
2. Croatia
3. Bosnia-Herzegovina
4. Serbia
5. Montenegro
6. Macedonia

Figure 4.1 Eastern Europe

4 Regional integration in eastern Europe

Wieslaw Michalak
Ryerson Polytechnic University

Intraregional cooperation and integration are not pursued with great enthusiasm in eastern Europe (Figure 4.1). Preferred slogans are a 'return to Europe' or 'joining Europe', both of which mean the same thing: a speedy integration with western European markets and the European Community (EC) in particular. The roots of this ambition, examined in this chapter, lie buried deep in the history of eastern Europe or, indeed, in the very origin of the term 'Eastern Europe' habitually used in the West to describe that part of Europe squeezed between the former Iron Curtain and the ex-Soviet Union. Until the Second World War, eastern Europe in its present form did not exist. Instead, there was a multiplicity of terms such as central Europe, the Balkans, the Habsburg Empire or Bohemia, all of which attempted to capture the sense of territory in an area where for centuries many different peoples mingled in a constant flux of changing political, social and ethnic circumstances. It was that territory which shielded western Europe from the many Asian and Russian invasions. The cost was a constant battle for identity and independence which significanlty delayed the industrial and social (but not political) development of this part of Europe. The Soviet takeover after 1945 put the straightjacket of 'Eastern Europe' on the many cultures, histories and geographies of this melting pot. The collapse in 1989 of the last imperial order in Europe – the Soviet Union – reopened the Pandora's box of conflicting national, ethnic, cultural, economic and political aspirations. For a great majority of east Europeans, past attempts at integration between the ex-Soviet Union and eastern European countries seem 'unnatural'. Without doubt most Czechs, Hungarians, Croats or Poles would consider their 'natural' focus to be western Europe and Roman Christianity.

The east European dream of a 'return to Europe' coincided with the end of the East/West confrontation and the collapse of the post-war order. At the same time, regionalism and the potential fragmentation of the GATT-orchestrated trading system polarised the West (Michalak and Gibb, 1992b). The deepening integration in the European Community has raised fears among east Europeans

Continental Trading Blocs: The Growth of Regionalism in the World Economy
Edited by R. Gibb and W. Michalak

that they will be left behind a new 'Golden Curtain' dividing the rich West from the poor East (Gibb and Michalak, 1993a, 1993b). Ironically if one looks at Europe as a whole, cross-border interaction, the movement of people, trade and transactions are far greater between western Europe and other regions of the world than with eastern Europe (Dawson, 1993; Wallace, 1992). In fact, economic, political and cultural interaction is far greater with the United States. Moreover compared to eastern Europe, western European states belong to a far greater number of international organisations, such as the Organisation for Economic Cooperation and Development (OECD), the Group of Seven (G7), the General Agreement on Tariffs and Trade (GATT), or the North Atlantic Treaty Organisation (NATO). To the extent that functional relationships and economic transactions induce the process of integration, western Europe can be expected to integrate more strongly with other parts of the world economy than with the newly democratic states of eastern Europe.

Despite this rather pessimistic scenario, even a casual look at a map (Figure 4.1) reveals that the dismantling of the Iron Curtain changed Europe beyond recognition. This chapter examines the prospects for regionalism in eastern Europe. First, the Soviet-style attempts at integration which lasted until 1990 are analysed and the implications of these attempts for the present conditions of intraregional integration are examined. Second, numerous blueprints for 'small' intraregional integration (i.e. integration between east European states) that have emerged since 1989 are evaluated. Third, the objectives and rationales associated with the most successful of these integrative efforts are explored and assessed. Finally, the discussion shifts to an enquiry into the relationships between eastern Europe and the EC. Focus here is concentrated on the prospects of 'large integration' between the two regions (i.e. integration between at least some east European states and the EC). It is postulated that the chances of east European membership of the EC are small until the process of economic transformation in the region is advanced further. Unfortunately, the optimistic 'return to Europe' scenario coincided with a turbulent period of internal disunity within the EC over the future of European integration; a disunity instigated in part by the events of 1989 in eastern Europe. An examination of the issues involved demands a brief look into post-war Soviet attempts to create the first 'socialist' trading bloc.

THE COUNCIL FOR MUTUAL ECONOMIC ASSISTANCE

In January 1949, Joseph Stalin announced the creation of the Council for Mutual Economic Assistance (CMEA) which was to include, in addition to the Soviet Union, Bulgaria, Czechoslovakia, Hungary, Poland and Romania (Figure 4.2). It was clear from the start that the reasons for establishing the CMEA were political rather than economic (Hamilton, 1990; Van Ham, 1993). The principal impulse behind this development was the perceived threat to the Soviet Union

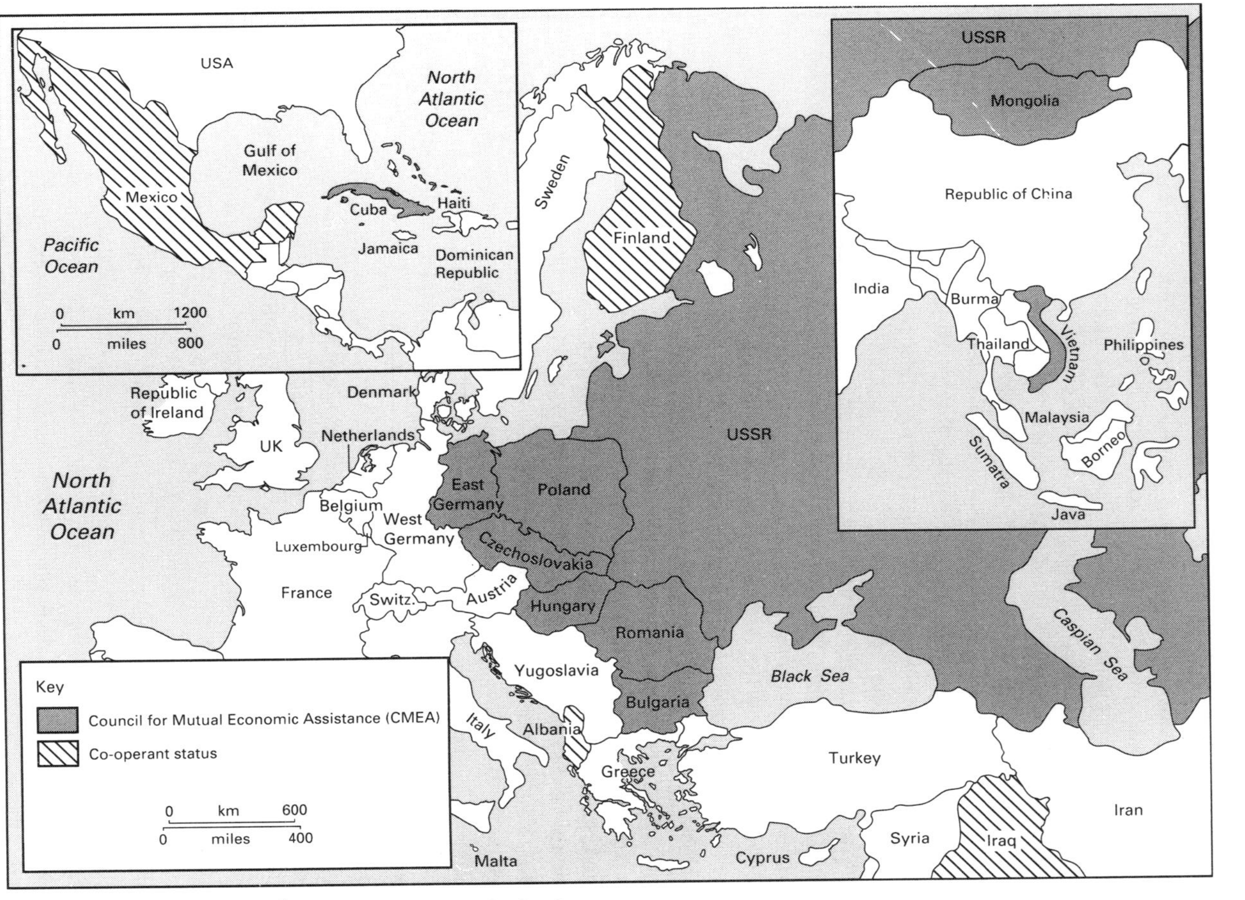

Figure 4.2 The Council for Mutual Economic Assistance

of the 1947 Marshall Plan, which offered aid not only to western Europe but also to eastern Europe (Korboński, 1990). The term 'assistance' in the name reflected the fact that the CMEA was set up primarily as a counterpart to the Organisation for European Economic Cooperation (OEEC) which monitored the Marshall Plan. In the view of the Soviet leadership, the economic integration efforts which followed the Marshall Plan in western Europe demanded a countermeasure in eastern Europe. It was argued that without some sort of framework the economies of eastern Europe would be difficult to control; as the example of Yugoslavia had amply demonstrated. Unfortunately, analysis of the factual material concerned with the creation of the CMEA is difficult since much of the evidence and relevant documents from 1949 were never made public (Van Brabant, 1974). As a result, a considerable difference of opinion exists about the origins of the CMEA. For example, some authors argue that the CMEA was not fully dominated by the Soviet Union and that some east European countries were able to use the organisation for their own national objectives (Van Ham, 1993).

The CMEA differed significantly in many institutional and conceptual aspects from Western institutions of economic integration. Unlike in western Europe, the integration process in the East was dominated, both economically and politically, by the Soviet Union. The Council was never intended as an evolving organisation like the EC, with its own political institutions and supranational powers transcending the local jurisdictions of member states. It was a committee, or rather a series of committees whose main function, at least superficially, was to promote industrialisation and a high rate of economic growth, coordination of medium- and long-term economic plans, specialisation in production and the maximisation of regional trade. The CMEA Secretariat in Moscow produced elaborate blueprints for economic cooperation and integration which had very little to do with the economic realities of east European countries. Moreover, it manifestly failed to deliver a platform for the multilateral coordination of economic plans and trade. Consequently, the CMEA never produced a uniform set of procedures enforcing the GATT-like rules of trade and cooperation between its members (CMEA members were not members of the GATT although some had 'observer' or 'associate' status – Haus, 1992). The integration of eastern Europe and the Soviet Union can therefore be described as '*radial*' since it was the Soviet Union and its bilateral links with each individual member state which defined the nature of the relationship. The linkages and consequently integration between smaller CMEA members was at best minimal. As a result, in the 1950s most CMEA participants fell back on autarchic economic policies supplemented by a network of bilateral agreements with the Soviet Union. All of this led to a very slow start, illustrated by the fact that the Council met only once between 1949 and 1954. In fact, the political functions of the CMEA took precedence over its economic objectives to such an extent that the organisation did not even have its own Charter until 1960.

The most significant impact of the CMEA on eastern Europe was the radical reorientation of traditional economic links from western and southern Europe in favour of the East and the transformation of the economy along the Soviet model of industrial development. The establishment of the CMEA put a sudden brake on the long-standing commercial ties between eastern Europe and Germany, Austria, Italy and France (Reynard, 1950). The increased volume of trade with the Soviet Union established the post-war pattern of dependence of east European economies on the resources of that vast country. Economic and political dependence was the result of a deliberate policy pursued by the Communist Party of the Soviet Union (CPSU), anxious to consolidate its political and military grip on eastern Europe. The Soviet leadership clearly intended to establish a distinct and defensible Soviet European bloc capable of defending Soviet imperial ambitions in Europe. Specifically, it ensured that the Soviet economic model would be enforced effectively in eastern Europe, making trade between eastern and western Europe nearly impossible. Consequently after 1950, east European countries embarked on 'catch-up industrialisation' programmes designed along the Stalinist model which, in the end, made their economies almost totally dependent on the raw materials and markets of the Soviet Union. In effect the CMEA contributed ultimately to the division of Europe into two distinct and hostile blocs rather than enabling the development of multilateral economic relationships.

The member states of the CMEA remained cautious toward 'socialist integration' throughout the existence of the organisation, despite numerous attempts by the Soviet leadership to expand trade (Kaser and Radice, 1986; Hamilton, 1990). Only at the end of the 1960s did the development of the European Economic Community (EEC) trigger renewed efforts at integration in eastern Europe. Especially after 1968, in response to the political disunity within its regional system demonstrated vividly by the Czechoslovakian crisis, the Soviet Union put renewed effort into the CMEA. It could be argued that something of a revival ensued as a result of the increasingly apparent success of western European economic integration. Although even east European planners and *apparatchiks* were cautious about the possibility of closer cooperation between Soviet-style economies, for ideological reasons it was argued that only the centrally planned model would be able to deliver a rational and well-balanced trading bloc. Despite those reservations, the Kremlin's enthusiasm for integration plans was motivated essentially by the desire to strengthen its hold over its satellites. Moreover, the Soviet leaders believed that a strong and coherent 'socialist trading bloc' would boost Moscow's negotiating position with the EC and the United States. The resulting stronger control over contacts between eastern Europe and western Europe would provide the additional benefit of closer 'brotherly' cooperation and integration. Another way of maintaining control over East/West trading relations was the Soviet Union's (and CMEA's) refusal to recognise the EC. It was not until June 1988, that a Joint Declaration was finally signed between the CMEA and the EC. Until then the Soviet Union attacked every aspect of the EC concept, especially that of political unity between 'imperialist forces of capitalism'. The political integration of the EC, and the remilitarisation of West Germany coupled with the NATO Treaty, were perceived as a direct threat toward Soviet interests in Europe and the Soviet Union itself. Although there were unofficial attempts to establish dialogue between the CMEA and the EC throughout the 1970s and 1980s, the tangible results of these diplomatic efforts were small. As a result, trade between the CMEA and the EC was relatively small (Gibb and Michalak, 1993a).

To stimulate flagging 'socialist economic integration', the Soviet Union organised a meeting of the CMEA Executive Committee in January 1969. The meeting produced a grand blueprint that not only envisaged the creation of a 'genuine socialist common market', but went so far as to plan for the political integration of member states and the creation of a supranational executive and planning institutions. In July 1971, the Comprehensive Programme for Socialist Integration (CPSI) was adopted. Most east European countries, however, politely declined to participate in the scheme which was quite unashamedly designed not only to increase Soviet control but also, in effect, advocated a gradual incorporation of eastern Europe into the Soviet Union. Consequently, the final text of the Programme included specific assurances against any supranational institutions with executive powers. Instead, the CPSI stressed improved coordination of intra- and extra-regional trade to the advantage of the Soviet Union and east European integration (Korboński, 1964; Wallace and Clarke, 1986). Several huge joint projects were carried out over the years as a tangible result of the CPSI, including the 'Friendship Oil Pipeline', the 'Peace Electric Power Grid' and the 'Brotherhood Gas Pipeline'.

The CMEA did not make the progress in economic integration experienced in western Europe. Only in energy and transportation did effective cooperation between member states take place. Throughout its existence until 1990 there was no consensus and a considerable degree of scepticism toward plans to embark on an integration process which might have developed the CMEA along EC lines. In fact, one could argue that between 1950 and 1970 a measure of disintegration occurred (Wallace and Clarke, 1986; Korboński, 1990; Marer, 1991). The reasons for this decline lie in the nature of the integration mechanism introduced by the centrally planned economic model. The regime for facilitating trade among CMEA members through bilateral agreements eliminated market impulses from trading relations. As a result, the CMEA reproduced in the sphere of international cooperation all the anomalies of bureaucratic planning typical of socialist economies (i.e. low quality, short supply, cost inefficiency, overmanning, etc.). Moreover, the isolationist and autarchic tendencies which prevailed throughout most of the post-war period in eastern Europe led to serious deficiencies in research, development and technological innovation. A realistic calculation of the costs and benefits of economic integration and trade was, in practice, impossible. Nevertheless, according to some estimates, the volume of intraregional trade was greater than it would have been in the absence of the CMEA (Brada, 1992; Hewett, 1976; Biessen, 1991). However, since there were no readily available measures to evaluate the degree of integration in eastern Europe it is extremely difficult, even after 1989, to determine how post-CMEA integration should proceed. For example, it is not even clear which of the past flows of goods, services and resources ought to be retained and which eliminated.

Another aspect of the CMEA which eventually contributed to its collapse was the promotion of specialisation in production among member countries. On the face of it this policy had some potential benefits for members. In theory, it was to lead to the promotion of efficiency and technological innovation, especially in industry, by stressing the areas of expertise, experience and resources of member countries which would not have to be duplicated throughout the bloc. This specialisation was to be implemented by promoting the central planning of production and investment throughout the region. Predictably, this aspect of 'socialist' integration was not successful. Almost all member countries sought their 'just' share of specialised production regardless of their ability to develop basic competence in the areas of specialisation. Thus industries and technologies perceived to be strategic, such as the electronics industry, computer engineering and manufacturing, chemicals and a whole array of high-technology research, were usually fought over in order to secure access to advanced engineering and design. The specialisation policy degenerated into a process of parcelling out new areas of production among all members, in effect broadening rather than narrowing the range of production (Brada, 1992). Moreover, the high technologies with military applications nearly all ended up in the Soviet Union or its most trustworthy ally, East Germany. Most other members of the 'socialist family' were simply shut off from any developments in these areas. For example, a very promising research and design effort in 'Odra' computers engineered and designed in Poland was closed down because it was the Soviet Union which was to 'specialise' in this industry.

The economic collapse of the Soviet economy in the 1980s wholly discredited the 'socialist' model of integration. In 1989, the introduction of hard currency payments for intra-CMEA transactions replacing the so-called 'transferable ruble' (which was neither transferable nor convertible) dealt a terminal blow to the process of CMEA integration (Palankai, 1991). The departure of East Germany and subsequent German unification removed one of the most important contributors to intraregional trade. By the end of 1990 the CMEA became largely irrelevant and on January 5 1991 the member states decided unanimously to dismantle the bloc.

Because of the lack of any reliable data it is extremely difficult to assess the legacy of the CMEA to east European economies. If analysed in a Vinerian sense (see Chapter 1) the trade within the bloc probably did not markedly raise its members' welfare. Most of the intraregional trade was the result of trade diversion after the division of Europe. This diversion was created first and foremost by political and military interference from Moscow. However, considerable differences in resource endowment and industrial base also played a role. None of the east European countries could reasonably be viewed as low-cost suppliers of industrial goods to the Soviet Union. Conversely, in spite of its vast natural resource base the Soviet Union could not be regarded as a relatively low-cost supplier of any type of fuel or raw materials to eastern Europe. Technological backwardness, overmanning and cost inefficiencies greatly increased the real production costs in the Soviet Union. Both the Soviet Union and eastern Europe benefited considerably from trade with other regions of low-cost production especially in Asia (Bradshaw, 1991). Thus the current hostility of east European countries to the idea of an organisation replacing the CMEA is a reflection of the *political* changes that took place in the Soviet Union as well as sombre economic analysis.

'SMALL INTEGRATION' IN EASTERN EUROPE

The failure of the CMEA and 'socialist integration' in eastern Europe had predominantly political causes. Immediately after the collapse of the totalitarian Communist regimes, the abolition of the remnants of CMEA-style integration became a priority for nearly all governments in the region. Since the introduction of hard currency in trade between ex-CMEA members exposed a whole range of structural problems in their economies, establishing a new organisation in place of the CMEA was not a priority. Indeed, outright hostility to the idea of such an organisation was a reflection of the revolutionary changes that had taken place in eastern Europe and the ex-Soviet republics. The many blueprints for regional integration which have emerged since 1989 can be classified in terms of two principles: that of trade and cooperation or deeper integration.

Initially the plans for economic integration in eastern Europe also included the Soviet Union. The significant assumption here was that the Soviet Union would continue to exist as a single economic entity undergoing market transformation. However, the disintegration of that country coupled with continuing uncertainty about the future course of political and economic reforms made such plans totally irrelevant. A much more pressing issue for the ex-Soviet republics continues to be the management of their own internal economic and fiscal relations rather than integration with eastern Europe (see Chapter 5). Moreover, it would be unrealistic to assume that in the climate of continuing political and economic upheavals the Russian Federation would be inclined to participate in any scheme with eastern Europe, which is going through a similar, if somewhat less intense, economic crisis. Consequently, the idea of a new integration scheme with the ex-Soviet republics would have hindered the emerging reorientation of trade and restructuring of industry (Brada, 1992). In short, any integration with the East would have been totally inconsistent with eastern Europe's basic aims of 'joining Europe' and the EC.

Brada (1992) argues that, at least in the short- to medium-term, proposals for creating formal organisations, such as a customs or payments union, are unlikely to yield any significant benefits for eastern European countries or ex-Soviet republics. To start with, the remnants of central planning and state monopolies still in place make integration virtually impossible because of their systematic incompatibility (see Chapter 5). Any forced policies to revert trade between eastern Europe and the Commonwealth of the Independent States (CIS) would simply reintroduce many of the inefficiencies that existed in the past. After all, the CMEA-style integration was based on a Johnsonian (Johnson, 1965) concept of regional integration promoting greater industrialisation, rather than static efficiency as seen by Viner. Thus it is likely that factors other than the lack of regional integration are responsible for the economic crisis in eastern Europe. Past intraregional trade was artificially high, and therefore uneconomically high. Consequently any efforts to prevent the decline and restructuring of trade between ex-CMEA members would be futile and counterproductive. As Brada put it, any rational discussion of the benefits of regional integration in eastern Europe:

> depends critically on the economic rationality of the trade that has been eliminated, on the actual relationship between intra-regional and extra-regional trade of the East European countries, and the sources of the decline in intra-regional trade. (Brada, 1992, 8)

The conversion to hard currency accounts, market prices and convertible currencies among ex-CMEA countries has led to confusion about the geographical distribution of trade between these countries. There is no doubt that in some cases very harsh reforms aimed at introducing market economies, such as the abolition of state subsidies, restrictive monetary and fiscal policies, price reforms and other austerity measures, contributed to the decline of output and trade. Since nearly all the ex-CMEA countries of eastern Europe are undergoing similar restructuring more or less at the same time, a foreign trade multiplier effect transmitted a deflationary shock throughout the region. On top of the serious consequences associated with the market reforms, German unification eliminated not only an extremely important trading partner but also West Germany as a major source of badly needed foreign investment. A *de facto* East German membership of the EC caused a series of shock waves throughout the east European economies who could not count on a generous influx of capital and instant access to the entire EC and EFTA market.

It is unclear whether the decline of trade in eastern Europe is the straightforward result of the demise of the CMEA. On the one hand, according to some recent estimates (Havrylyshyn and Pritchett, 1991; Brada, 1992), the collapse of the CMEA should, at least theoretically, lead to a 60 to 75 per cent decline in intraregional trade. However, Wang and Winters (1991) indicate that even without the CMEA the market reforms alone would have reduced trade by at least 50 per cent since roughly half of it could not in any case be justified on economic grounds. It appears, therefore, that the demise of 'socialist' integration did not have a great effect on intraregional trade and that market reforms contributed far more to its decline than regional disintegration. Brada (1992) and Köves (1992) argue that the restructuring of intraregional trade, including the decline of its exaggerated volume, and the accompanying redirection of this trade toward the West, will in the long-run improve the welfare of east Europeans regardless of the eventual scheme adopted for integration within the region. Because of the lack of comprehensive data it is not entirely clear whether increased trade with the West is trade diverting or trade creating: however, preliminary data suggests that the latter is the case (Brada, 1992). For example, the declining share of manufacturing imports from the ex-Soviet republics and the parallel increase of such trade with western Europe suggests a greater degree of rationalisation. Conversely, the greater proportion of fuels and raw materials imported from Ukraine and Russia indicates a much more realistic structure of trade with the East. This type of trade restructuring is necessary before any formal attempt at economic integration in the region.

The Czech Republic, Hungary, Poland and Slovakia have made the greatest efforts to establish some degree of regional integration based on the market economy (often referred to as 'small integration'). Particularly important were the currency and price reforms successfully pushed through the parliaments of the four countries. As a result, regional economic integration has a much better chance of success between these countries than in any other part of eastern Europe or the ex-Soviet Union. In fact, these four countries have already established the first eastern European economic bloc – the Central European Free Trade Agreement (CEFTA) – which came into force on March 1, 1993. The agreement calls for mutual tariff reductions in three phases over an eight-year period. In the first phase tariffs will be eliminated on industrial goods. The four countries had different tariffs ranging from 30 to 60 per cent on their exports before the agreement. Border controls of goods traded within the region are to be phased out by 1997 and all remaining tariffs are to be eliminated by 2001.

The agreement is modelled after the European Free Trade Association (EFTA) which accepted Bulgaria, the Czech Republic, Hungary, Poland, Romania and Slovakia as associate members earlier in 1992 and 1993. The new trading bloc – CEFTA – creates a sizeable market of 64.2 million people. Its principal aim is to reverse years of decline in trade between member states, this time on the basis of market forces rather than command economies. The agreement to reduce protective customs duties means that present tariffs among the four countries will be matched to the level of duties imposed on trade between the member states of the EC. Despite these developments in east European integration, serious differences of opinion about the purpose of any further deepening of relationships remain. For example, some argue that any increase in trade, especially between uneconomic state monopolies, will hamper the long-term prospects for industrial restructuring and comparative advantage. Vaclav Klaus, the Czech Prime Minister, repeatedly argued as late as January 1993 that any further integration between these economies makes little sense since it will merely delay the badly needed restructuring of uncompetitive industries. Moreover, increased trade in the absence of a market mechanism may result in further unemployment and adjustment costs, in effect slowing down genuine integration in the region.

Interestingly it is usually western, and especially EC, politicians who insist on further integration of eastern Europe not only in the economic but also in the political sphere. The insistence of the Community on some form of integration aroused a degree of suspicion among east Europeans about the true motives behind such proposals. In response to such pressures the so-called 'Visegràd Group' was established in order to coordinate policies between the Czech Republic, Hungary, Poland and Slovakia. The idea was also popular among some well-known 'pro-European' intellectuals from eastern Europe, Adam Michnik, Gyorgy Konrad, George Schöpflin, Vaclav Havel and Milan Kundera among others, who argued that such an alliance could act as a guarantee of the region's independence both from the East and West. Moreover, there was very considerable interest in such a grouping in the West even before 1989. A few months before the 'velvet revolution', Zbigniew Brzezinski – a well-known American policy maker and former security adviser to President Carter suggested a revival of the idea of a Polish-Czechoslovak federation. This was an old idea proposed during the Second World War by Sikorski and Benes, the leaders of the Polish and Czechoslovak governments in exile. Although nothing came out of these plans at the time because of the Soviet takeover, the collapse of the Communist regimes introduced an intriguing new possibility of a renewed East Central European Federation which would include Hungary. In fact, Brzezinski went as far as to suggest that western economic and political assistance should be made *conditional* to some degree on the willingness of east Europeans to engage in multilateral cooperation. As Brzezinski explained:

> In light of the Benelux experiences, it would be worthwhile to encourage it [integration in Eastern Europe], because that way such international cooperation abilities could develop in all three countries, which will be greatly needed in Western Europe. (Brzezinski, 1990)

The idea found vigorous support in western Europe and the EC. In particular, such cooperation would calm western fears about the stability of the region and the possible 'balkanisation' of east-central Europe.

Despite pressure from the West and significant support for the idea by

Poland, Vaclav Havel initially rejected the plan and directed his first steps as the President of the newly democratic Czechoslovakia toward Germany and Austria and not to Poland or Hungary. Western Europe, and Germany and Austria in particular, figure much higher on the political agenda of the Czech Republic. Nevertheless, after months of pressure the three governments of Czechoslovakia, Hungary and Poland met in February 1991 in the old capital of Hungary, Visegràd. After a few days of negotiations Josef Antall, Vaclav Havel and Lech Walesa signed the Declaration of Cooperation. Thus the 'Visegràd Triangle', renamed the 'Visegràd Group' after the break-up of Czechoslovakia, was established.

Although the 'Visegràd Group' is clearly a step toward east European integration, the Declaration fell short of expectations. Despite Polish proposals to establish a formal institutional structure. Hungary and Czechoslovakia insisted on defining the cooperation only in terms of regular consultations. During the organisation's next meeting in Cracow, the four respective governments committed themselves to coordinating their efforts in pursuing the objective of integration with the EC. Particularly significant was the commitment to consultations and the harmonisation of policies aimed at developing linkages with western European security organisations such as NATO and the Western European Union (WEU). The 'Visegràd Group' is therefore evolving as a transitionary umbrella organisation focusing on issues of security. Especially high on its agenda is preparation of the four countries for eventual membership of NATO. Until the Russian warning in 1993 directed toward the West, some NATO functionaries expressed support for some form of association status for the east European countries. Unfortunately, further security arrangements in eastern Europe and some ex-Soviet Republics ran into serious difficulties with the Russian leadership. The Russians argued against integration and regional blocs such as the 'Visegràd Group', CEFTA or Hexagonale, on the grounds that they were harmful to Russian interests and might eventually lead to an anti-Russian *cordon sanitaire*. This is why Russia warned Poland in 1992 not to contemplate any kind of economic or military alliance with the Ukraine; a move almost unanimously perceived as an attempt to weaken Russia. Indeed, in its new military doctrine published in October 1993, Russia identified Poland as a higher threat to its security than China or Iran. Thus Russia prefers a kind of 'finlandisation' of eastern Europe: bilateral agreements and a non-aligned political orientation. In short, rightly or wrongly, Russian politicians perceive eastern European regional integration as a direct threat to Russian interests. Needless to add, few in eastern Europe would seriously contemplate a move endangering their relations with Russia, even if such a move would please western Europeans.

The West and the EC insist, however, on some kind of eastern European integration. The motive behind this insistence seems to be predominantly political not economic. The usual line of argument is that without good-neighbourly relations among eastern European countries, they cannot enjoy good relations with the West. Another argument often used by EC politicians is that the West should avoid encouraging any kind of 'race' for Community membership which could harm 'small' regional integration. However, Köves (1992) argues that there is absolutely no evidence to support the belief that rivalry for western aid and an acceleration of the process of adjustment to EC laws, regulations and norms would endanger relations among eastern European or EC countries. Probably the opposite is true, as adjusting to EC rules, defining the conditions and producing the timetable for joining the EC would help not only to shape democracy and economic reforms but also to promote cooperation and integration between the east European countries. There is no guarantee that establishing an institutional facade of multilateral integration in eastern Europe would prevent conflicts. As Köves puts it:

> if such cooperation were to be established because of Western pressures alone it would be very difficult to avoid concluding that the whole idea of Central European integration was but a convenient way of preventing the Central European countries from becoming integrated into Europe. As practically desirable as closer cooperation among these countries is, it is unrealistic to expect significant changes in the region's economic situation as a result of any conceivable attempt at 'small' integration. (Köves, 1992, 91)

The simple truth is that eastern European countries are far from being each other's most important trading partners. For example, in 1992 trade between the Visegràd partners amounted to a mere $2.2 billion, that is less than 10 per cent of their respective total trade for that year. In contrast, EC exports to eastern Europe have increased by over 20 per cent annually since 1990. At the same time exports from eastern Europe to the EC boomed to an annual average of 41 per cent for Poland, 53 per cent for ex-Czechoslovakia and 18 per cent for Hungary between 1990 and 1992 (Business Central Europe, 1993).

The limits to trade development among moderately developed countries are well known and should be taken into consideration in any discussion about the relative merits of east European integration (see Chapter 7). This is why all of the east European countries pursue the expansion of trade with the developed West and not with similarly developed or, as some would prefer, underdeveloped regions. Finally, the political and economic integration of eastern Europe faces quiet but determined opposition on the part of the Czech Republic and, to a lesser extent, Hungary. In a statement given during an interview with *Le Figaro* in January 1993, Vaclav Klaus insisted that cooperation in eastern Europe was an artificial process created by the West. Klaus argued that the 'Visegràd Group' was not created from below as the expression of the political will of east Europeans, but from above as a political decision reflecting an insistence of the West to create another alternative bloc in eastern Europe. Nearly all east Europeans seem convinced that regional integration at this stage of transition would be counterproductive. The only reasonable option is a quick 'large integration' with western economies and the EC.

'LARGE INTEGRATION' AND EC MEMBERSHIP

From the beginning of the revolutionary changes in 1989, several newly democratic governments in the region declared their intention to 'join Europe'. In practice this translated into a policy which established full Community membership as a prime political and economic objective. The benefits associated with EC membership and so-called 'large integration' were recognised in most east European countries long before the collapse of the Soviet economy and the demise of the CMEA (Michalski and Wallace, 1992). Moreover, EC membership symbolises both the re-creation of a link between what after the Second World War became known as eastern Europe, and the more pragmatic intention of economic and political integration with the West. Perhaps the most important step in this direction to date was the decision to abandon CMEA-style integration in favour of resurrecting trade with the West. This dramatic reversal of the post-war patterns of economic relations led to a considerable controversy both in eastern Europe and the EC over the means and consequences of future enlargement (see Chapter 3). East European politicians consider trade with the West and the EC in particular as a source of economic and fiscal discipline, stability and the basis for economic recovery achieved through greater productivity, efficiency, foreign investment and transfers of technology (Michalak, 1993). The dramatic political changes in eastern Europe have allowed individual countries for the first time since the Second World War to negotiate directly with the EC. Perhaps even more importantly, the reorientation toward the West is an expression of a desire to re-establish a common European culture and assure future security by establishing a common defence and military infrastructure against external threats. The President of Hungary, Arpad Goncz, expressed these aspirations in the following way:

> Working out good relations with our neighbours is a matter of coming to terms with our past and present, but we want especially to come to terms with our future and establish ourselves as part of a common Europe. It is the western, or European standards of social, political and economic activities, not to speak of the western standards of living, that we want to see in Hungary, so we turn our eyes and mind to the West above all. (quoted in Miall, 1993, 92)

EC membership is seen as the best way of *de facto* east European integration through common participation in the EC's political, economic and social institutions. In September 1989, initial exploratory steps were taken to formalise the

Community's relations with eastern Europe. Unfortunately, the resulting 'first generation' Agreements did little to liberalise trade with the West and included only timetables for the removal of some trade restrictions by 1995 (Horowitz, 1991; Grabska, 1992). However, as the political barriers between eastern Europe and the West began to disappear after 1989, these agreements quickly became irrelevant (De Michelis, 1990; Brzezinski, 1991).

The Czech Republic, Hungary, Poland and to a lesser degree Slovakia have been at the forefront of the new round of negotiations with the EC because of the radical package of reforms which these countries adopted at the outset of the 1990s. In response to increasing pressure from the 'Visegràd Group' and after long and difficult negotiations, the Community concluded new 'second generation' Association Agreements (also referred to as 'Europe Agreements'). Under these Agreements, the terms of which are nearly identical for all four countries (considerably different Association Agreements were later concluded with Bulgaria and Romania), the EC accepted an asymmetric tariff reduction regime in favour of eastern Europe, recognising the latter's much weaker economic position. Accordingly, the EC will reduce tariffs on its imports of industrial goods from eastern Europe during the five-year period following the Agreements, whilst the eastern European signatories will reciprocate with tariff reductions in the following five years. However, the principle of an asymmetry of concessions should not be confused with the actual benefits of such an asymmetric relationship. It is instructive to examine briefly the negotiations leading to the formal accords.

Eleven months of often difficult and acrimonious negotiations were marked by three conflicts over trade issues between the 'Visegràd Group' and the EC. All three conflicts have their roots in the EC's trade policies and especially its recent protectionist tendencies (see Chapter 3). The first conflict was over agriculture. In September 1991, France assisted by Ireland and Belgium vetoed an agreement leading to increased quotas for meat from the three east European countries. In response Poland withdrew from the negotiations threatening the entire process. Since the Polish protest had broader implications for the international prestige of the EC, the Commission of the EC applied intense pressure on both France and Poland to return to the bargaining process. In fact, French objections were purely symbolic and designed for domestic political consumption. A vivid example of just how acrimonious the negotiations were is reflected in the fact that the rejected agreement would have increased EC quotas for all four countries by 500 tonnes of beef and 900 tonnes of lamb per year; by EC standards an entirely insignificant amount of trade. In the end, a compromise was reached and the EC agreed to increase the meat quotas by 10 per cent per year over a five-year period. However, the planned liberalisation of trade in other agricultural products will begin only after 1995. Thus, in effect, the EC continues to restrict access for eastern European agricultural products.

The two other areas of disagreement concentrated on the manufacturing and service sectors. In particular, textiles, steel and coal were singled out by the Commission as 'sensitive' trade areas. The less developed members of the Community, such as Portugal and Spain, expressed concerns over textiles competition in EC markets. In response to these concerns, the EC insisted on extending the tariff liberalisation process until at least 1998. Furthermore, the Portuguese textile industry is to be subsidised through direct financial assistance from Brussels. Finally, Germany, France, Portugal and Italy have voiced concerns over the elimination of EC quotas and duties on steel products and coal. Initially, these countries lobbied the Commission for a voluntary export restraint (VER) provision to be imposed on eastern Europe as part of the Association Agreements. Luckily for the eastern European producers the Commission rejected these proposals. Unfortunately, the EC introduced instead a safeguard clause permitting Community members to restrict steel imports by invoking anti-dumping procedures. In effect, the EC gave its members a free hand to decide arbitrarily to what extent east European steel products are to be allowed on their markets. Predictably EC steel producers, who for years have been struggling to compete, convinced their respective governments to reduce import quotas for steel products from Poland, Slovakia, the Czech Republic and Hungary, accusing them of dumping practices. Consequently, in the face of increased protectionist pressure from within the EC, in 1993 the Commission imposed additional tariffs on imports of steel, cutting off the only markets on which thousands of jobs in eastern Europe depend, in direct contradiction of the GATT.

According to a recent study which examines the costs and benefits of enlarging the EC to include members from the East, the combination of low incomes, large populations and even larger agricultural sectors would translate into a net transfer of EC money from both the structural funds and the Common Agricultural Policy (CAP) (CEPR, 1992). The economies of eastern Europe are poor by EC standards (Table 4.1). Only Slovenia and ex-Czechoslovakia had in 1991 GDP per capita income levels approaching the poorest member of the EC, Portugal. Poland and Romania, the most populous east European countries, had GDP per capita levels almost three times lower than Portugal. In addition the dramatic fall in production levels, which for nearly all countries in the region averaged around 10 per cent per year or more during the 1990–1991 period, increased the potential troubles for the EC. Only Poland achieved a significant improvement in growth rate, which for 1993 is expected to reach a respectable 4.5 per cent. Most other countries in the region are expected to do much worse, with either a further contraction in their industrial output or very slow growth. As a result, recent estimates point unambiguously to a large net transfer of funds from the EC in the event of enlargement (CEPR, 1992).

The CEPR study estimates that net annual transfers to the five potential east European members would amount to over $15 billion per annum (Table 4.2). As expected, the most populous countries in the region with the largest agricultural sectors would receive the largest amounts of money; over $6 billion to Poland

Table 4.1 Key economic indicators for east European countries in 1992

Country	*GDP per capita (in $)*	*Production change average between 1991 and 1992 (% of change)*
Albania	500	–15.5
Bulgaria	2300	–17.5
Croatia	2900	–16.5
Czechoslovakia	3400	–9.5
Estonia	2600	–7.5
Hungary	2600	–8.0
Latvia	2800	–4.0
Lithuania	2400	–7.5
Poland	1900	–9.5
Romania	1400	–10.5
Slovenia	5200	–10.0
Portugal	3800	+2.7
Germany (western part)	20400	+2.0
EC Average	18000	+2.2

Source: CEPR, 1992, 61 (modified).

Table 4.2 Estimated EC contributions and receipts of selected east European countries (all values in millions of $)

Country	*Average trade with the EC between 1988 and 1990 (% of change)*	*Contributions to the EC*	*Receipts from structural fund*	*Receipts from CAP*	*Net contributions to the EC*
Bulgaria	41	310	1422	609	–1721
Czechoslovakia	46	728	1605	526	–11403
Hungary	50	402	1481	642	–1721
Poland	42	964	5428	1663	–6127
Romania	34	467	3764	955	–4252
Total	–	2871	13700	4395	–15224

Source: CEPR, 1992, 72 (modified).

and $4 billion to Romania per annum. However, even Bulgaria, Hungary and ex-Czechoslovakia would present a serious financial drain on the EC budget. According to CEPR estimates, even if eastern European incomes and agricultural output doubled relative to their 1989 levels, the net receipts of funds from the EC would not diminish. On the contrary, the net transfer of money would increase. These estimates suggest strongly that it will take a long time before eastern European economies develop into a position where, under existing EC rules, they would be acceptable candidates for full EC membership. Clearly, such large transfers of capital from the EC's already strained budget would be politically unacceptable to current member states. Consequently, considering the current political difficulties facing EC decision making, full membership is not a realistic objective for eastern European countries for at least two decades (CEPR, 1992).

However, CEPR argues that a clear signal committing the EC and EFTA to a full-scale integration of eastern Europe into their internal markets is required if a successful transformation is to be achieved. A firm commitment to the speedy dismantling of trade barriers and tariffs is needed if east European countries are ever to be in a position to enter the EC. There are substantial mutual benefits in the immediate large-scale liberalisation of trade and 'large integration' between eastern and western Europe. In fact, the liberalisation of trade between these two regions in Europe is arguably the single most important factor capable of promoting the successful transformation of eastern European economies and societies in preparation for full EC membership. Moreover, the expansion of trade with eastern Europe including agriculture, textiles, steel and coal producers is in the self-interest of the EC. Such a liberalisation along GATT lines would substantially benefit Western producers by providing 'export market opportunities to western economies on a scale and speed unprecedented in modern history' (CEPR, 1992, 8). Unfortunately the current Association Agreements do not encourage positive liberalisation of trade in these crucial areas.

CONCLUSIONS

The difficulties involved in establishing formal trade links between eastern and western Europe reflect, in part, the internal tensions within the EC itself. The EC's agricultural policy and relative decline in sectors such as textiles, steel and coal have long bedevilled the Community's trade policies (see Chapter 3). Moreover, the hesitant policies adopted by the Community expose an even deeper disunion in the EC over the speed and direction of its future integration. Thus the protectionist measures introduced by the EC against eastern Europe are, of course, not new. However, it is doubtful whether an East/West trade confrontation would have any significant effect on EC trade policy. The Community negotiates from a position of unchallenged strength, making any balanced negotiations virtually impossible. This is why it is highly likely that the issue of East/West economic integration will be determined in international fora such as the GATT or bilateral negotiations between NAFTA (see Chapter 6) and the EC; in other words, totally outside east European influence. It is also unclear how long the EC can continue these tactics while expanding the number of east European associates. Significantly, the text of the Association Agreements explicitly recognises full membership as the ultimate objective of the accord. However, no timetable for membership negotiations is specified.

Despite this recognition, it is not at all clear whether Association Agreements are designed for the countries which may be able, in the end, to join the EC as full members. In other words, EC politicians will have to think hard whether to restrict Association Agreements to those countries that they are prepared in principle to admit as potential full members. Perhaps in recognition of the weakness of this position, Hans van den Broek, the EC Commissioner for External Affairs, recommended in April 1993 that the EC should commit itself to membership of the Visegràd Group. At the same time, however, he also ruled out membership of the EC for Russia and other ex-Soviet republics, arguing that the EC should draw the line at the states with which the EC has Association Agreements.

The largely unexpected collapse in 1989 of the old order in Europe caught most by surprise. Despite considerable progress, the future of the ensuing east European integration is still uncertain. The reluctance of the West to open-up their markets and the Russian attitude toward the region raise the serious possibility of the rebirth of the infamous *cordon sanitaire* designed by west Europeans to shield themselves from eastern upheavals. Indeed, western insistence on 'small' integration is perceived, at least by some, as another way of re-creating this traditional role. Once again eastern Europe finds itself squeezed between two powerful neighbours, Russia and the West. Although there are no imminent threats, some of the signals coming from Moscow are all too familiar to east Europeans. Equally important however, are relations with the EC and EFTA, or more specifically Germany and Austria.

Albeit the largest economy in the Community, Germany has remained dependent on other, especially European, markets. For example, in 1991 no less than 54 per cent of German exports went to the EC, and a further 16 per cent to EFTA countries. For the German and Austrian governments, participation in regional and multilateral integration has been a means to influence trading relations with its partners. Expansion of German and Austrian trade into eastern Europe is of paramount importance to these countries (Morgan, 1992). This is probably one of the principal reasons why Germany has been an ardent supporter of early integration between the EC and the Visegràd Group and of establishing timetables for membership negotiations with the other east European countries. Austria has played a similar role with respect to EFTA. Moreover, investment by German and Austrian corporations and multinationals in the Visegràd Group exceeded those of other countries (Michalak, 1993). Not surprisingly, the unbalanced nature of investment flowing from the West again raised fears of German expansionist policies which were, in substantial part, responsible for the present state of eastern Europe. To be fair, at least at present, it appears that German investment is motivated by purely market considerations. Indeed, east Europeans compete for that investment. Simply put, Germany has a huge comparative advantage over the depressed east European economies. It is also both geographically and economically ideally placed to service the vast needs of this region for capital, technology, marketing skills, etc. There is also no serious evidence that German foreign policy aims to re-establish a national hegemony or political dominance in eastern Europe.

For the immediate future, the relationship between the EC and eastern Europe will be based on existing Association Agreements, which provide the first significant step toward 'large' integration. However, interaction can and should be strengthened in the economic sphere by improving market access, including the notorious 'sensitive' sectors, and by expanding the political and security framework for future integration. Eastern Europe can ill afford to remain outside the EC at a time of increasing protectionism amongst the principal trading blocs (see Chapter 2). Although intraregional 'small' integration is important, the reluctance of east Europeans to go along this path is in no small part dictated by a fear of being locked outside the gates of the largest existing trading bloc, well-known for its protectionist actions. It is an almost universally shared opinion in the East that the GATT rules alone will not guarantee access to the EC market.

It is likely that the future of integration in eastern Europe lies with a multi-polar and multi-speed Europe into which the EC is likely to evolve (see Chapter 3). In principle, the tri-polar design for the European Political Area proposed by the Commission enables EC enlargement to include eastern Europe. This form of integration could be achieved without any significant costs being incurred by the EC. Regardless of the final blueprint adopted, it is clear that the old paradigm of western European integration cannot be sustained without modification. The old order has been tailored to the practical needs of western Europe which cannot be replicated in eastern Europe. Clearly, the new paradigm of integration in Europe must include eastern Europe if the old division of Europe between 'haves' and 'have nots' is ever to be overcome.

References

Biessen, G. (1991) Is the impact of central planning on the level of foreign trade really negative?, *Journal of Comparative Economics*, 15, 22–44.

Brada, J.F. (1992) *Regional Integration in Eastern Europe: Prospects for Integration Within the Region and With the European Community*, paper presented at World Bank and CEPR Conference on New Dimensions in Regional Integration, April 2–3, Washington, D.C.

Bradshaw, M.J. (ed.) (1991) *The Soviet Union: A New Regional Geography*?, Belhaven, London.

Brzezinski, Z. (1990) For Eastern Europe: A $25 Billion Aid Package, *The New York Times*, March 7.

Brzezinski, Z. (1991) To Strasbourg or Sarajevo?, *European Affairs*, 5, 20–24.

Business Central Europe (1993) Open it up, *Business Central Europe*, May 1, 7–9.

CEPR (Centre for Economic Policy Research) (1992) Is Bigger Better? The Economics of EC Enlargement, *Monitoring European Integration*, CEPR Annual Report, London.

Dawson, A.H. (1993) *A Geography of European Integration*, Belhaven, London.

De Michelis, G. (1990) Reaching out to the East, *Foreign Policy*, 79, 44–55.

Gibb, R.A. and Michalak, W.Z. (1993a) The European Community and Central Europe: prospects for political and economic integration, *Geography*, 78, 16–30.

Gibb, R.A. and Michalak, W.Z. (1993b) Foreign debt in the new East-Central Europe: a threat to European integration?, *Environment and Planning C*, 11, 69–85.

Grabska, W. (1992) Realizacja Unii Europejskiej a problem terytorialnego rozszerzenia Europy (The European Union and a territorial expansion of the Community), *Sprawy Międzynarodowe*, 45, 35–50.

Hamilton, F.E.I. (1990) COMECON: dinosaur in a dynamic world?, *Geography*, 75, 244–246.

Haus, L.A. (1992) *Globalizing the GATT: the Soviet Union's Successor States, Eastern Europe and the International Trading System*, The Brookings Institution, Washington, D.C.

Havrylyshyn, O. and Pritchett, L. (1991) *Trade Patterns After the Transition*, Working Paper, The World Bank, Washington, D.C.

Hewett, E.A. (1976) A gravity model of CMEA trade, in: J.F. Brada, (ed.) *Quantitative and Analytical Studies in East-West Relations*, Indiana University Press, Bloomington, Ind., 35–46.

Horowitz, D. (1991) The impending 'second generation' agreements between the European Community and Eastern Europe: some practical considerations, *Journal of World Trade*, 25, 55–80.

Johnson, H.G. (1965) *The World Economy at the Crossroads: A Survey of Current problems of Marey Trade and Economic Development*, Clarendon Press, Oxford.

Kaser, M.C. and Radice, E.A. (eds.) (1986) *The Economic History of Eastern Europe, 1919–1975, Vol. 2*, Clarendon Press, Oxford.

Korboński, A. (1964) COMECON, *International Conciliation*, September, 12–28.

Korboński, A. (1990) CMEA, economic integration, and perestroika, 1949–1989, *Studies in Comparative Communism*, 23, 47–72.

Köves, A. (1992) *Central and East European Economies in Transition: The International Dimension*, Westview Press, Oxford.

Marer, P. (1991) Foreign economic liberalization in Hungary and Poland, *AEA Papers and Proceedings*, 81, 329–333.

Miall, H. (1993) *Shaping the New Europe*, Royal Institute of International Affairs, Pinter, London.

Michalak, W.Z. (1993) Foreign direct investment and joint ventures in East-Central Europe: a geographical perspective, *Environment and Planning A*, 25, in press.

Michalak, W.Z. and Gibb, R.A. (1992b) The debt to the West: recent developments in the international financial situation of East-Central Europe, *Professional Geographer*, 44, 260–271.

Michalski, A. and Wallace, H. (1992) *The European Community: The Challenge of Enlargement*, Royal Institute of International Affairs, London.

Morgan, R. (1992) Germany in the New Europe, in: C. Crouch and D. Marquand (eds.) *Towards Greater Europe? A Continent Without an Iron Curtain*, Blackwell, Oxford, 105–117.

Palankai, T. (1991) *The European Community and Central European Integration: The Hungarian Case*, Institute for East-West Security Studies, Occasional Paper Series, Westview Press, Boulder, Col.

Reynard, P. (1950) The unifying force for Europe, *Foreign Affairs*, 28, 39–55.

Van Ham, P. (1993) *The EC, Eastern Europe and European Unity: Discord, Collaboration and Integration since 1947*, Pinter, London.

Van Brabant, J.M. (1974) On the origins and the tasks of the CMEA, *Osteuropa Wirtschaft*, 19, 192–193.

Wallace, W.V. (1992) From twelve to twenty-four? The challenges to the EC posed by the revolutions in Eastern Europe, in: C. Crouch and D. Marquand (eds.) *Towards Greater Europe? A Continent Without an Iron Curtain*, Blackwell, Oxford, 34–51.
Wallace, W.V. and Clarke, R.A. (1986) *COMECON, Trade and the West*, Pinter, London.
Wang, Z.K. and Winters, L.A. (1991) *The Trading Potential of Eastern Europe*, Discussion Paper No. 1610, Centre for Economic Policy Research, London.

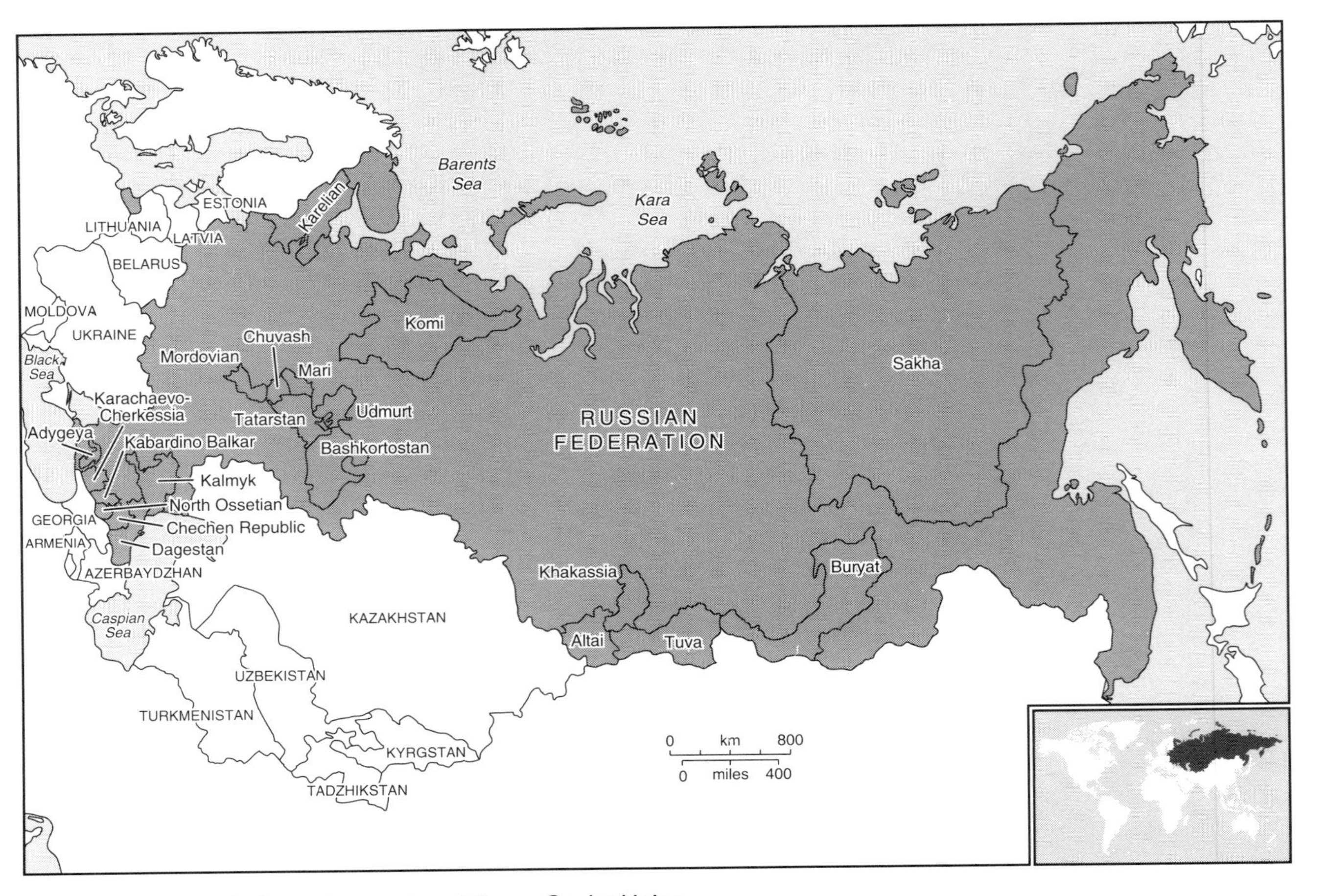

Figure 5.1 Newly independent states of the ex-Soviet Union

5 The Commonwealth of Independent States

Michael Bradshaw
University of Birmingham

INTRODUCTION

To some it might seem strange that a book on continental trading blocs should contain a chapter dealing with the republics of the former Soviet Union. After all, had the Soviet Union still been in existence it would have been considered a single sovereign state which played a relatively minor role in the global economy. In 1988, before the current economic crisis, the Soviet Union's share of world exports stood at 3.9 per cent and its share of imports was 3.6 per cent. However, the Soviet Union is no longer in existence and the Commonwealth of Independent States (CIS), established on 8 December 1991, *is* a supranational organisation. On that date an agreement creating the Commonwealth of Independent States was signed by the three slavic republics, Russia, Belarus and Ukraine. Its membership later grew to include 12 of the republics of the former Soviet Union. The only republics which have declined membership are the Baltic States of Latvia, Lithuania and Estonia.

The CIS is not at present a regional trading bloc; but, as this chapter reveals, there is every intention that the CIS will provide the basis for some form of economic union. Thus one could justify the inclusion of the CIS on the grounds that some of its member states are likely to form a continental trading bloc in the near future. There is also a second reason for considering the CIS and it is that the disintegration of the Soviet Union highlights many of the issues that are central to the question of economic integration.

The Soviet Union was a single economic space. Although a federal state, political power in the Union of Soviet Socialist Republics (USSR) was concentrated within the central organs of the Communist Party of the Soviet Union (CPSU) and the Soviet Government in Moscow. All Union, or federal, institu-

Continental Trading Blocs: The Growth of Regionalism in the World Economy
Edited by R. Gibb and W. Michalak

tions dominated the decision-making process and the 15 constituent Union Republics had very little control over their economies. The Soviet Union was a free trade area in the sense that there were no duties on trade between the Union Republics; it was a customs union in that there was a single policy governing trade outside the Union (in fact the Soviet Government, through the Ministry of Foreign Trade, exercised a state monopoly over foreign trade); it was a common market in that there was free factor mobility across republican boundaries, although there was control over the mobility of labour and the central planning authority, Gosplan, directed capital investment; it was an economic union in the sense that there was a single monetary system with a central bank (Gosbank) and a common currency (the ruble); finally, it was a political union in that it had a common foreign policy and unified defence forces. Although this may be an unusual way of looking at the Soviet Union, it is useful because it highlights many of the problems created by the rapid disintegration of the Soviet economy. In a very short time all these components of an integrated 'Soviet economic space' have been dismantled as one state has disintegrated into 15 Newly Independent States (NIS) (see Figure 5.1). Furthermore, there are signs that the centrifugal forces that led to the collapse of the Soviet Union are now gaining strength in Russia, itself a federation with 21 Autonomous Republics (recently three regions, Svedlovsk, Chelyabinsk Oblasti and Primosrkiy Kray, declared themselves to be republics).

As the leaders of the CIS contemplate the costs and benefits of economic (re)integration they find themselves in an unusual situation. Until recently they were part of a single state, they shared a common currency and were part of a unified economic space. In the short period since the final demise of the Soviet Union, in December 1991, the governments of the NIS have sought to impose control over their domestic economies and to establish individual state identities. Now the prospect of economic (re)integration between the republics of the former Soviet Union raises many of the same fears as the Maastricht Treaty does in the minds of many in the European Community: the erosion of the power of national governments, loss of sovereignty and the possibility of domination by stronger members. At the same time, just as integration in western Europe was seen as a means of reducing the likelihood of conflict, so the creation of an economic union is seen as reducing the potential for conflict between member states of the CIS. Thus the disintegration of the Soviet Union and the emergence of the CIS touch many of the issues that are central concerns of this book.

In order to understand the problems now facing the member states of the CIS and to assess the prospects for economic (re)integration, the first section of this chapter examines how the Soviet central planning system influenced the structure of republican economies. The second section describes the pattern of inter-republican economic relations that was a consequence of central control. Section three examines the economic consequences of the collapse of the Soviet economic space and documents the collapse of the ruble zone. The concluding section assesses the progress made by the CIS and prospects for economic (re)integration.

SOVIET ECONOMIC SPACE

The source of many of the economic problems now facing the governments of the Newly Independent States is that they have inherited economies that are, in most cases, the product of almost 70 years of Soviet economic planning. A key feature of the Soviet economic planning system was the dominance of sectoral planning (planning by industrial branch or sector) over horizontal planning (planning by republic or economic region). Apart from Khrushchev's Sovnarkhoz reforms (which replaced industrial ministries with regional economic councils), which lasted from 1957 to 1964, the planning and development of the Soviet economy had been organised by industrial sector, not by republic or region. As is shown later, this is having a profound influence upon the economic viability of the Newly Independent States. This is not the place to describe in any detail the nature of the Soviet central planning system (see Nove, 1977; Pallot and Shaw, 1981; Dyker, 1983, 1992; and Gregory and Stuart, 1990), rather the concern here is with those aspects of the planning system that promoted the integration of economic activity and created a single economic space.

The parallel system of Party and State in the Soviet Union ensured that the policies decreed by the Politburo were implemented by the Soviet Government. At the republican level, party organisations and union and republican government ministries and state committees operated to fulfil Moscow's plans. It would be wrong to think that the republican and local authorities were powerless to influence central policy, however the key elements of Soviet economic power were closely controlled by party officials in Moscow. The industrial ministries which controlled the Soviet economy were of three types: union, union-republic and republican. Union ministries operated only at the union level and controlled enterprises in the most important sectors of the economy, including the defence industry. In 1984, before Mikhail Gorbachev came to power, there were 33 all-union ministries. Enterprises under the jurisdiction of union-republic ministries were under '*dual subordination*' which meant that they were responsible to the ministry in Moscow and also to the republican government. In reality Moscow often exerted a stronger influence than the local authorities. Union-republic ministries tended to be concerned with local natural resources, such as forestry, petroleum refining and the coal industry (Dellenbrant, 1986). In 1984 there were 28 union-republic ministries. Finally, there were the republican ministries which controlled activities of local importance, such as service provision.

The European Commission (1992) reported that in early 1989 the average share of industry under republican control in the Soviet Union was only 5 per cent and that in no republic was it more than 10 per cent. Clearly, these figures do not refer to industries under dual subordination. Dellenbrant (1986) provides information about the Turkmen Soviet Socialist Republic (SSR) where 44 per cent of production was under the jurisdiction of union ministries, 50 per cent was under union-republic ministries and about 6 per cent was under republic ministries. Under this system the degree of central control was strongly influenced by the economic structure of the various republics. Those republics and regions dominated by heavy industry which were (particularly defence production) under the jurisdiction of union ministries were subject to far greater central control than republics whose economies specialised in light industry and agriculture. Thus, as Table 5.1 illustrates, the Baltic States which specialised in light manufacturing and the Central Asia States which specialised in agriculture had greater influence over their economies than Russia or Belarus whose economies were dominated by heavy industry. This variation in the degree of central control may, in part, explain why Estonia, for example, has been able to break free from Soviet economic space.

The industrial ministries, together with Gosplan (the State Planning Committee) and various other state committees were the key agents guiding the development of the Soviet economy. In theory at least, Gosplan devised the Five-Year Plans and Annual Plans to coordinate the activities of the industrial ministries. Gossnab (the State Committee for Material and Technical Supply) organised the exchange of material among enterprises and the industrial min-

Table 5.1 Industrial production controlled by the federal (Union) and regional (Republican) governments, 1985

Republic	*Per cent Union*	*Per cent Republican**
USSR	58	42
Russia	71	29
Ukraine	37	63
Belarus	51	49
Uzbekistan	29	71
Kazakhstan	35	65
Georgia	30	70
Azerbaijan	34	66
Lithuania	39	62
Moldova	19	81
Latvia	41	59
Kyrgyzstan	30	70
Tajikistan	25	75
Armenia	42	58
Turkmenistan	40	60
Estonia	28	72

*Includes union-republic production.
Source: Goskomstat SSSR, 1991b, 140.

istries and their subordinate enterprises fulfilled the plans. In an environment of scarcity and with ever-increasing demands to increase production, ministries and enterprises resorted to various means to maintain their supply lines. The economist Alec Nove (1977) has used the term 'centralised pluralism' to describe the division of the Soviet economy into quasi-autonomous units, the industrial ministries. Shaw (1985) has examined the spatial consequences of ministerial behaviour. He described how the predominance of administrative controls promoted vertical integration rather than horizontal coordination and created an environment within which cost and profit were secondary to plan fulfilment. Security of supplies was a key factor ensuring plan fulfilment, thus there was a tendency towards 'ministerial autarky'. To ensure that their enterprises got the inputs required, industrial ministries developed their own supply systems, producing many products that might otherwise have been supplied, often more efficiently, by enterprises subordinate to other industrial ministries. This behaviour led to wasteful duplication of production facilities and placed an added burden upon the transportation system. In reality, much of what will later be described as inter-republican trade was in fact intra-ministerial trade, trade between enterprises subordinate to the same ministry, and inter-ministerial trade, trade between enterprises subordinate to different ministries (IMF et al., 1991). Thus, if the central planning system and the industrial ministries were the key actors orchestrating inter-republican trade, it is easy to see how the abolition of those organisations would result in a collapse of trade and would leave enterprises, cut off from their traditional patterns, unable to produce goods.

There are two further aspects of Soviet economic development that have contributed to the economic crisis caused by the disintegration of a single economic space: the high level of monopoly in Soviet industry and the high degree of industrial concentration. Nove (1981) has ascribed the Soviet tendency towards very large enterprises not only to the stalinist penchant for 'gigantomania' and the desire to have the 'world's biggest tractor plant', but also to the fact that for central planners it is actually easier to manage a small number of very large plants, than a large number of small enterprises. Analysis in the ill-fated Shatalin Plan and the IMF report on the Soviet economy provides details on the level of monopoly and concentration in Soviet industry. The Shatalin Plan (Shatalin, 1990, 77–78) reported that almost 2000 products with a value of 11 billion rubles were only produced at one enterprise. The level of monopoly was particularly high in the machine-building industry, where monopoly production accounted for 80 per cent of output, with 166 monopoly enterprises in 180 different products. However, the problem of monopoly production was widespread; in 209 of the 344 major commodity groups in industry, one large enterprise accounted for more than 50 per cent of output, and in 109 of those cases the share of the total output of that product was more than 90 per cent. Table 5.2 provides some examples of monopoly production (for a more detailed list of monopoly producers see Snyder, 1993, 207–243). However, these figures are

Table 5.2 Examples of monopoly production in the former Soviet Union

Product	*Plant*	*Location*	*Per cent of output*
Die-casting machines	Tiraspol' Plant for Die-Casting Machines named for S.M. Kirov	Tiraspol' (Moldova)	99
Trolley Buses	Trolley Bus Factory named for Uritsky	Engels, Saratov Obl. (Russia)	97
Corn Harvesters	Kherson Combine Plant Assoc. named for G.I. Petrov	Kherson, Kherson Obl. (Ukraine)	100 100
Potato Harvesters	Ryazan' Combine Plant	Ryazan', Ryazan Obl. (Russia)	100
Cotton Harvesters	Tashkent Agricultural Machinery Associations	Tashkent (Uzbekistan)	100
Tramway rails	Kuznetsk Metallurgy Combine	Novokuznetsk, Kemerovo Obl. (Russia)	100
Plastic-lined pipes	Pervoural'sk New Pipe Plant	Pervoural'sk, Sverdlovsk Obl. (Russia)	
Sewing machines	Podol'sk Sewing Machine Association	Podol'sk Moscow Obl. (Russia)	100

Source: Soviet Geography, 1991, 190–193.

somewhat misleading. As already noted, Soviet enterprises tended to internalise their linkages making them less vulnerable to disruptions from monopoly supplies, so the high level of monopoly in the production of end products may not necessarily be paralleled in the supply of inputs. Nonetheless there are cases of extreme dependence. In 1993 a fire at the Kama engine production facility brought Russian truck production to a virtual standstill as the plant supplied all the other truck producers with engines.

A further consequence of this monopoly production is the tendency to concentrate production in very large enterprises. Thus one very large plant would produce a particular product for the entire Soviet market. The supposed benefits of economies of scale often failed to take into account the transportation costs associated with supplying such a geographically large market. According to the IMF (IMF et al., 1991, Vol. 3, 287), in the late 1980s there were about 46 000 industrial enterprises in the Soviet Union. However, the 600 largest enterprises had 25 per cent of the book value of assets, produced one-third of the industrial output and employed 20 per cent of the industrial workforce. In 1987, only 16.4 per cent of enterprises had a workforce of more than 1000, but those enterprises employed 73.1 per cent of the industrial workforce and accounted for 73.9 per cent of the volume of production (Goskomstat SSSR, 1988, 14). An international comparison helps to put these figures into perspective. In West Germany in 1985, 2.6 per cent of enterprises employed more than 1000 people and those enterprises accounted for 39.7 per cent of the workforce (Yakovlev, 1991, 4). The level of concentration is particularly high in the defence industry. Cooper (1993, 2) reports that in the core enterprises of the military-industrial complex, for which military production was the basic activity, the average workforce at the beginning of 1992 was 10 000. Clearly the high levels of monopoly and concentration present major obstacles to privatisation and military conversion, and also have profound geographical implications.

In 1990 the four key members of the CIS – Russia, Ukraine, Belarus and Kazakhstan – accounted for 78.9 per cent of the population of the former Soviet Union, 81 per cent of employment and 90 per cent of industrial production (Bradshaw, 1993, 11). Russia is by far the most important republic. In 1990 Russia produced 90.4 per cent of the former Soviet Union's oil production, 78.6 per cent of natural gas production, 56.2 per cent of coal production and 58 per cent of steel production (Goskomstat SSSR, 1991a, 358–61). It is only in the agricultural sphere and light industry that the other republics are of significance. A 1985 data set on industrial employment provides further evidence of Russia's dominance; of the 14 industrial branches identified, the only branch where Russia did not account for more than 50 per cent of employment was the fuel industry, which is explained by the Donbas coalfield in the Ukraine (Bradshaw, 1993, 13–14). The same data set provides information on employment in strategic industries, the military-industrial complex and production of strategic minerals; Russia and Ukraine accounted for 71 and 17.5 per cent of employment in these branches respectively. Cooper (1993, 5) reports that Russia accounts for 70 per cent of the enterprises of the former Soviet defence complex, responsible for three-quarters of former Soviet military output.

Not only is there a concentration of production within the slavic republics, but within the republics themselves industry is concentrated in particular regions. Utilising the same 1985 data set, Sagers (1992a, 499) examined the regional industrial structure of the former Soviet Union. His analysis revealed a considerable spatial concentration of industrial production:

> In terms of absolute amount of industrial production, just a few oblast-level units contain a very large share of national output. For example, the top five account for 14.2 per cent of production; the largest 10, 24.2 per cent; the largest 20, 40.0 per cent; and the largest 30, 51.4 per cent.

In many regions there is also a high level of specialisation, for example several regions have more than 50 per cent of production coming from a single sector. The areas where the defence sector is of primary importance include, Novgorod (39.2 per cent), Kaluga (46.9 per cent), Republic of Mari-El (45.7 per cent), Voronezh (40.2 per cent), Republic of Udmutia (57.0 per cent), Novosibirsk

(45.3 per cent), Omsk (42.5 per cent) and Magadan (41.4 per cent) (Horrigan, 1992, 36–37, using the same 1985 data set). Just as the large monopoly producers that dominated Soviet industry are now vulnerable to disruptions in supply, so the regions that depended on them are equally exposed.

It should be apparent that the Soviet central planning system produced a particular pattern of economic activity that reflected the priorities of policy makers in Moscow and the peculiarities bred by the system itself. The policy of maximum self-suffiency at the all-union level resulted in substantial trade diversion, with enterprises being forced to reply on domestic suppliers regardless of cost. The patterns of sectoral and regional specialisation were not the result of comparative advantage based on market prices, therefore the pattern of inter-republican trade, or the form of economic integration, generated by the Soviet system is likely to be radically different from that generated in market-oriented economies. The next section examines the nature of trade relations between the republics.

ECONOMIC RELATIONS BETWEEN THE REPUBLICS

During 1990, continuing conflict between the Soviet Government and the Republican Governments heightened interest in the nature of economic relations between the republics. Economists at the Institute of Economics and Industrial Production at Novosibirsk have long been interested in the nature of inter-regional relations in the Soviet Union (see Granberg, 1990, 1991; Seliverstov, 1991); however, the prospect of economic disintegration prompted numerous studies by western institutions and academics (Belkindas and Sagers, 1990; European Commission, 1991; Gros, 1991; Havrylyshyn and Williamson, 1991; McAuley, 1991a, 1991b; Slay 1991; IMF, 1992; Michalopoulos, 1993; Michalopoulos and Tarr, 1992; Senik-Leygonie and Hughes, 1992). Most of these analyses examine the role of inter-republican trade and foreign trade in the economies of the republics. Because many of them use the same data sets they reach similar conclusions. (During 1990 data on inter-republican economic relations were published by Goskomstat SSR in the annual economic handbook *Narodnoye Khozyaystvo* and the journal *Vestnik Statistiki.*) Some of these data are presented in this chapter.

What is the value of inter-republican trade? Table 5.3 provides an evaluation of the value of republican trade (inter-republican and foreign) in 1988, before trade was affected by economic decline and rapid inflation. Two sets of values are provided, one in domestic prices and another in so-called 'world prices'. On the basis of the domestic pricing system, all republics, with the exception of Azerbaijan and Belarus, show a deficit in their trading relations. When the value of trade is recalculated on the basis of world prices all republics, except Russia and Turkmenistan, record a deficit; however, Russia's terms of trade are trans-

Table 5.3 Republican trade (inter-Republican and foreign), 1988 (billion rubles)

Republic	*Imports*		*Exports*		*Balance*		*Per cent*
	Domestic	*World*	*Domestic*	*World*	*Domestic*	*World*	*[GNP]*
Russia	135.86	101.90	102.54	132.70	−33.32	30.80	22.3
Ukraine	49.86	47.40	46.94	44.50	−2.92	−2.90	34.0
Belarus	17.84	18.50	19.92	16.40	2.08	−2.10	52.0
Uzbekistan	12.32	10.50	10.49	8.00	−1.83	−2.50	39.7
Kazakhstan	16.40	15.60	9.10	9.00	−7.30	−6.60	34.2
Georgia	6.49	5.30	5.90	3.40	−0.59	−1.90	43.8
Azerbaijan	5.70	5.10	6.80	4.60	1.10	−0.50	41.3
Lithuania	7.49	7.80	5.96	4.10	−1.53	−3.70	54.5
Moldova	6.10	5.10	5.06	2.50	−1.04	−2.60	52.2
Latvia	5.60	5.00	4.90	3.70	−0.70	−1.30	54.1
Kyrgyzstan	3.77	3.20	2.56	2.10	−1.21	−1.10	45.6
Tadjikistan	3.49	2.80	2.33	1.70	−1.16	−1.10	46.7
Armenia	4.88	3.60	3.76	2.20	−1.12	−1.40	53.7
Turkmenistan	2.90	2.40	2.60	2.40	−0.30	0.00	42.2
Estonia	3.70	3.20	3.00	1.90	−0.70	−1.30	58.9

GNP based on EC calculations
Source: Vestnik statistiki (1990b, 49) and European Commission, (1991, 154).

formed from a deficit of 33.3 billion rubles in domestic prices to a surplus of 30.8 billion rubles. The main reason for this dramatic change is the difference between the domestic and world prices for energy resources and raw materials. The domestic price is a fraction of the world price, thus Russia's exports are systematically underpriced. In terms of inter-republican trade, this price distortion could be seen as an implicit subsidy by Russia (in 1991 Russia produced 90 per cent of the former Soviet Union's oil and 79 per cent of its natural gas). In 1991 this implicit subsidy was estimated to be worth 33.5 billion rubles (Economist Intelligence Unit, 1991a, 92).

Unfortunately there are numerous problems with these data on inter-republican trade: they include merchandise trade, no account is made of invisibles, they do not allow for cross-border shopping, the prices used include state subsidies (thus underestimating the true value of food imports, for example), but do not include turnover tax. All of these problems should lead us to treat the data with extreme caution. Various attempts have been made to compensate for these problems, for example McAuley (1991b) has explored the question of invisibles (earnings from transport services, tourism, financial services, investment incomes and transfers), while Goskomstat (Vestnik statistiki, 1990a, 1990b) and the European Commission (1991) have tried to account for the various technical omissions. The inclusion of these adjustments does very little to change the overall picture generated by the use of world prices: clearly the domestic pricing system is the most important factor distorting the true value of inter-republican trade relations.

How important is inter-republican trade? The final column in Table 5.3, based on European Commission calculations, provides an indication of the importance of foreign trade in the various republics, showing trade turnover as a percentage of GNP. As one would expect, the larger republics were the most self-contained, while the Baltic States and Belarus were particularly open. The distribution of energy production is a key factor in explaining this variation; Belarus, for example, has a large chemical and petrochemical industry but no oil or gas resources. Tables 5.4 and 5.5 (and Figure 5.2) make the distinction between inter-republican trade and foreign trade activity. These statistics are based on domestic prices: in the case of foreign trade domestic prices tended to undervalue exports and overvalue imports, hence all republics record a deficit. The final column of each table shows the share of inter-republican and foreign trade respectively, in total trade turnover. For all the republics, with the exception of Russia, trade with other republics was far more important than trade with the outside world. Figure 5.3 shows that for all the republics Russia was the most important trading partner (The data for Figures 5.3 and 5.4 were obtained from Michalopoulos and Tarr (1992, 42 & 44). (These data were supplied to the World Bank by Goskomstat and are based on domestic prices.) One set of bars (Russia's share of imports) indicates the role of Russia in the total imports of each republic; for Ukraine in

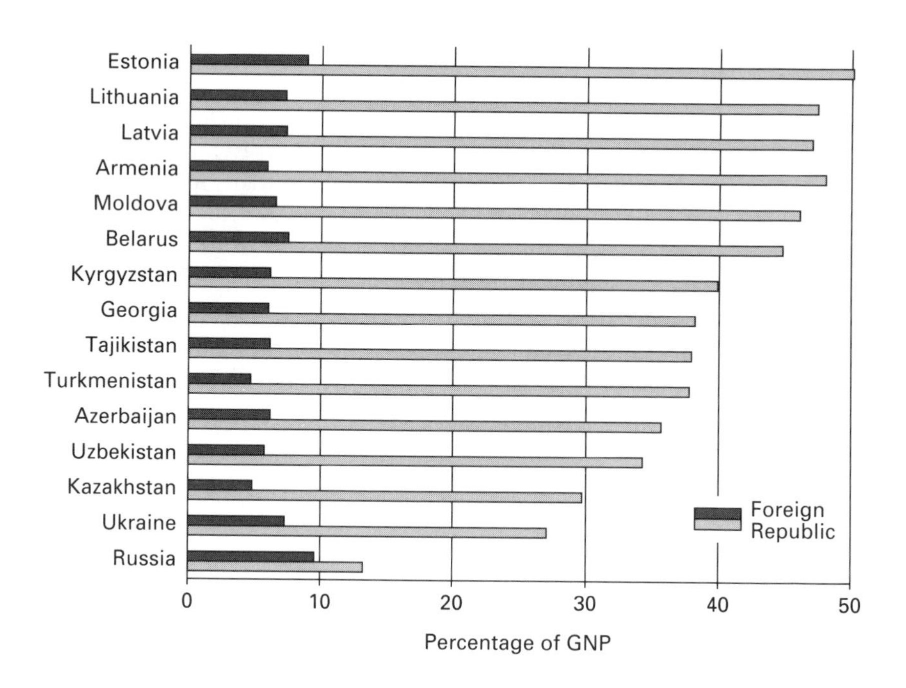

Figure 5.2 Role of trade in Republics as a percentage of GNP, 1988

Table 5.4 Turnover of trade between republics, 1988 (Billion rubles, domestic prices)

Republic	*Imports*	*Exports*	*Balance*	*Per cent [GNP]*	*[Share of] Total Trade*
Russia	68.96	69.31	0.35	12.9	58.0
Ukraine	36.43	40.06	3.63	26.9	79.0
Belarus	14.17	18.22	4.05	44.6	85.8
Uzbekistan	10.62	8.96	−1.66	34.1	85.8
Kazakhstan	13.70	8.30	−5.40	29.5	86.3
Georgia	5.22	5.50	0.28	37.9	86.5
Azerbaijan	4.30	6.40	+2.10	35.4	85.6
Lithuania	6.24	5.43	−0.81	47.3	86.8
Moldova	5.00	4.80	−0.20	45.9	87.8
Latvia	4.60	4.50	−0.10	46.9	86.7
Kyrgyzstan	3.00	2.50	−0.50	39.7	86.9
Tadjikistan	3.02	2.00	−1.02	37.7	86.3
Armenia	4.02	3.68	−0.34	47.9	89.1
Turkmenistan	2.50	2.40	−0.10	37.6	89.1
Estonia	3.00	2.70	−0.30	50.1	85.1

Source: Vestnik statistiki (1990b, 36) and European Commission, (1991, 154).

Table 5.5 Republican foreign trade, 1988 (billion rubles, domestic prices)

Republic	*Imports*	*Exports*	*Balance*	*Percent [GNP]*	*[Share of] Total Trade*
Russia	66.90	33.31	−33.59	9.4	42.0
Ukraine	13.43	6.88	−6.55	7.1	21.0
Belarus	3.67	1.70	−1.97	7.4	14.2
Uzbekistan	1.70	1.53	−0.17	5.6	14.2
Kazakhstan	2.70	0.80	−1.90	4.7	13.7
Georgia	1.27	0.40	−0.87	5.9	13.5
Azerbaijan	1.40	0.40	−1.00	6.0	14.4
Lithuania	1.25	0.53	−0.72	7.2	13.2
Moldova	1.10	0.26	−0.84	6.4	12.2
Latvia	1.00	0.40	−0.60	7.2	13.3
Kyrgyzstan	0.77	0.06	−0.71	6.0	13.1
Tadjikistan	0.47	0.33	−0.14	6.0	13.7
Armenia	0.86	0.08	−0.78	5.8	10.9
Turkmenistan	0.40	0.20	−0.20	4.6	10.9
Estonia	0.70	0.30	−0.40	8.8	14.9

Source: Vestnik Statistiki, (1990b, 36) and European Commission, (1991, 154).

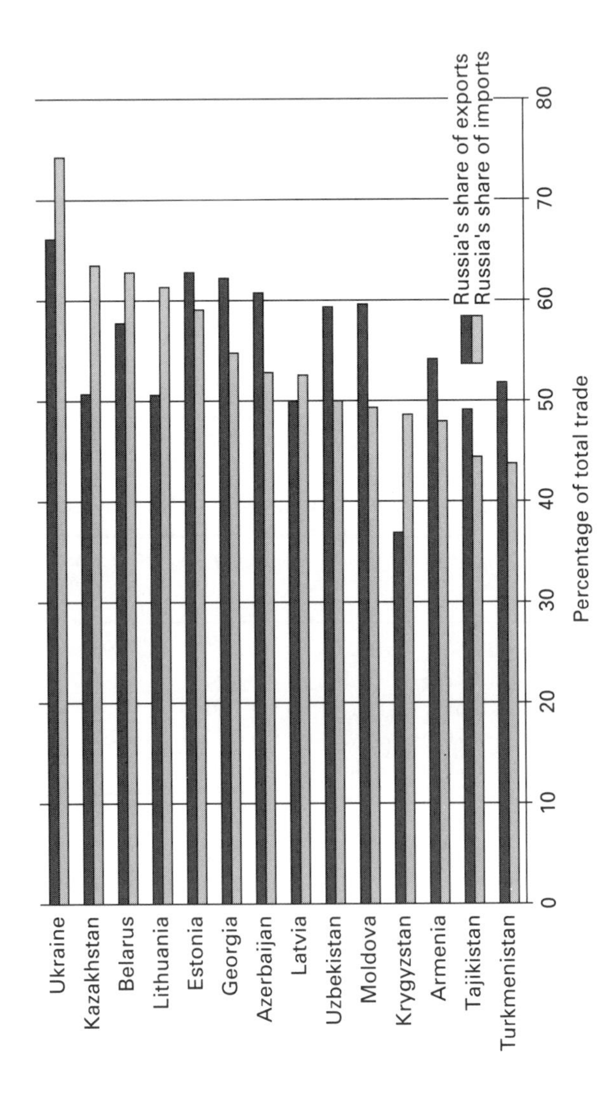

Figure 5.3 Russia's role in trade, 1990

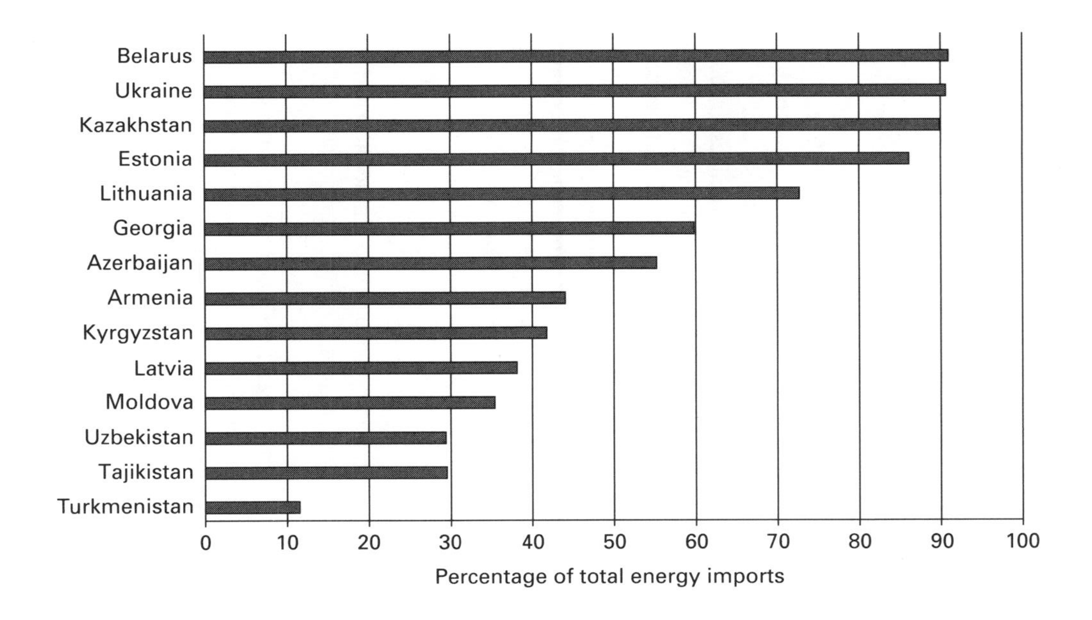

Figure 5.4 Role of Russian energy exports, 1990

1990, 74.1 per cent of imports came from Russia. The other set of bars (Russia's share of exports) indicates the share of each republic's exports that were destined for Russia; for example, Russia accounted for 65.9 per cent of Ukrainian exports. From these data it is clear that all the republics were far more dependent upon trade with Russia than upon trade with one another. For example, in 1990 10.8 per cent of Estonia's exports went to Latvia and Lithuania, and the other two Baltic states accounted for 9.8 per cent of Estonia's imports, while 59 per cent of Estonia's imports came from Russia and 62.6 of Estonian exports were destined for Russia. Figure 5.4 shows the share of Russia in inter-republican energy trade; the data include all forms of energy, oil, gas, coal and electricity. For the most part the Slavic and Baltic republics and Kazakhstan are the most dependent upon energy supplies from Russia. In the Caucasus Azerbaijan was an important regional supplier of energy, while Turkmenistan was an important supplier of energy to the other Central Asian republics.

Discussion so far has focused upon inter-republican trade. Table 5.5 provides information on republican participation in foreign trade. In 1988, Russia accounted for 76.2 per cent of Soviet exports and 68.4 per cent of imports. The reason for this dominance is apparent when one examines the commodity structure of exports. For example, in 1989 fuels and electricity accounted for 41.1 per cent of Soviet exports to other socialist countries and 55.1 per cent of exports to non-socialist developed countries (IMF et al., 1991, Vol. 2, 70). In 1988 Russia accounted for 91.3 per cent of Soviet exports of oil and gas (Smith, 1993, 202). Russia was also the most significant exporter of chemicals, petrochemicals and non-ferrous metals. As with inter-republican trade, the fragmentation of the Soviet economy has heightened interest in the foreign trade activity of the republics (see for example, Hewett and Caddy, 1992; Smith, A., 1993). More recent data for the CIS show that in 1991 Russia accounted for 80 per cent of CIS exports and 67 per cent of imports (Bradshaw, 1993, 31). In 1991 Russia recorded a sizeable foreign trade surplus while all the other CIS member states recorded trade deficits. The state monopoly of foreign trade meant that all decisions governing external trade were made in Moscow; consequently the republics had little control over foreign trade activity on their territory. Following the collapse of the Soviet Union, the governments of the Newly Independent States have to develop their own foreign trade strategies and enterprises have to establish an identity on world markets.

When one combines the above information on merchandise trade with data on income flows between republics, it is clear that the slavic states, particularly Russia, subsidised the economies of the Central Asian republics. On the basis of an analysis of income redistribution among the republics, Belkindas and Sagers (1990, 638) have concluded that:

> long-term donors consist of the RSFSR, the Ukraine, Belorussia and Azerbaijan. The long-term recipients include Uzbekistan, Kazakhstan and Tadzhikistan, while

the other republics appear to have recorded both positive and negative balances.

The economic development strategy pursued by Moscow was in large part responsible for this pattern of dependence. Investment policy in Central Asia produced a classical colonial pattern of development with the Central Asian republics exporting agricultural products and raw materials to the slavic core in return for imports of manufactured goods. This pattern of unequal exchange, and the compensating flow of income from the core, helped to maintain the high levels of interdependence among the republics of the Soviet Union. In assessing the level of economic integration between the republics, the IMF (1992) concluded:

> the share of inter-republican trade is considerably greater than the share of intra-regional trade in either the European Community or Canada. On average, for example, inter-republic trade accounted for 72 percent of total republic trade in the observation year [1988], compared with 44 percent and 59 percent for Canada and the European Community, respectively. However, trade with the rest of the world was relatively low, so that the *total* amount of trade undertaken by each republic was similar to that of countries and regions of similar size elsewhere. (IMF, 1992, 7)

In summary one can say that the majority of the republics of the former Soviet Union were highly dependent upon trade with one another, particularly trade with Russia, but were relatively isolated from the world economy. This pattern of interdependence was the result of a territorial division of labour produced by the central planning system, rather than the interaction of coherent 'republican economies'. As a result of distortions in the domestic pricing system and the dominance of sectoral planning, the actual pattern of trade obscured many subsidies and did not reflect specialisations based on comparative advantage. The collapse of the Soviet Union has resulted in the break-up of the system that produced this distorted pattern of economic interaction. It is therefore not surprising that the republics of the former Soviet Union have experienced a collapse in external trade and have resorted, at times, to extreme measures to protect domestic interests. The next section examines the economic consequences of the deintegration of the former Soviet republics.

THE COLLAPSE OF SOVIET ECONOMIC SPACE

Since the beginning of 1990, the republics of the former Soviet Union have suffered two trade shocks in close succession. First, the collapse of intra-CMEA trade (see Chapter 4) that resulted from the Soviet Union's unilateral decision as of January 1990 to trade on the basis of world prices in convertible currency.

Second, the reduction in inter-republican trade that has resulted from the disintegration of the Soviet economy.

In 1990, trade with other CMEA members accounted for 35.3 per cent of the Soviet Union's exports and 36.5 per cent of its imports. By the end of 1991, the CMEA's share of Soviet foreign trade had fallen to 23 per cent of exports and 24.5 per cent of imports. Between 1990 and 1991, as a direct result of the breakdown of the CMEA trade and payments system, Soviet exports to CMEA member-states declined 57 per cent and imports 63 per cent (Smith, 1993, 174). In 1991 Russia's total export earnings fell by 29.1 per cent (Langhammer et al., 1992, 619). During 1992 the decline of Russia's foreign trade continued. According to preliminary government statistics, in 1992 Russia's exports were worth $38.1 billion, down 21 per cent on 1991. First quarter statistics for 1993 shown no improvement: Russia's foreign trade turnover was down 36 per cent over the corresponding period in 1992. During 1992 only 20 per cent of Russia's exports were destined for the former members of the CMEA and only 15.3 per cent of exports originated from those states. Thus not only has the volume of trade declined, but the geographical distribution of trade has changed.

Unfortunately, information on the collapse of inter-republican trade is much more scarce. The IMF (1992, 7) reported a Russian Government estimate that suggests that during 1991 the total volume of inter-republican trade fell by at least 15 per cent. It was also estimated that deliveries from Russia to other republics in January 1992 were 25–30 percent lower than in January 1991 (Business Eastern Europe, 1993, 4). A recent World Bank study has concluded that between 1990 and 1992 inter-republican trade fell in real terms by 50 per cent (Michalopoulos, 1993, 4). While much of the decline in inter-republican trade can be ascribed to the abolition of the central planning system, it has been exacerbated by at least three other factors: the behaviour of the republican governments, the collapse of the ruble zone and the crisis in the Russian oil industry.

TRADE FRICTION BETWEEN THE REPUBLICS

The origins of the current crisis in inter-republican economic relations lie in the conflict between the republican governments on the one hand and Mikhail Gorbachev and the Soviet government on the other. In the process of negotiating a new Union Treaty (the signing of which prompted the ill-fated August Coup), the republican governments sought to gain control over their economies through declarations of economic sovereignty. These declarations asserted that republican legislation took precedence over Soviet legislation. The so-called 'battle of the laws' resulted in each republic trying to protect its own interests. A key aspect of this 'republicanisation' of the Soviet economy was control over trade, both foreign and inter-republican. As individual republics sought to implement their own economic reforms, others found it necessary to protect their economic space by controlling trade. The European Commission (1992, 73) reports how in September 1990, the Russian Government's decision suddenly to increase meat prices by 50 per cent led to an immediate shortage in the neighbouring republics of Belarus and Ukraine, as farms shipped meat to Russia to benefit from the higher prices. The governments of Ukraine and Belarus responded by controlling exports of foodstuffs. As the prospect of full independence became more realistic, so the republics sought to establish even greater control over cross-border trade, and some even established customs posts. Following the collapse of the Soviet Union, the individual republics have sought to conserve national resources by imposing quotas and export controls (McAuley, 1991b; Noren and Watson, 1992). Because of the high level of interdependence, detailed earlier in this chapter, such actions served only to accelerate the rate of economic decline. It has been suggested that between 50 and 80 per cent of the fall in industrial production is due to the collapse of inter-republican trade (Granberg, 1993, 70). However, it is not only the direct actions of the republics that have precipitated a crisis in inter-republican trade; the breakdown of the financial system and the collapse of the ruble zone are also major contributing factors.

THE COLLAPSE OF THE RUBLE ZONE

The Soviet Union had a unified monetary system with a central bank, Gosbank, and a single non-convertible currency, the ruble. Financial flows between the republics were managed by Gosbank and its republican branches. Thus the Soviet Government controlled the macroeconomic policies that influenced the financial environment in the constituent republics. However, in the Soviet Union financial interactions were secondary to the activities of the central planning system and the Communist Party. The system of taxes and price subsidies was often used as a means of redistributing income between the republics. The non-convertibility of the ruble isolated the domestic economy from the international financial environment and enabled the government to control the 'official' exchange rate. In the mid-1980s the ruble was worth one pound sterling; in late June 1993 there were 1569 rubles to the pound. Previously the official exchange rate overvalued the ruble, whilst the current market rate substantially undervalues it. The instability of the ruble undermines the whole economic system, as individuals and enterprises seek to obtain hard currency to protect themselves against inflation and further declines in the value of the ruble.

Since the abolition of the Soviet state and its financial institutions many republics have sought to establish monetary sovereignty (control over macroeconomic policy). Because most of the infrastructure of the Soviet financial system was located in Moscow, Russia inherited the Central Bank and Foreign

Trade Bank and assumed control of the printing presses that produced paper rubles. In 1992 rapid inflation and Russian control of the printing presses led to a shortage of rubles. Visiting Moscow in February 1992, it was impossible to buy rubles for hard currency because the banks did not have any rubles. The republics responded by providing enterprises with credits to pay for inter-republican transactions and introduced coupons and parallel currencies to provide the populace with 'money' with which to purchase goods and services. At the same time the centralised banking system was devolved as Gosbank republican branches became state banks in their own right. Inter-republican payments have to be made through these clearing banks with payments between enterprises in different republics taking months. With very high inflation, it is no surprise that enterprises avoid cash payments, preferring instead to engage in barter trade. All of these problems have undermined the ruble as a medium for exchange and have accelerated the collapse of the ruble zone.

Hansson (1993) suggests that there are three reasons why a republic might want to abandon the ruble and establish its own currency: enhancement of national sovereignty, economic stabilisation, and the promotion of structural adjustment. A national currency is an important symbol of sovereignty: even the Russian Government is printing new Russia rubles to replace Soviet banknotes. However, there is more to monetary sovereignty than iconographic value; the creation of a national currency enables the government to gain control over macroeconomic policy. It is very difficult to regulate the money supply if that money is supplied by another state. Monetary sovereignty enables governments to design macroeconomic policy to meet a specific economic and political situation. During 1992, other republics in the ruble zone were forced to keep pace with the reforms introduced by the Russian Government. Inflation generated by price liberalisation in Russia was quickly exported to the other republics. Monetary sovereignty also provides republican governments with greater control over the local tax base and balance of payments. In short, abandoning the ruble and introducing national currencies enables individual republics to distance themselves from the consequences of Russia's macroeconomic policy. Nevertheless, the high level of dependence upon trade with Russia means that none of the republics, the Baltic States included, can really isolate themselves from the effects of Russian price liberalisation.

For some republics, such as Estonia, the creation of their own currency can be a powerful tool for promoting industrial restructuring and privatisation. The creation of a convertible currency can encourage foreign direct investment, thus integrating the economy into the world economy and opening domestic enterprises to international competition. Of course, if mismanaged it can easily bring balance-of-payments problems: so far Estonia has succeeded in controlling the money supply through the use of a currency board (the kroon is tied to the Deutschemark). In a very short period of time Estonia has been able to reorient its foreign trade relations away from the former Soviet Union. In 1991, Russia accounted for 64 per cent of Estonia's foreign trade turnover, the rest of the USSR 33 per cent and the rest of the world a mere 3 per cent. At the end of the first quarter of 1993, 22.1 percent of Estonia's foreign trade was with Russia, 18 per cent with the rest of the former USSR, 23.4 per cent with Finland, 10.6 per cent with Sweden and 10.6 per cent with the rest of the world. This success may be short-lived as Estonia may soon have to pay for energy imports from Russia at world prices and in hard currency. Russia's attitude toward concessionary energy prices has not been helped by what is perceived as discrimination against ethnic Russians in Estonia. The creation of a credible stable currency also promotes privatisation, as people are more prepared to invest if they feel that having money and shares is worthwhile. Stability is vital if the republics are to develop a modern independent financial and banking sector.

At present it would seem that the ruble is failing to fulfil the major functions of money: to act as a store of value, a medium of exchange and unit of account. This being the case, there is little incentive to stay within the ruble zone. On the basis of his analysis of the monetary reform in the former Soviet Union, Hansson (1993) concludes:

> The most important step in ensuring a positive outcome is a recognition by sides that the present Ruble zone is dead. Attempts to resurrect it with only minor modifications would be as futile as breathing life into the USSR, Yugoslavia or the CPSU. These efforts can lead only to a more chaotic and confrontational breakup, and a shattering of expectations on all sides. (Hansson, 1993, 177)

In their attempts to gain monetary sovereignty, the republics of the former Soviet Union have pursued a number of different strategies. It is possible to classify these into three groups: coupons, ruble substitutes and bona fide currencies. Some republics have moved through all these 'stages'; others, such as Estonia, have leapt straight from the ruble zone to bona fide currencies. Many republics introduced coupons as a means of establishing local monetary policy, while remaining within the ruble zone. Frequently, workers received part, or all, of their wages in coupons which could then be exchanged for goods and services. At the local level ration coupons were also introduced to control goods in short supply. Visiting Novosibirsk in Russia in May 1991, a handful of coupons was presented as a souvenir: they had proved useless as goods were unavailable. Coupons are also a means of stopping the flow of scarce goods out of a republic or region, as only those employed locally have access to them. However, a market in exchanging coupons quickly develops, and in many instances the local coupons have failed to hold their value against the ruble.

A number of republics introduced coupons in late 1991 and early 1992 and have since replaced them with parallel currencies. The introduction of parallel currencies enabled the republican governments to influence the money supply, compensating for the shortage of cash rubles emanating from Moscow. For example, Belarus issued coupons in November 1991 and in June 1992 moved to special 'accounting notes', known as the Belarus ruble (1 Belarus ruble = 10 Russian rubles). Since November 1992, these special notes have become the sole legal tender. At present they are not a bona fide currency. Following Russia's own currency reform in late July 1993, Russia is now pressuring Belarus to leave the ruble zone or accept Russia's monetary policies. Because of Belarus' economic dependence on Russia, it is likely to form some kind of currency union based on the Russian ruble. A more conservative strategy has been pursued by Georgia, which although not part of the CIS remained part of the ruble zone. Georgia introduced coupons and as of June 1993 the coupons were declared the sole legal tender. Russia's currency reforms forced the Georgian government to outlaw the use of 'Soviet' rubles. Eventually a national currency, the 'lary' will be introduced.

As of June 1993, all three of the Baltic States had established their own bona fide currencies. Estonia was the first republic to leave the ruble zone: the kroon was introduced in June 1992. Latvia first introduced a parallel ruble (the rublis) then in March 1992 the 'lat' was introduced as the national currency. In October 1992 Lithuania introduced the 'talonas' (coupon) as a parallel means of payment. In June 1993 the 'litas' was introduced as the national currency. Elsewhere, the Ukraine introduced coupons in January 1992, and in December 1992 these were replaced by a new temporary monetary unit, the 'karbovanets'. Eventually, in late 1993 or early 1994, the 'hryvnia' will be introduced as Ukraine's national currency. In a surprise move, in May 1993 Kyrgyzstan introduced its own currency, the 'som'. Kyrgyzstan failed to notify its neighbours of the introduction of the som and Uzbekistan and Kazakhstan have been hesitant to accept payment in it. A physical shortage of som caused some technical difficulties as the populace was unable to obtain sufficient money with which to carry out transactions.

In late July 1993, the Russian Central Bank declared that all ruble notes printed before the beginning of 1993 were invalid. Initially citizens were given two weeks to swap up to 35 000 old rubles for new, but these restrictions were later lifted. This rather clumsy currency reform led to widespread panic and further undermined the authority of President Yeltsin. Russia's unilateral decision to render invalid all 'Soviet' rubles in circulation has had a negative impact on the ruble zone. Within a week of the Russian currency reforms, four republics announced that they would leave the ruble zone – Azerbaijan, Georgia, Moldova and Turkmenistan. This would leave a core ruble zone of Russia, Armenia, Belarus, Tajikistan. Uzbekistan and Kazakhstan. The current chaos in the ruble zone is likely to continue as all the republics are making plans to introduce some form of national currency. The creation of national currencies and the uncertain status of the Russian ruble will further complicate republican trade relations and will increase the need for multilateral institutions to enable payments to be made between the member states of the CIS. It may be that the death of the ruble zone will serve as a catalyst for new, more meaningful, forms of economic cooperation between the Newly Independent States.

Confusion over the future of the ruble has led Williamson (1992) to suggest a distinction between the 'ruble zone' and the 'ruble area'. The ruble zone refers to the region where the ruble is the sole currency: this region is shrinking fast and in the future may not include all of Russia, since some of the Autonomous Republics in Russia have talked of introducing their own money. The ruble area refers to a group of countries with separate currencies that continue to use the ruble for international purposes. Thus, a convertible Russian ruble would be used for foreign trade and trade between states in some form of post-Soviet economic union. This begs two questions: in the near future, is the ruble likely to be strong enough to perform such a task: and are the republics likely to cooperate? At present the prospects do not look good: despite good intentions trade frictions are continuing to grow. Russia's decision to liberalise energy prices threatens to bankrupt many of the republics who are already substantially indebted to Russia. Of course, this may force those who cannot afford to leave, or who have no one to pay their bills for them, into creating some form of ruble payments area.

RUSSIA'S ENERGY CRISIS

Energy trade lies at the heart of inter-republican economic relations and is the source of Russia's dominance over many of the republics. Changes in Russia's energy industry therefore have the potential to change radically the scale and structure of inter-republican economic relations. At present the Russian oil industry is facing a crisis. Mismanagement during the early 1980s, aggravated by the policies of perestroika, has led to a rapid decline in oil production in West Siberia (Gustafson, 1989; Dienes, 1993). During the 1990–91 period Soviet oil production fell by 9.6 per cent, from 570 million tons to 515 million tons (Sagers, 1992b). During the same period Russian oil production fell 10.6 per cent, from 516 million tons to 461.1 million tons. During 1992 oil production fell a further 14.1 per cent to 395.8 million tons (Sagers, 1993, 207). Forecast production for 1993 is around 350 million tons. This represents a 37 per cent decline in Russian oil production since 1989. According to the news agency Interfax, after the first five months of 1993 Russian oil production stood at 146.5 million tons, down 12.5 per cent against the same period in 1992. Oil exports to non-CIS countries reached 33.28 million tons, 90 per cent came from Russia. Fortunately, natural gas production has remained relatively stable; at the end of the first five months gas production totalled 269.1 billion cubic metres, which is comparable to 1992 production. The problem is that oil is a key export commodity for Russia, it is the single most important source of hard currency and a key commodity in Russia's trade with the other republics of the former Soviet Union.

In 1992, Russia exported about 66 million tons of oil, approximately 15 per cent of total production; this was actually an increase of 17.2 per cent over exports in 1991. During 1992 domestic oil consumption, as measured by deliveries to Russian oil refineries, fell by 12.7 per cent to 250 million tons (Sagers, 1993, 207). Nevertheless, with domestic production continuing to fall, Russia can no longer expect to meet domestic demand, expand exports to hard currency markets and supply the other republics at previous levels. Given that Russia is only receiving a fraction of the world price in non-convertible currency for its exports to the other republics, it is not surprising that it has already cut back oil deliveries to the other states. According to the Russian State Statistical Committee, in 1992 Russia supplied the other republics with 75.5 million tons of oil and 106.4 billion cubic metres of natural gas. In 1993 Russia plans to supply the former union republics with 55.8 million tons of oil (this is 26 per cent less than 1992 and represents about 15 per cent of Russian production) and 96 billion cubic metres of natural gas (Rossiyskiye vesti, 1993). Many republics are unhappy about the reductions in oil supply. For example, under an intergovernmental trade agreement, Russia agreed to supply Ukraine with 20 million tons of oil in 1993, but Ukraine maintain that it needs 40 million tons.

While the Russian government has raised energy prices on a number of occasions, inflation and the decline of the ruble mean that the ruble price is still nowhere near the world price. In 1992, the average price of oil sent to the former Soviet Republics was R18 500 per ton (34 per cent of world price). According to Russian figures, in 1992 Russian supplies of oil and gas to the other republics of the former Soviet Union amounted to $16 billion, for which Russia received only $800 000 worth of equipment for the fuel and energy industry, less than one-third of the total envisaged in intergovernmental agreements. A number of the republics have already experienced difficulties in financing energy imports from Russia. By the spring of 1993, Lithuania owed Russia $40 million for natural gas deliveries. In early July Russia cut off gas supplies to Lithuania and Estonia for non-payment of bills. The Russian Government also threatened to cut off oil supplies to Estonia if it failed to improve its treatment of ethnic Russians. Thus, dependence upon Russian energy is not only a financial problem, it leaves the republics vulnerable to political leverage. At the end of the first quarter of 1993 the other republics owed Russia 530 billion rubles, equivalent to 6.6 per cent of Russian GDP (Ivanter et al., 1993, 19).

Russia is now adopting a tough stance in its trade negotiations with the other republics. It has suggested that debtor countries repay debts accumulated in 1992 and 1993 either with hard currency or by transferring state property (interest and blocks of shares). In future, credits offered by Moscow will be translated into dollar rates and a potential borrower will be requested to deposit $5 billion in Russia's foreign trade bank as insurance. These measures threaten to have the same impact on inter-republican trade as the Soviet Union's unilateral decision to move to world prices and hard currency had on intra-CMEA trade. Meanwhile, Russia is introducing new investment programmes to increase domestic oil production, and western investors and international aid agencies, such as the EBRD, are being encouraged to help modernise the oil and gas industry. An Oil and Gas Council has been created to try to coordinate the energy policies of the Newly Independent States. At the same time, the other republics are seeking foreign help to develop their own energy resources and construct new pipelines to reduce their dependence on Russia. For example, Ukraine hopes to obtain oil from Iran and Armenia is to be supplied with natural gas from Turkey. In the long run, the stabilisation of Russian oil production and the development of new non-Russian oil and gas fields will generate a new pattern of energy trade; in the short term many of the republics remain dependent on Russia. The current situation seems intractable. On the one hand, Russia cannot afford to continue to subsidise its energy exports to the other republics; on the other hand, the republics cannot pay world prices and cannot pay in hard currency. Noren and Watson (1992) have highlighted the problem:

> Energy trade at concessionary prices is probably the most striking aspect of past commerce between the republics. At world prices the Russian surplus in energy trade with the other republics amounts to the equivalent of $24 billion; Ukraine's deficit would approach the equivalent of $15 billion.' (Noren and Watson, 1992, 92)

The problems of financing inter-republican trade are further complicated by the fact that Russia does not have sufficient oil to satisfy the needs of the republics. In response to the problems of the energy industry, 13 of the 15 states of the former Soviet Union have created a Council of Heads of Government on Oil and Gas. The council has its secretariat in Tyumen, West Siberia. Its aim is provide finance to halt the decline in Russia's oil production. Perhaps this initiative is recognition that the problems the republics have inherited from the Soviet Union can only be resolved by collective action. The final section of this chapter considers prospects for cooperation and reintegration among the republics of the former Soviet Union.

CONCLUSIONS: PROSPECTS FOR REINTEGRATION

This chapter has shown how the Soviet system produced a high level of economic interdependence among its constituent republics. Following the collapse of the Soviet Union, the actions of the republics have resulted in a dramatic reduction in inter-republican trade which has been a major factor promoting industrial decline. Now, in the face of an energy crisis and the collapse of the ruble-based payments system, the republics are seeking to promote economic reintegration in an attempt to reduce the economic costs of transition. However, at present the republics seem to lack the political will to create meaningful multilateral or supranational institutions. The CIS, which emerged from the ruins of the Soviet Union, is unlikely to form the basis of a new regional trading bloc. At best, the CIS represents a loose coalition of states who share a common interest in reducing the costs of Soviet disintegration. Agreements within the CIS are not multilateral, instead all member states have to reach bilateral agreements with one another. Consequently, CIS summits have resulted in numerous ambitious declarations, but very little real action.

In early 1993, 10 member states of the CIS agreed to create an interstate bank. The aim of this bank is to establish an institutional mechanism for multilateral clearing and settlement of interstate payments in Russian rubles or hard currency. The accelerated collapse of the ruble zone and the introduction of new national currencies make the creation of such a bank essential if monetary fragmentation is not to promote a further decline in inter-republican trade. However, although all the member states have signed an agreement the bank is yet to begin operation. In March 1993, five CIS states – Russia, Kazakhstan, Kyrgyzstan, Tajikistan and Uzbekistan – agreed to form a customs union. In July 1993, the three slav republics – Belarus, Russia, and Ukraine – announced their intention to create an economic union. This slav union is in part a reaction to the lack of real progress by the CIS and aims to form a nucleus of deeper economic integration within the wider framework of the CIS. It can be expected that other states, such as Kazakhstan, will soon become part of this nucleus. This may result in a two-speed CIS with a more deeply integrated core, surrounded by a free trade area comprised of the remaining CIS states. Such an arrangement would recognise the different economic conditions that exist in the peripheral states of Central Asia. The result may be similar to the European Economic Area (EEA) comprised of the European Community and the European Free Trade Association (see Chapter 3). At present, the focus of these efforts is to provide a framework for interstate trade, much of which is based on bilateral intergovernmental trade agreements. However, if economic integration is to promote transition, then state-controlled trade between republics must soon give way to free trade between enterprises in the various states. In other words, economic integration must provide a mechanism for nurturing market relations, rather than a means of preserving state control over economic relations.

Outside the framework of the CIS, new regional associations are emerging. The three Baltic States have formed the Baltic Union. Kazakhstan and the four Central Asian states have formed a loose coalition to manage common problems, such as the Aral Sea. At the same time, new regional associations are being joined that link former Soviet republics with states in neighbouring regions. For example, two new regional trade and development banks are being created in the Black Sea region and Central Asia. The Black Sea Economic Cooperation Bank includes: Turkey, Bulgaria, Romania, Greece, Albania,

Moldova, Ukraine, Russia, Georgia, Armenia and Azerbaijan. The Economic Co-operation Organisation involves: Turkey, Iran, Pakistan, Afghanistan, Azerbaijan, Kazakhstan, Turkmenistan, Kyrgyzstan, Tajikistan and Uzbekistan. There are hopes that a similar regional association can be developed in the Baltic. For the former Soviet republics such associations provide a means of gaining much-needed capital investment to promote industrial restructuring. These initiatives also suggest that the collapse of Soviet power is resulting in a new geopolitical and geoeconomic configuration, as the states of the former Soviet Union are able to control their own economic and political relations with their neighbours.

In the short term the focus of effort has been upon measures to reduce the economic costs of the collapse of inter-republican trade. A lack of political trust and a short-sighted desire to hoard scarce resources have undermined attempts to create new institutions to manage trade and payments. The non-Russian republics have been happy to provide credits to finance large trade deficits with Russia. Perhaps the demise of the ruble zone and the consolidation of the Russian ruble will provide renewed impetus for the creation of an interstate bank or some form of payments union. International institutions such as the World Bank and the IMF may have a role to play in supporting new trading arrangements and encouraging free trade between the republics of the former Soviet Union. Overall, any new initiatives to promote the economic reintegration of the former Soviet republics must focus upon providing an environment that promotes the development of new trading relations, rather than perpetuating the inefficiencies of the Soviet system.

References

Belkindas, M.V. and Sagers, M.J. (1990) A preliminary analysis of economic relations among the Union Republics of the USSR: 1970–1988, *Soviet Geography*, 31, 629–656.

Bradshaw, M.J. (1993) *The Economic Effects of Soviet Dissolution*, Royal Institute of International Affairs, London.

Business Eastern Europe (1993) The business outlook: Russia, *Business Eastern Europe*, 15 March, 4–5.

Cooper, J. (1993) *The Conversion of the former Soviet Defence Industry*, Royal Institute of International Affairs, London.

Dellenbrant, J.A. (1986) *The Soviet Regional Dilemma*, M.E. Sharpe, New York.

Dienes, L. (1993) Prospects for Russian oil in the 1990s: reserves and costs, *Post-Soviet Geography*, 34, 79–110.

Dyker, D.A. (1983) *The Process of Investment in the Soviet Union*, Cambridge University Press, Cambridge.

Dyker, D.A. (1992) *Restructuring the Soviet Economy*, Routledge, London.

Economist Intelligence Unit (1991a) *USSR: Country Report No. 4*, The Economist Intelligence Unit, London.

European Commission (1992) *European Economy No. 45*, European Commission, Brussels.

Goskomstat SSSR (1988) *Promyshlennost' SSSR* (Industry USSR), Finansy i Statistika, Moscow.
Goskomstat SSSR (1991a) *Narodnoye Khozyaystvo v 1990g* (The National Economy of the USSR in 1990), Finansy i Statistika, Moscow.
Goskomstat SSSR (1991b) *Soyuznyye respubliki: osnovnyye ekonomicheskiye i sotsial'nyyee pokazateli* (Soviet Republics: Basic Economic and Social Indicators), Goskomstat SSSR, Moscow.
Granberg, A.G. (1990) Ekonomicheskiy mekanizm mezhrespublikanskikh i mezhregional'nykh otosheniy (Economic mechanism of interrepublican and interregional relations), in: V.I. Kuptsova (ed.) *Radikal'naya eknonmicheskaya reform*, Vysshaya Shkola, Moscow, 310–326.
Granberg, A.G. (1991) Inter-republican integration: Russia's position, *Business in the USSR*, December, 48–49.
Granberg, A.G. (1993) The economic interdependence of the former Soviet Republics, in: J. Williamson (ed.) *Economic Consequences of Soviet Disintegration*, Institute for International Economics, Washington, D.C., 47–77.
Gregory, P.R. and Stuart, R.C. (1990) *Soviet Economic Structure and Performance*, 4th Edition, Harper & Row, London.
Gros, D. (1991) Regional disintegration in the Soviet Union: economic costs and benefits, *Intereconomics*, September/October, 207–213.
Gustafson, T. (1989) *Crisis Amid Plenty: The Politics of Soviet Energy Under Brezhnev and Gorbachev*, Princeton University Press, Princeton, NJ.
Hansson, A.H. (1993) The trouble with the Ruble: monetary reform in the former Soviet Union, in: A. Aslund and R. Layard (eds.) *Changing the Economic System in Russia*, Pinter, London, 163–182.
Havrylyshyn, O. and Williamson, J. (1991) *Open for Business: Russia's Return to the Global Economy*, The Brookings Institution, Washington, D.C.
Horrigan, B. (1992) How many people worked in the Soviet defense industry, *RFE/RL (Radio Free Europe/Radio Liberty) Research Report*, 1, 33–39.
IMF, WB, OECD, EBRD (International Monetary Fund, The World Bank, Organisation for Economic Cooperation and Development, European Bank for Reconstruction and Development) (1991) *A Study of the Soviet Economy, Vol. 1*, OECD, Paris.
IMF (International Monetary Fund) (1992a) *Common Issues and Interrepublic Relations in the former USSR*, International 169h Monetary Fund, Washington, D.C.
Ivanter, A., Kirichenko, N. and Khoroshavina, N. (1993) First-quarter CIS statistics indicate severe drop-off in GDP, *Commersant*, 2 June, 17–19.
Langhammer, R.J., Sagers, M.J. and Lücke, M. (1992) Regional distribution of the Russian Federation's earnings outside the former Soviet Union and its implications for regional económic autonomy, *Post-Soviet Geography*, 33, 617–634.
McAuley, A. (1991a) Costs and benefits of de-integration in the USSR, *Moct-Most*, 2, 51–65.
McAuley, A. (1991b) The economic consequences of Soviet disintegration, *Soviet Economy*, 7, 189–214.
Michalopoulos, C. (1993) *Trade Issues in the New Independent States*, World Bank, Washington, D.C.
Michalopoulos, C. and Tarr, D. (1992) *Trade and Payments Arrangement for States of the Former USSR*, The World Bank, Washington, D.C.
Noren, J.H. and Watson, R. (1992) Interrepublican economic relations after the disintegration of the USSR, *Soviet Economy*, 8, 89–129.
Nove, A. (1977) *The Soviet Economic System*, Allen & Unwin, London.
Nove, A. (1981) An overview, in: I.S. Koropeckyi and G.E. Shroeder (eds.) *Economics*

of Soviet Regions, Praeger, New York, 1–8.
Pallot, J. and Shaw, D.J.B. (1981) *Planning in the Soviet Union*, Croom Helm, London.
Rossiskiye vesti (1993) 5th March.
Sagers, M.J. (1991) Regional aspects of the Soviet economy, *PlanEcon Report*, 7, No. 1–2.
Sagers, M.J. (1992a) Regional industrial structure and economic prospects in the former USSR, *Post-Soviet Geography*, 33, 238–268.
Sagers, M.J. (1992b) Review of the energy industries of the former USSR in 1991, *Post-Soviet Geography*, 33, 487–515.
Sagers, M.J. (1993) Russian crude oil exports in 1992: who exported Russian oil?, *Post-Soviet Geography*, 34, 207–211.
Seliverstov, V. (1991) Inter-republican economic interactions in the Soviet Union, in: A. McAuley (ed.) *Soviet Federalism: Nationalism and Economic Decentralisation*, Leicester University Press, London, 108–127.
Senik-Leygonie, C. and Hughes, G. (1992) Industrial profitability and trade among the former Soviet Republics, *Economic Policy*, October, 354–386.
Shatalin, S.S. (1990) *Perekhod k rynku* (Transition to the Market), Ministry of Publishing and Mass Information. Moscow.
Shaw, D.J.B. (1985) Spatial dimensions in Soviet central planning, *Transactions of the Institute of British Geographers*, 10, 401–412.
Slay, B. (1991) On the economics of interrepublican trade, *RFE/RL (Radio Free Europe/Radio Liberty) Research Institute: Report on the USSR*, 3, 1–8.
Smith, A. (1993) *Russia and the World Economy: Problems of Integration*, Routledge, London.
Soviet Geography (1991) News notes, *Soviet Geography*, 32, 190–193.
Snyder, T. (1993) Soviet Monopoly in Williamson, J. (ed) *Disintegration* Institute for International Economics, Washington DC, 175–243.
Vestnik statistiki (1990a) Ekonomicheskiye vzaimosvyazi republik v narodnokhyaystvennom komplekse (Economic interrelations of the republics in the national economic complex), *Vestnik statistiki*, 3, 36–353.
Vestnik statistiki (1990b) Ob'em vvoza i vyvoza produktsii po soyiznym respublikam za 1988g vo vnutrennykh i mirovykh tsenakh (Volume of imports and exports of products by the Soviet republics in 1988 in internal and external prices), *Vestnik statistiki*, 4, 49–60.
Williamson, J. (1992) *Trade and Payments after Soviet Disintegration*, Institute for International Economics, Washington, D.C.
Yakovlev, A. (1990) Monopolizm v ekonomike SSSR (Monopoly in the economy of the USSR), *Vestnik statistiki*, 1, 3–6.

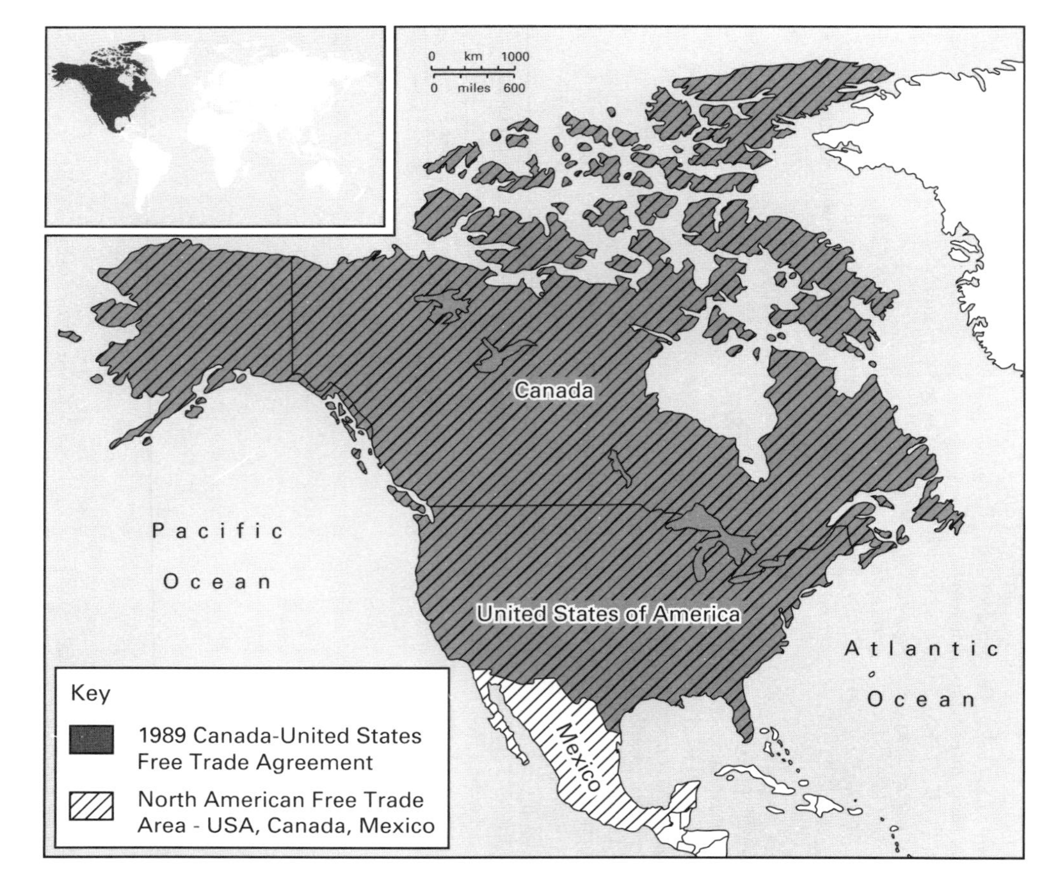

Figure 6.1 North America

6 The North American Free Trade Area: an overview of issues and prospects

James McConnell and Alan MacPherson

State University of New York at Buffalo

INTRODUCTION

The leaders of Canada, Mexico and the United States (U.S.) signed a historic trade accord on December 17 1992, designed to create a trinational market area of over 360 million people with a combined purchasing power approaching $6 trillion. The North American Free Trade Agreement (NAFTA, see Figure 6.1), if adopted by the three countries, will be the most comprehensive free trade pact (short of a common market structure) ever negotiated between regional trading partners. Moreover, it will represent the first reciprocal free trade accord between industrial countries and a developing nation, which explains why the ongoing negotiations are of special interest to Third World nations, particularly those located in Latin America.

The prospect of strengthened economic ties among these three nations has created a wide diversity of viewpoints and a growing polarisation regarding the merits and disadvantages of such a union. For example, many proponents view the accord:

> as an integral part of a national competitiveness strategy, one that complements domestic economic reforms designed to improve productivity and promote the ability of local industries to compete more effectively against foreign suppliers at home and in world markets. (Hufbauer and Schott, 1993, 116)

In contrast, opponents argue that such a trade alliance will exploit human and environmental resources, benefit only large multinational corporations, result in major job losses across the trilateral market area, reduce wages and standards of living in the two northern-most members of the union, and even threaten consumers in the

Continental Trading Blocs: The Growth of Regionalism in the World Economy
Edited by R. Gibb and W. Michalak

U.S. and Canada because of unsanitary agricultural practices in Mexico (including the use of sewage-tainted irrigation water and unsafe pesticides).

The goal of this chapter, therefore, is to examine the pressures for regionalism in North America in the light of several economic and political irritants that constrain widespread acceptance of the NAFTA concept. Three main questions are addressed in the sections that follow. First, what are the principal factors encouraging greater regionalism among the three nations? Second, what types of regional economic interests and arguments most strongly support and/or oppose the trilateral arrangement? And third, given the key provisions of the NAFTA text and the prevailing economic and political conditions within North America, what are the long-term prospects for the proposed trade accord?

MOVEMENTS TOWARD CLOSER INTEGRATION

Despite major economic and political misgivings about each other, the three North American nations began to move closer together in several important commercial spheres during the late 1980s. While trade between Canada and the U.S. more than doubled to some $170 billion over the 1980s, important advances in trade were also being forged between Mexico and its two neighbours to the north. By 1991, Mexico's two-way trade with Canada had increased to over $2 billion, while the value of its shipments to and from the U.S. had risen to some $50 billion – a 1987–1991 increase of well over 100 per cent in both cases (International Monetary Fund, 1993). As the three nations moved into the 1990s, the most important export and import market by value of shipments for both Mexico and Canada was the U.S.; the first- and third-ranked export markets for the U.S. were Canada and Mexico, respectively; and the second- and third-ranked import markets for the U.S. were Canada and Mexico, respectively.

In addition to bolstering trade ties, the three nations also increased their ownership of assets in each other's territory. For example, of the total stock of foreign direct investment (FDI) in Mexico, over 65 per cent is owned by the U.S. and 2.4 per cent by Canada; approximately 70 per cent of all Canada's foreign direct investment is in the U.S.; while recent estimates of the value of Mexican assets in the U.S. range from $40 billion to over $80 billion (Investment Canada, 1991). To put these Mexican direct investments in perspective, however, it should be noted that less than 2 per cent of all FDI in the U.S. comes from Mexico (United States General Accounting Office, 1992) – a rather small proportion when compared with the 65 per cent share of FDI in Mexico that is owned by U.S. companies.

A synopsis of recent economic growth trends for the three NAFTA nations is presented in Table 6.1. Here the sheer size of the U.S. economy stands out quite clearly. Even so, Mexico stands out as a fast-growing participant in international

Table 6.1 Economic trends among NAFTA nations

Variable		*1970*	*Year 1990*	*% Change 1970–1990*
GNP (US$B)	Canada	83.8	572.7	583.4
	USA	1016.8	5685.8	459.1
	Mexico	35.1	200.5	471.29
Population (millions)	Canada	21.3	26.9	26.2
	USA	205.0	252.7	23.2
	Mexico	50.7	84.5	66.6
GNP per capita (US$)	Canada	3929	21220	440.1
	USA	4959	22501	353.7
	Mexico	692	2374	243.1
Exports (US$M)	Canada	16707	127459	662.9
	USA	42469	416510	880.7
	Mexico	1348	27121	1911.9
Imports (US$M)	Canada	13659	121529	789.7
	USA	39866	490110	1129.4
	Mexico	2236	38184	1607.6
FDI Balance (US$M)	Canada	+508	+239	–52.9
	USA	–6130	–7300	19.1
	Mexico	+323	+4742	1368.1

Source: International Monetary Fund (1992).

trade and investment. Despite a remarkably high rate of population growth, per capita incomes in Mexico are rising fairly quickly. In net terms, moreover, it is important to note that Mexico has recently been absorbing an annual average of $4 billion in FDI, as well as some $4 billion worth of imports. While Table 6.1 is intended to provide a general snapshot only (a more disaggregated and time-sensitive picture would require a separate study), the data clearly indicate that NAFTA is a proposed alliance between unequal partners. Having said this, however, the degree of integration and complementarity between the three nations is already substantial. For example, over 65 per cent of Mexico's exploding import bill is the result of merchandise shipments from the U.S., while a significant proportion of Mexico's positive balance on FDI is the result of incoming U.S. investment.

While a detailed description of the commercial connectivity between the NAFTA nations goes beyond the scope of this chapter, Table 6.2 provides a brief illustration of the extent to which the three countries are tied together. In terms of exports, for example, the U.S. currently derives almost 30 per cent of its foreign earnings from Canada and Mexico, making these two destinations more important than any other single export market. Mexico imports fully 71 per cent of its merchandise needs from the U.S., and exports 74 per cent of its foreign shipments to U.S. customers. Canada is also highly dependent upon the U.S. market for both exports and imports. In 1991, for example, 62 per cent of

Table 6.2 Trilateral trade patterns among the NAFTA nations (Per cent of total trade among NAFTA nations*)

		1980		*1991*	
		Imports	*Exports*	*Imports*	*Exports*
Canada	USA	67.5	60.6	61.9	75.1
	Mexico	0.5	0.6	1.6	0.3
USA	Canada	16.3	16.0	18.4	20.1
	Mexico	5.0	4.5	6.2	7.9
Mexico	Canada	1.7	0.7	0.8	5.4
	USA	58.8	64.7	70.7	74.5

*These percentages have been calculated on the basis of total international trade patterns rather than trinationally exclusive trade shares.

Source: International Monetary Fund, 1993.

Canada's imports came from the U.S., compared to only 1.7 per cent from Mexico. On the export front, over 75 per cent of Canada's foreign shipments are destined for U.S. customers, compared to less than 1 per cent in the case of Mexico. Clearly, the U.S. functions as a strategic hub between two commercial partners that are relatively peripheral to each other (both geographically and economically). From a trading perspective, Canada and Mexico have a clear interest in pursuing a trilateral agreement, if only because stable access to the large U.S. market is a commercial necessity for both countries. In 1991, for example, roughly 17 per cent of Canada's gross national product (GNP) came from merchandise exports to the U.S. (the comparable figure for Mexico was approximately 12 per cent), suggesting a major role for the U.S. as a source of income for the two smaller NAFTA countries (International Monetary Fund, 1993). From a global trading perspective, few other nations are as absolutely and proportionally dependent upon a single export market (and import source) as this.

FACTORS FAVOURING THE CREATION OF A TRILATERAL TRADE AGREEMENT

The proponents of a North American trade alliance tend to rally around one or more of three fundamental arguments. These basic principles are that the trilateral accord enables the partners to achieve important geopolitical and commercial objectives within the global community, it enhances the long-term economic well-being of the nations, and it creates uniform standards and relatively stable political conditions within the regional market. These arguments are discussed in more detail below.

Global geopolitical and commercial objectives

One of the most popular arguments in support of a trilateral agreement is that such an alliance will enable the three countries to achieve important geopolitical and commercial objectives. First, NAFTA is expected to create a counterbalance to the rising tide of regional bloc formation that is occurring in Europe and Asia. In contrast to the multilateral non-discriminatory philosophy of the General Agreement on Tariffs and Trade (GATT), regional trading blocs are founded upon the principle of preferences. Such moves toward regionalism in international commerce tend to generate reciprocal action by other nations as a mechanism of defense. Hence, the creation of NAFTA is viewed as a way to counter bloc formations in Europe and Asia. Support for this perspective is provided by a recent report from the government of Canada in which it is noted that:

> Regional responses to the internationalisation of economic activities in Europe and Asia, as well as in the America's, have presented Canada with a direct challenge. The FTA (i.e., the Canada – U.S. Free Trade Agreement) was the first response to that challenge. The Canada – U.S. – Mexico negotiations built on that achievement. (External Affairs and International Trade Canada, 1993, 3).

Similarly, two Canadian economists argue that 'NAFTA is a strategic response by the United States to global trends such as the increasing economic threat arising from an expanded and unified Europe and a more assertive Japan' (Grinspun and Cameron, 1993, 16). Additional support is advanced by Hufbauer and Schott in their claim that:

> Juxtaposed against the European Community, which is broadening and deepening the integration of its twelve member states and expanding the scope of its trade preferences with both the EFTA and the eastern European countries, the NAFTA is often seen as North America's response to the European trading bloc. (1993, 6).

In short, the agreement is seen as a defensive mechanism that places the governments of the three countries in a more favourable position to bargain against threats of increased protectionism that might be put forth by trading blocs in Europe, Asia and elsewhere.

Another objective that underlies current enthusiasm for NAFTA is based upon the adage from geopolitics that the physical proximity of nations plays an important role in shaping the foreign policy relationships among those same nations (Balassa, 1962, 39–40; Russett, 1967, 158). By working toward the harmonisation of rules, regulations and standards, the economic heartlands of these three countries – extending some 2000 air miles from southern Ontario to Mexico City – can be integrated by a dense and relatively low-cost network of transportation and communication that is unequalled anywhere else in the world. These countries not only stand to benefit economically from their physical juxtaposition, but their similar location relative to other countries in Latin America provides an opportunity to broaden the geographical boundaries of the trading bloc to embrace nations whose trade linkages are already strongly oriented to the North American market-place. It would be a mistake to underestimate the importance, particularly to the U.S., of eventually extending the North American trade bloc southward to include Central and South American nations. Currently, for example, the countries of Latin America have a combined population of over 445 million (which is some 8.4 per cent of the world's total population), a combined GDP of more than $825 billion, and almost 60 per cent of the merchandise trade of these nations is with the three North American countries. To take advantage of the rapidly growing market potential in Latin America, in June 1990 former U.S. President Bush launched the 'Enterprise for the Americas Initiative', which is perceived by many as a blueprint for the eventual creation of a hemispheric free trade area that extends from northern Canada and Alaska to Tierra del Fuego.

A North American trading bloc is also seen as helping the members achieve a third global objective. Specifically, the creation of NAFTA provides a fall-back position should the GATT multilateral negotiations fail to achieve their desired objectives. Hence a key benefit of creating regional trading blocs in North America, Europe and elsewhere is in 'easing the way back to a full-blown multilateral trading system, which we currently do not have' (Belous and Hartley, 1990, 32). In other words, the three nations are working within the rules and spirit of the GATT to create, in a stepwise manner, a model of economic cooperation that can be followed by other trading partners. This model is especially important in demonstrating that industrialised nations can form a mutually beneficial trade alliance with a developing nation. Hence a successfully negotiated NAFTA underscores the merits of free trade as a bulwark for economic development, and it provides the impetus for a more open, multilateral trading system under (hopefully) a revitalised GATT 'umbrella'.

NAFTA enhances economic well-being

A second set of arguments is designed to demonstrate that the creation of a North American trade accord will contribute to the long-term economic growth and development of the three participants. This argument is based upon several basic propositions. The first of these is that a free trade arrangement, with established rules of operating and resolving disputes, creates a relatively stable and predictable environment for commerce. A primary benefit of such an atmosphere is that business executives can contemplate longer-term decisions with a greater degree of certainty, which ideally serves to strengthen the efficiency and global competitiveness of their manufacturing and service-based establishments.

Within a stable and relatively predictable business environment, the three countries are also in a better position to benefit from the strong economic complementarities that exist among them. For example, the economic recovery programme for Mexico requires large infusions of FDI, much of which can be provided by Canada and the United States. Moreover, manufacturers in both Canada and the U.S. need offshore production capabilities to minimise labour costs and thereby to enhance their global competitiveness. The prime choices for locating such activity would appear to be Mexico and certain countries in East Asia and eastern Europe. The relative advantage of Mexico over other options for foreign production lies not only in Mexico's relatively low-cost, literate labour force, but also in its close geographic proximity to U.S. and Canadian markets (which significantly reduces transportation costs).

The trade accord is also expected to create new jobs, especially in technology-intensive industries for Canada and the US, and in more labour-intensive production for Mexico. Although many forecasts have been made of net jobs likely to be gained or lost in the intermediate term, a conservative estimate, prepared by the Institute for International Economics in Washington, D.C., is that the agreement will create about 170 000 new jobs by 1995 just in the US, five years after the agreement was first proposed in 1990 (Hufbauer and Schott, 1993, 14). In fact, Hufbauer and Schott point out that if the proposed accord is rejected, the US could actually lose jobs as investment capital leaves Mexico. This, in turn, could force Mexico to contract its imports, thereby reducing exports to Mexico from the United States. At the same time, however, these authors make a compelling argument that regardless of the intermediate gains and losses associated with NAFTA, the long-term impact of the accord on employment throughout the region will not be significant. They believe that macroeconomic conditions and policies, such as global business cycles and the fiscal and monetary policies of the three countries, are much more significant in determining employment outcomes than are microeconomic events, such as defence conversion or the creation of NAFTA (Hufbauer and Schott, 1993, 20).

One of the most cogent expectations in support of NAFTA is the belief that by creating an expanded market-place and permanent rules for trade and investment, the alliance will promote long-term efficiencies of production, expand opportunities for capital investment, and increase commodity trade. In turn, increased productivity and trade, along with a more rational allocation of resources within the region, will increase economic growth, improve the global competitiveness of businesses, and reduce prices for consumers within the region. The eventual outcome expected, therefore, is rising standards of living for all three countries.

Rising wage levels in Mexico are of particular interest because approximately 70 per cent of Mexico's imports are from its northern neighbours. More economic development in Mexico means more purchasing power for Mexicans and increased demand for goods from Canada and the United States. In addition, however, rising income in Mexico also has implications for the country's ability to address its environmental problems adequately. To counter the argument that no trade accord should be approved until Mexico more effectively enforces its environmental laws, NAFTA proponents have noted that when countries become wealthier and more industrialised, they spend a far greater percentage of their GDP on environmental protection. For example, of 42 countries analysed in a recent Princeton University study, pollution levels began to decrease in those with per capita GDP rising above $4000 to $5000 (Smith W.R., 1993, 13). Mexico, with a current per capita GDP of $3700, has already begun to devote larger resources to environmental matters.

Uniform standards and stable political conditions

NAFTA proponents also argue that closer integration will harmonise various standards and strengthen democratic institutions across the region. In addition to reducing most of the existing market-access barriers to the movement of commodities and capital throughout the region, for example, NAFTA also establishes rules against distortions to investment (e.g. local content and export performance requirements), protects intellectual property rights, and establishes uniform procedures for resolving commercial disputes. Moreover, the agreement stands as a landmark accord in that it attempts to ensure that existing standards for handling environmental issues are maintained, although it does not contain provisions to upgrade the enforcement of existing standards, nor does it adopt more rigorous environmental standards. These efforts to harmonise standards across the region are accompanied in the trade accord by the concept of 'national treatment', in which Canadian, Mexican and US businesses must be treated at least as well as domestic-based firms when conducting business within the trilateral market area.

In addition to the issue of standards, proponents point out that it is in the interest of each of the three nations to have political stability and to strengthen democratic institutions within Mexico. The argument is made that economic growth and development in Mexico under NAFTA could provide that stability; however, for this transformation to occur, Mexico needs sustained assistance from its two northern neighbours. Mexico's present economic condition puts enormous pressures upon the country's public officials to balance ongoing macroeconomic policy reforms with internal economic and political realities and constraints. The country's overwhelming reliance upon the US market for its exports and imports can be bolstered by the new trade deal. In addition, Canadian and US companies can provide financial and entrepreneurial support for the efforts of the Mexican government to integrate the achievements of *maquiladora* production (cross-border investments along the Mexico-US border) more thoroughly with national industrial development efforts in other parts of the country. Moreover, a trilateral alliance is expected to help Mexico restore confidence among both Mexican and foreign investors that the country's economic recovery is relatively secure. This assurance is essential if the Mexican government is to succeed in its efforts to attract the capital resources that are essential to reduce the country's debt payments and reinvigorate its national economy.

A relatively stable economic and political climate in Mexico is also seen by some as a way to contain the problem of illegal immigration into the US from the south. NAFTA proponents see the illegal immigration issue, which has become an area of increased political contention between the two neighbouring countries, as one that is primarily related to the harsh economic disparities that exist between the two nations. The trade accord is seen as a way of strengthening the Mexican economy, creating new jobs in Mexico, and thereby significantly reducing the illegal migration of people across the border. For example, the consulting firm CIEMEX-WEFA calculates that over the ten-year period ending in 2002, the gross number of illegal immigrants into the US could decrease by 600 000 as a result of the economic growth stimulation of NAFTA (Hufbauer and Schott, 1993, 25).

RESISTANCE TO NAFTA

Voices in opposition: an overview

In general, the opponents of a trilateral accord support one or more of the following four themes. The first position is that a union between two industrialised nations and a developing nation will eventually work to the detriment of the more developed partners. For example, the 'pauper labour' argument advances the notion that imports by a rich country from a poor country must inevitably reduce the standard of living in the rich country. This view, which is held by many labour unions in Canada and the US, is based upon the assumption that the lower wages in Mexico will further depress wage levels in the two northern neighbours once closer economic integration occurs. Others express concern about the downward pressure NAFTA is likely to place upon the working conditions of labour, environmental standards, health and safety regulations, and cultural 'industries'. Still others argue that Mexico's legal system is not compatible with those that currently exist in the US and Canada. In short, the argument is that the vast economic, political and legal differences that exist among the three are simply too great to be dealt with effectively in a commercial alliance.

A second set of views in each of the three countries reflects a range of economic insecurities regarding the expected outcome of a comprehensive trade alliance. In addition to forecasting major job losses, critics argue that only large corporations will benefit from the accord, many small and medium-sized establishments will be forced out of business by the expected increase in competition, the costs of worker-adjustment assistance and environmental clean-up along the Mexican-US border are likely to be much more than predicted, the overall standard of living in the two northern-most countries will be lowered, and the likelihood of currency devaluation by Mexico threatens expected gains in trade from closer integration. Many of these beliefs, which are especially common among critics in Canada and the US, reflect in part the circumstantially poor timing of the NAFTA negotiations. In short, the legislative bodies of the two countries are being asked to accept the uncertainties associated with the trade accord at a time when both are attempting to sustain their recovery from a severe recession. To make matters worse, businesses in both nations continue to confront stiff opposition from foreign competitors at home and in the global market-place, and both countries are trying to reduce their enormous federal budget deficits.

A third viewpoint in opposition to the accord points to the differential effects that are likely to occur across geographical boundaries if NAFTA is created. For example, heated political debates have already occurred in the US between those representing border states and regions and those from more interior areas of the country. Government leaders of the states of Texas, New Mexico, Arizona and California are advancing the argument that because of their location along the 1000-mile border with Mexico, they will need billions of federal dollars to meet the environmental and infrastructure costs that will inevitably be associated with expanded trade with that country. Included in their proposed budgets are funds to expand already clogged border crossings and airport facilities, build new roads and water-treatment plants, upgrade waterways and health-care facilities, improve education, and reverse the degradation of the regional environment. At the same time, however, many legislators from states located away from the border are complaining that they are being asked to support the allocation of federal subsidies to assist the border region in implementing a trade agreement that by many estimates threatens the employment base of their states. Moreover, critics in Canada and the US argue that low-wage, unskilled workers will tend to be the principal ones displaced by the trade accord; hence manufacturing establishments and communities located in the rural areas of the trilateral region will be likely to suffer at the expense of those in more urban populated areas.

A fourth line of reasoning against NAFTA takes the position that the accord threatens the autonomy of one or more of the three partners. For example, Mexico is concerned about maintaining control over its petroleum reserves, while Canada voices concern about controlling its water resources and cultural industries. In addition, Canada and Mexico are strongly resisting efforts by the US to negotiate supplemental agreements that would enforce environmental laws, harmonise labour rights and standards, provide stronger safeguards against unexpected and overwhelming surges in imports, and coordinate exchange rate policies.

These 'voices' of opposition to NAFTA tend to vary in intensity and importance when comparisons are made between the three North American countries. To examine this spatial variation, some of the major points of resistance to the agreement in each of the three nations are presented below.

Opposition in Mexico

Although the Mexican government has spent millions of dollars lobbying in support of NAFTA, internal opposition continues to persist. For example, the Mexican political Left, gathered around Cuauhtemoc Cardenas and the Party of the Democratic Revolution, has shown an ambivalent interest in free trade. Cardenas has accused the President of Mexico, Carlos Salinas, of 'unprecedented subordination' to US interests in the areas of drugs, immigration and industrialisation, and has also accused him of gambling that 'he can buy off' Mexico's middle and lower classes with a large influx of resources from the United States. According to Cardenas, Salinas is hoping that recent economic reforms will bring about sufficient prosperity for the official party to remain in power. He further contends that the government is delaying political reforms in the hope that prosperity will make them irrelevant for the power structure.

For many union leaders in Mexico, a free trade deal would flood the country with imports, create structural unemployment, open the nation to increasing control from external interests and, ultimately, act as a vehicle for the exploitation of Mexican workers – few of whom enjoy legally guaranteed standards when it comes to working conditions, health care and retirement planning (among other things). Interestingly, even labour organisations in the US have supported their Mexican counterparts by claiming that NAFTA will 'do little to improve the lives of Mexican workers because their low wages must be maintained to attract continued US investment' (International Trade Reporter, 1991a, 19). Not only are many Mexicans fearful of external control over vital resources and sectors (Mexico has a long tradition of hostility to foreign bosses), they are also concerned that their relatively small manufacturing and service establishments – long protected by high tariffs – will be unable to compete against the larger multinational corporations from Canada and the United States.

Mexicans from all sides of the political spectrum are concerned that the Salinas government will abandon Mexico's traditional control of its oil industry, railways and electricity. While they are already alarmed over concessions made in the petrochemical industry, the oil industry is a particularly sensitive issue. Mexico nationalised all foreign-owned oil companies in 1938, and a state monopoly has existed ever since. However, there is widespread agreement that the oil industry has long been hampered by inefficiency and corruption, and few critics would deny that this sector badly needs new investment (especially foreign technology). It is estimated, for example, that Mexico requires $20 billion in investment just to the end of 1995 to meet domestic demand, which is increasing at 5 per cent a year, and to keep exports constant at 1.3 million barrels a day (Hufbauer and Schott, 1993, 120). The US has a vested interest in expanding Mexico's oil production, and it would seem logical for US trade negotiators to press this point to the limit. It would appear, however, that many of these concerns are somewhat unfounded because the proposed trade accord does not remove Mexico's historic restrictions on foreign investment and private participation in the country's primary petroleum industry; it does not ensure foreign investment in oil exploration, production or refining; and it does not permit Canadian or US firms to enter Mexico's retail gasoline market.

Others vigorously opposed to the agreement include various business interests in Central Mexico, especially in the highly protected areas of textiles and leather goods. The financial and service sectors in Mexico City also feel threatened by the expected increase in competition. Most Mexican firms are simply too small to compete with US-based multinationals. In the area of financial services, for example, negotiations have resulted in the proposal that by 1 January 2000, all Mexican restrictions on entry into the financial services market will be eliminated. Furthermore, temporary safeguard measures may be applied in the banking and securities sector only until January 2007, and only if foreign market shares reach 'high' levels. In addition, US exports could do serious damage to Mexican agricultural interests, specifically in grain, cattle, beef, most fresh fruits and vegetables, nursery products and soyabeans because, once NAFTA goes into effect, Mexican tariffs on these products will be eliminated immediately. The Mexican petrochemical industry could also be negatively affected because currently the average Mexican duty on chemical imports is 20 per cent. Under NAFTA, Mexico has agreed to lift investment restrictions on 14 of its 19 basic petrochemical industries and on 66 secondary petrochemicals. Many are concerned, therefore, that the Mexican chemical industry may have difficulty remaining competitive.

Another problem concerns negotiations over liberalisation of US laws governing movement of labour. The Mexican coordinator general of the trade accord has claimed that:

> Mexico does not want to export workers, but wants to export goods .... We want to provide them the opportunity to work here. We would not put on the table the idea that the FTA [foreign trade agreement] should lift all immigration barriers. (Mishra, 1990)

Yet more than one million Mexicans work in the US illegally, and some US officials believed at the outset that Mexico would push in the final stages of the negotiations to classify labour as a service. However, the agreement does not directly address the issue of illegal migration.

Perhaps the most contentious issue for Mexico (and to a somewhat lesser extent for Canada) is the US proposal that so-called side-bar (supplemental) agreements be negotiated in parallel to the NAFTA talks. In particular, the Clinton administration is calling for the creation of trinational commissions on labour and the environment with the power to investigate abuses and to recommend appropriate trade sanctions. Both Mexico and Canada have rejected such proposals, and in fact the Mexican ambassador to the US, Jorge Montano, has indicated that Mexico is prepared to pull out of the trade agreement if the Clinton administration continues to push for such commissions. While Mexico and Canada are willing to have commissions created to oversee the accord, they do not want the commissions to have the ability to impose sanctions, if only because this would be seen as an infringement upon their sovereignty, as well as a potential device for trade harassment. In response to Mexican resistance (and to the strong opposition of most Republicans in the US Congress), President Clinton has indicated that he is asking for sanctions only for 'repeated and persistent' violations of agreements made by NAFTA countries. Clinton notes that: 'I don't think that any of us should make agreements and expect there to be no consequences to their repeated and persistent violation' (International Trade Reporter, 1993, 813).

Opposition in the United States

Although a bilateral agreement with Mexico and a trilateral agreement involving Mexico and Canada had strong support from the Bush administration (and from most US businesses), equally strong objections to such accords have been voiced from several quarters, including President Clinton. For example, many critics are reluctant to move into negotiations which, if multilateral negotiations fail, could subsequently be viewed as attempting to create a 'fortress North America' rather than a non-discriminatory global environment for trade.

Other US complaints come from regional businesses and labour constituents who see free trade with Mexico as a threat to their livelihood. Because of the huge differences in the size and level of economic development of the US and Mexican economies, many are worried that the US will be forced to accept an asymmetrical deal in which Mexican tariff cuts and other trade-related matters will be phased in more slowly than on the US side. Moreover, the AFL-CIO (i.e. the American Federation of Labour and Congress of Industrial Organisations, which consists of affiliated labour unions) claims that 'a free trade agreement will only encourage greater capital outflows from the US and bring about a further increase in imports from Mexico, further harming the US industrial base' (New York Times, 1990). Even US automakers are worried about 'made-in-Mexico Japanese cars heading north to compete with Canada- and US-made vehicles' (Baker, 1990, 42).

Well-organised opposition is also being heard from various industrial groups and labour unions in the 'rustbelt' states. For example, groups such as the Religion and Labour Council of Kansas City and the Federation for Industrial Retention and Renewal in Chicago are gearing up for a fight in Congress against free trade with Mexico. The basic position of such groups is that a free trade pact with Mexico would spark a second wave of US investment in Mexico, beyond the *maquiladoras* clustered along the border, which would then result in the further relocation of production and assembly operations southward. As argued by an economist for the United Auto Workers, a US-Mexican trade accord 'is an extension of the southern strategy to move manufacturing to the low-wage, non-union Sunbelt' (Holstein and Borrus, 1990, 113). Other unions argue that not only will a trade deal with Mexico cost US jobs, but it will also put downward pressure on US industrial wages. Parenthetically, it is worth noting that similar arguments were used in 1987 and 1988 by groups in Canada that were opposed to the Canada–US Free Trade Agreement (FTA). Despite claims to the contrary, most of these critics argue that low-cost labour in Mexico will cause some southward shifting of economic activity and jobs (particularly blue-collar jobs). The unknown question is how many such jobs are likely to be lost.

The differential effects that are likely to occur across geographical boundaries have received growing attention in the light of a recent report from the governor of Texas that her state will need between $8 billion and $10 billion from the Federal Government over the next 16 years to adapt to the impacts anticipated from NAFTA. The border states of California, New Mexico and Arizona are expected to present similar claims in the near future. For many representatives and senators from northern industrial states, such requests are what one Congressman referred to as 'adding insult to injury' (Brown, 1993). Specifically, some view the job bases of industrial states located away from the border as seriously threatened by a trade alliance with the low-wage economy of Mexico. Yet the border states, which are expected to benefit substantially under NAFTA, are asking the Federal Government, and thus other states, to subsidise their start-up costs.

Further political frictions are expected between rural and urban areas of the United States. For example, a position paper prepared by the Federal Reserve Bank of Chicago suggests that the negative effects of NAFTA are most likely to be experienced in the non-metropolitan areas of the United States. The reasoning is that because average wages and labour skills in Mexico are lower, those US manufacturing jobs most likely to flee to Mexico would be the lower-wage, lower-skill, non-service-type jobs, which are those that have historically located in rural areas of the country (Testa, 1992, 12). This argument is somewhat problematic given the conclusions reached by the Institute for International Economics that no overall tendency exists for US exports to Mexico to support high-skilled US jobs, nor for US imports from Mexico to displace low-skilled US jobs; hence the total displacement expected over the next five years under NAFTA is predicted to be less than 2 per cent (Hufbauer and Schott, 1993, 21). This conclusion was based upon the authors' findings that the median weekly wage rates in 1990 of US exports to Mexico and US imports from Mexico were approximately the same: about $420 to $425 per week.

Worries about job migration have even caused some to wonder how much political support will exist for a trilateral agreement in northern US border cities, such as Buffalo, New York, which have gained in several ways from the Canada–US Free Trade Agreement. The logic is that legislators from these border locations, who may have supported a trade accord with Canada, are likely to be opposed to a similar pact with Mexico for fear of losing newly won jobs from Canada to lower-cost locations in Mexico. However, a recent analysis of these Canadian-owned subsidiaries and their parent corporations suggests that such fears are unfounded (MacPherson and McConnell, 1992). The study reports that most of these establishments were located directly across the Ontario-New York State border where a physical presence could be established in the U.S. and where, at the same time, the parents could remain in close geographic proximity to their subsidiaries. Maintaining close geographic ties with these subsidiaries is considered to be an important objective because most of the parent firms are relatively small and new to international business. At this time, therefore, it is unlikely that these corporations would be attracted to locations some 1500 miles away in Mexico.

In addition, many environmentalist groups in the U.S. are concerned that trade liberalisation will undermine their attempts to impose rigorous environmental standards on U.S. industry. With free trade, industries obeying these new standards would be operating under a disadvantage relative to foreign competitors that do not need to meet the same standards, or to North American firms that locate their production facilities in Mexico to avoid environmental restrictions. It should be noted that Mexico has constructed a comprehensive legal framework for protecting the country's environment; however, strong enforcement of the legislation has not been forthcoming. Hence environmental groups have been joining with labour unions to put pressure on the Clinton administration to negotiate supplemental agreements that will require Mexico to enforce its environmental laws rigorously, or face externally imposed sanctions.

Two other contentious arguments in opposition need to be addressed. The first is that the primary beneficiaries of NAFTA will be large multinational corporations. While such companies will certainly have positive gains from the trade accord, such statements are misleading. For example, as co-directors of the University at Buffalo's Canada-U.S. Trade Center which provides export-marketing assistance to small manufacturing establishments, the authors of this chapter have been working with a number of executives who expect to benefit significantly from expanded export sales to Mexico. The second point concerns a recent proposal from a U.S. legislator calling for yet another supplemental agreement to NAFTA that would require the three countries to coordinate exchange-rate policies. The concern is that Mexico may soon devalue its peso substantially to help counter soaring trade and payments deficits. If this were to occur soon after NAFTA went into effect, it could put downward pressure on wages in both the U.S. and Mexico and, the critics argue, provide even more incentive for U.S. companies to relocate south of the border. As others have pointed out, however, a sudden devaluation of the peso is highly unlikely because it would reignite inflation and create turmoil in Mexico's capital markets. Mexico has been successful in devaluing the peso at a modest rate of 4.6

per cent annually over the past seven years. Moreover, it has reduced its inflation from a high of 160 per cent in 1987 to a current rate of around 11.9 per cent. It is unlikely, therefore, that the Mexican government would risk jeopardising these achievements by suddenly reducing the value of its currency.

Opposition in Canada

Early on in the NAFTA negotiations, many in Canada who were initially opposed to broadening their country's trade alliance with the U.S. to include Mexico found themselves in an awkward position. For instance, if the U.S. and Mexico were to be successful in working out a bilateral deal, this could benefit the U.S. at the expense of both Mexico and Canada. In other words, a 'hub-and-spoke' system, with the U.S. hub having separate bilateral accords with Canada and Mexico, would allow the U.S. to get preference in the Mexican market in competition with Canada, and preference in the Canadian market in competition with Mexico (Wonnacott, 1990, 6–7). For some Canadian critics, therefore, the worst possible outcome for Canada was a U.S.–Mexican bilateral pact. The following potential problems could arise if Canada were to be excluded from a U.S.–Mexico agreement:

1. Canada's share of the US market in a number of areas is protected by US protectionist policies, specifically through quotas and tariffs on Mexican imports. Because Mexican exports to the US bear an increasing resemblance to those from Canada, particularly in the automotive sector, Canadian exports to the US could be hurt.
2. Canada does not offer duty-free status to goods purchased in the US if they are manufactured in a third country, but goods assembled in the US with a sufficiently low proportion of intermediate content from a third country could qualify as US goods. Canada could expect more imports of US finished products containing intermediate components made in Mexico.
3. Greater penetration of the US market by Mexican imports could force US producers to shift a portion of their output to Canadian markets, putting even greater competitive pressure on Canadian producers.
4. Lower labour costs and geographic advantages could divert investment from Canada to Mexico by both North American and overseas firms looking for a production base in North America. The US could also be the preferred investment location since it would be the only country with free access to markets in all three countries. (International Trade Reporter, 1990a, 1447)

As a result of these considerations, many critics of the agreement in Canada reluctantly argued that Canada must be present at the bargaining table for a trilateral arrangement so that it could protect (and perhaps even improve) the gains it had won under the FTA with the United States.

In contrast, other Canadian critics, many of whom were opposed to the original Canada–US FTA, remained in opposition to any negotiations that would broaden the alliance to include Mexico. Their fears of a trilateral accord arose from several basic perceptions about such an agreement. For example, New Democratic Party trade critic David Barrett warned that: 'This whole concept is a nightmare of US continentalists come true: Canada's resources, Mexico's labour, and US capital' (International Trade Reporter, 1990b, 1468).

Maude Barlow, chairperson of the Council of Canadians, provides further criticism by noting that Canadian jobs have already been lost because of Canadian manufacturers moving their operations to the *maquiladora* region. Under a trilateral accord, she argued, 'there will be no reason for manufacturing to take place in Canada' (International Trade Reporter, 1990a, 1448). Supporting this position, the Canadian Labour Congress claims that Canada will lose more than 120 000 jobs in the textile industry alone if NAFTA is approved. The position of Canadian labour (as well as the Liberal Party), therefore, is that recent job losses in Canadian industry have been caused mainly by the FTA, and not by the types of competitive problems highlighted by authors such as Rugman (1991) and Porter (1991). The latter argue that Canada's economic difficulties stem from structural and investment deficiencies that are themselves not closely connected with bilateral or multilateral regulations of commerce.

How successful has Canada been in the NAFTA negotiations in protecting and/or improving the gains it won under the FTA with the United States? As the trilateral agreement is presently configured, Canada has been successful in obtaining clearer and more predictable rules of origin, an extension of duty drawback provisions, an improved mechanism for consultation and dispute settlement, a strengthened sideswipe exemption from US safeguards, and a reduced US capacity to retaliate in dispute-settlement cases. These gains are significant because they have been used as the basis for criticism by opponents of the original FTA with the United States. It is also clear that under the proposed trilateral accord Canada has managed to retain exemption for the country's cultural industries and any interbasin diversion of water in its natural state. Canada also maintains its right to screen and deny, if desired, potential FDI projects exceeding $150 million. These are considered by some in Canada to be important issues of sovereignty.

Another concern voiced by many Canadian NAFTA critics is that efforts on the part of the US to increase the percentage of North American content in order for automobiles to receive duty-free treatment would jeopardise existing and future Japanese investment in the Canadian auto sector. Under the Canada–US FTA this percentage was fixed at 50 per cent, and Canada was concerned that if the percentage were to be increased significantly, foreign auto investors, particularly those from Japan, would have less incentive to maintain or expand automobile production within Canada. Under the currently negotiated NAFTA, automotive goods must contain a specified percentage of North American content based on a net-cost formula (certain sales-related expenses are excluded). For vehicles, engines and transmissions, the percentage is 50 per cent for the first four years of the agreement, 56 per cent for the second four years, and 62.5 per cent after eight years. This new method for calculating regional content traces the value of foreign components, replacing the 'roll-up' method under the Canada–US FTA (which counts a part or component as 100 per cent regional if its domestic value exceeds 50 per cent). In the final analysis, the NAFTA talks resulted in a higher percentage of domestic content than Canada had originally desired, but, according to Canadian government officials, the accord improves the calculation of the rules of origin and thereby enables Canada to avoid future disputes, such as those involving the production of Honda motor cars in Ontario (External Affairs and International Trade Canada, 1993, 14). In this particular case, engines manufactured by Honda (US) were shipped duty-free to a Honda assembly plant in Ontario (Canada) – yet these same engines were classified as having less than 50 per cent North American content when they returned to the US embodied within fully manufactured cars. Canada Customs had applied a roll-up procedure in its rules of origin calculations (thus permitting the engines to enter Canada duty-free), whereas the US Customs Service had applied a somewhat different accounting method (roll-down). This dispute came to a head when the US attempted to apply retroactive tariff penalties against Honda Canada, leading to much bitterness on both sides of the border. While there is a good deal more to the Honda dispute than this, the important point is that the proposed trilateral accord will eliminate cross-border discrepancies in the calculation of North American content. The net-cost formula endorsed by NAFTA is designed to eliminate costly and time-consuming disputes of this nature.

Finally, Canadian critics of Mexico's environmental policies and working conditions for labour supported the US Administration's call for the creation of special commissions to urge higher environmental protection standards and a more rigorous protection of workers' rights within the three-country region. However, as further protection of its sovereignty, Canada joins Mexico in refusing to grant such commissions the right to impose sanctions upon those who violate such standards.

FACTORS LIKELY TO INFLUENCE THE PROBABILITY THAT NAFTA WILL EMERGE

Six interrelated factors are likely to influence the probability that a trilateral trade agreement will eventually be fully approved by the respective legislative bodies in Canada, Mexico and the United States. Three of these factors are dependent upon conditions and events that are internal to the three-country area. These 'internal' factors are: 1) the general assessment of President Salinas' efforts to transform Mexico's economic policies and to assure stronger enforcement of environmental standards, workers' rights laws, and voting practices; 2) economic conditions and political considerations in Canada and the US over the short term; and 3) reactions to key provisions of the trilateral agreement that have been negotiated. Another set of factors relates to conditions and events outside the North American arena. These 'external' concerns include: 1) the success of the GATT-sponsored multilateral free trade negotiations; 2) the rising tide of sentiment in Europe and Asia that regional solutions are necessary to resolve global trading issues; and 3) mounting pressures elsewhere in the western hemisphere to pursue formal trade negotiations with the United States. Each of these concerns is discussed briefly below.

Internal factors

Whether or not Canada, Mexico and the US create a formal trade accord will depend upon how successful the Salinas government continues to be in transforming the Mexican economy. The task of convincing potential investors within as well as outside Mexico that economic policy reforms and political stability will continue is another major factor. Since joining the GATT in 1986, Mexico has significantly opened its economy to global competition; it has slashed maximum tariff rates from 100 per cent to 20 per cent (reducing its average tariff rate to approximately 11 per cent); it has maintained efforts to keep inflation down; the government has encouraged entrepreneurial investment, privatisation and deregulation; it has concluded debt-relief agreements with official and private creditors; and it has eased foreign investment rules.

While these are major accomplishments within a relatively short time-span, the new challenge lies in continuing this transformation while at the same time balancing short-term gains with long-term objectives. As pointed out by Sidney Weintraub, the major long-term goals of the Salinas government are to keep inflation down, promote non-oil exports, continue to privatise and deregulate the domestic economy, stimulate foreign and domestic investment, and lower the federal debt (Weintraub, 1990, 41–42). Weintraub also points out that not all of these objectives can be achieved simultaneously. They are dependent upon the continued restructuring of the Mexican economy. In the short run, this policy involves slow increases in real wages, a decline in government subsidies, export-led growth, and increasing trade and investment ties to the US (Weintraub, 1990, 42–48). Over the next few years, many of these policies may become increasingly unpopular, especially if inflation is not curbed and if income distribution problems are not seen to be given sufficient attention by the federal government. Mexico must also determine how best to address the estimation that its currency is overvalued by approximately 10 per cent. The expectation that a sudden change is likely to occur in the country's exchange-rate policy could strengthen resistance to the agreement in Canada and the United States.

Achieving long-term economic development goals is also dependent upon convincing potential domestic as well as foreign investors that the Salinas experiment will succeed. It is estimated, for example, that some $85 billion of Mexican capital is invested outside the country (*Wall Street Journal*, October 1, 1990). In fact, since 1982, Mexico has transferred more than $40 billion abroad (Weintraub, 1990, 42). These financial assets must be retrieved if long-term economic recovery is to be successful. Mexico must also address more effectively the continued concerns of many opponents in Canada and the US that it is violating workers' rights and failing to enforce its own environmental laws. Despite a broad and expensive lobbying effort on the part of the Mexican government, these issues continue to be used effectively by NAFTA critics.

Another 'internal' consideration that is likely to be important in determining whether or not a NAFTA emerges relates to current economic and political conditions in Canada and the United States. For a variety of reasons, the present economic environment in these two countries is less than ideal to support further efforts to liberalise trade policies on the continent. For example, both countries are trying to recover from economic recessions, and weekly news of plant closures and rising unemployment does not bode well in some circles for talking about reducing trade barriers with Mexico. Moreover, the economies of both countries are under severe strain from very high federal budget deficits. Adding to the negative economic picture is the belief by an increasingly large number of people, particularly in Canada, that the Canada–US FTA is not as economically beneficial as it was supposed to be, and that a trilateral alliance would only compound existing problems.

Characteristically, deteriorating economic conditions in Canada and the US have enabled a variety of opposition groups in both countries to mobilise political support for their positions. For example, organised labour is leading a broad coalition of agricultural, consumer and environmental groups opposed to a free trade pact with Mexico, and even some business groups, such as the American Textile Manufacturers Institute, warn that such an agreement threatens the domestic workforce (Maggs, 1991). In short, as the current governments of the three countries continue to promote the acceptance of the negotiated accord by their respective legislative bodies, the atmosphere surrounding such efforts will be significantly influenced by the economic and political conditions in the two northern neighbours.

A third factor that will be likely to determine whether or not the trilateral accord is eventually adopted involves how strong the reactions are to key provisions of the accord. It is an important observation that, at least at the present time, most of the contentious issues surrounding NAFTA appear to be associated with a debate over how many jobs will or will not be lost in each country if the accord is approved, and concern over the welfare and working conditions of Mexican workers and whether or not Mexico is likely to enforce its laws protecting the environment. It is unlikely that the first issue will ever be resolved;

opinions and estimates are many and diverse. On the other hand, all three countries are attempting to address the issues of workers' rights and the environment by negotiating 'side deals' that would create special commissions to monitor conditions in Mexico. The issue that remains to be worked out is how much autonomy is given to these special commissions to regulate conditions within the region. As Canada's International Trade Minister, John Crosbie, has stated: 'This is a trade agreement. Therefore we don't intend to dictate to Mexico what their social, cultural, or labour policies should be' (International Trade Reporter, 1991b, 184). Nevertheless, various political factions in the US Congress indicate that if the special commissions do not have the authority to impose trade sanctions for labour and environmental abuses, they will not support the agreement.

External factors

Prospects for NAFTA are also dependent upon economic and political events external to the area. The interplay between external and internal forces is important because of the growing interdependence of the global economy. Thus how successful the leaders of the three nations are in moving a final agreement through their respective legislative bodies will also depend upon conditions and events outside the North American arena. Three external factors are especially relevant at present.

First, the extent to which it is perceived that the Uruguay Round of the GATT talks can be brought to a successful conclusion is likely to be an important factor in the timing and contents of the trilateral accord. While the GATT has been successful over the past 40 years in cutting tariff levels, its members are still faced with difficult issues relating to non-tariff barriers, trade in services, intellectual property rights, direct investment issues and, perhaps most controversial of all, the issue of agricultural subsidies. For many members of the GATT, therefore, negotiating a bilateral or regional trade agreement seems to be a quick solution to many of the problems that arise in international business. By pursuing a more geographically limited goal, the three North American nations remain in compliance with the 'escape clause' of Article XXIV of the GATT charter. This provision permits GATT members to form regional blocs if the accord covers most of the goods/services traded among members, and if the bloc does not form new trade barriers to nations outside the bloc (see Chapter 1). Nevertheless, the basic principle of a regional trading bloc is preferences, whereas the goal of the multilateral GATT system is non-discrimination. Thus, at least in the short term, regional trade blocs violate the non-discriminatory principle of the GATT. If the GATT succeeds in promoting multilateral free trade, pressure in North America to pursue a regional accord may wane. On the other hand, if the GATT talks continue to falter, resistance to the regional approach could fade, and the three North American countries could feel compelled to move ahead with the trilateral accord.

Another external matter relates to the momentum of market integration already under way in Europe and East Asia. Increasingly, it appears that the European Community perceives its long-term economic interests to be in eastern Europe. With its relatively high wage scale, the Community may need a high level of protectionism to defend its standard of living while it integrates eastern Europe into its regional economy. At the same time, Japan and several other East Asian countries (notably Taiwan, Singapore, Malaysia and Indonesia) are creating a new pattern of horizontal integration. Fostering this closer orientation is the strong export performance of these countries, the relative strength of the yen, and the threat of increased protectionism in North American and European markets (Nanto, 1990).

Not only does the growth of such blocs create uncertainty for businesses that depend upon global markets, it also disrupts corporate planning and frequently presents government leaders with difficult policy problems. Such an unstable world environment can lead to a fragmented trading system where groups of countries respond in kind to perceived, if not actual, protectionism elsewhere. Such posturing within the present global community comes at an awkward time for the US, if only because many believe that the best hope that country has in the short term of pulling out of the current economic downturn is to follow an export-led growth strategy (Bergsten, 1991, 2–6). However, for this strategy to work at the global level, the European Community and East Asian markets must remain economically healthy and open. If, as many believe, these markets are likely to become increasingly difficult to penetrate, then a North American free trade area could become a viable alternative.

A third source of external pressure for a North American regional market comes from other countries in the western hemisphere. Most of the members of the Central American Common Market, the Andean trade pact and the recently formed Southem Cone trade bloc have responded positively to the proposal former President Bush made on 27 June 1990, for the 'Enterprise for the Americas Initiative'. The plan includes reducing part of the official debt owed by Latin American countries to the US, developing a new programme to stimulate increased FDI, and giving stronger emphasis to environmental protection in the hemisphere. In anticipation of this initiative, the US has already signed framework agreements with Bolivia, Costa Rica, Chile, Colombia, Ecuador, Honduras and Mexico. Similar agreements are expected with Argentina, Brazil, Uruguay, Paraguay and Venezuela (International Trade Reporter, 1991a, 65).

Most of the nations in Latin American seem eager to begin discussing a free trade pact with the United States. Despite their enthusiasm for such an accord, however, the governments of many of these nations are watching closely what happens to NAFTA. It is clear that the negotiated trilateral accord is not a carbon copy of the Canada–US agreement. In other words, various components of the alliance reflect the significant differences among the economic, political and even legal systems of the three countries. Of concern to some countries in Latin America are perceived flaws in the NAFTA provisions for non-member accession. For example, several of the industry-specific provisions of the pact (e.g. auto and textile origin rules) were designed without regard to their possible application to other countries, and the agreement does not specify the application procedures or the criteria that new members would have to meet to join NAFTA (Hufbauer and Schott, 1993, 114–115). Nevertheless, the accession clause of the accord does not contain geographical limitations for future members. In effect, therefore, the trilateral regional market is open to all countries, and those in Latin America are likely to be first to test the membership rules once NAFTA becomes a reality.

CONCLUSIONS

As the current century draws to a close, political and commercial irritants between the three North American countries are likely to weaken in the face of integration efforts that are already well under way. With or without a formal trade agreement, recent trends in trilateral trade are unlikely to reverse themselves. For optimists, the NAFTA concept promises a regional reallocation of resources on the basis of comparative advantage. Thus, for example, both Canada and the US are expected to increase their penetration of the Mexican market in areas such as telecommunications equipment, advanced electronics and capital goods. Southbound flows of technology-intensive merchandise are expected to be complemented by northbound flows of lower value-added (labour-intensive) products. Given that Mexico has become a fast-growing market for basic consumer durables, moreover, Canadian and US subsidiaries with FDI interests in Mexico are likely to source many of their more sophisticated inputs from home bases in the north. Looking further into the future, many optimists see the NAFTA experiment as a harbinger of regional industrial rationalisation on a large scale, bringing substantially enhanced incomes to all three nations over the long run. This type of optimism resembles a Schumpeterian perspective with a distinctly regional twist, in that gales of creative destruction are expected to revitalise declining areas as they move toward more advanced and more specialised sectors of activity.

On a darker note, opponents of the NAFTA concept invoke the neoclassical logic of factor mobility to illustrate the wage-depressing effects of integration upon the two more northerly countries. Voices of dissent have also emerged with regard to the social costs of adjustment, especially in regions that contain large numbers of trade-sensitive workers. Many of these workers are locked into sectors that have lost their competitive advantage in world markets. Examples include Mexico's textile and leather industries, Canada's printing and durable goods assembly industries, and virtually all the low-technology manufacturing sectors that pervade the 'rustbelt' regions of the United States. Opponents of the NAFTA accord have been quick to point out that trade adjustment is likely to be most painful in regions that contain the least capacity for worker retraining and attracting inward investment. In many areas of debate, these voices of dissent have become increasingly organised, articulate and convincing. From a multilateral perspective, for example, several opponents of the NAFTA negotiations have argued that a trilateral trade area will ultimately result in trade diversion, an inward-oriented commercial posture, and a redirection of international trade discussion from global to regional arenas. These types of outcome, should they come to pass, would detract from global welfare in fairly obvious ways.

Irrespective of which standpoint one happens to adopt, however, the fact remains that very rapid changes are taking place in the manner in which international business is being conducted in North America. As the international economist Jagdish Bhagwati has noted, profound commitments to trade policy are generally the result of 'a mix of ideological factors (in the form of ideas and examples), interests (as defined by politics and economics), and institutions (as they shape constraints and opportunities)' (Bhagwati, 1988, 17). The course of events in North America over the next few years will determine, at least for this part of the international community, whether or not Bhagwati is correct in his assertion that major new interests and forces, prompted by growing globalisation and interdependence in the world economy, are so compelling that 'the deck is not stacked in favour of protectionism' (Bhagwati, 1988, xii).

References

Baker, S. (1990) Along the border, free trade is becoming a fact of life, *Business Week*, 12 June.

Balassa, B. (1962) *The Theory of Economic Integration*, Allen & Unwin, London.

Belous, R.S. and Hartley, R.S. (eds.) (1990) *The Growth of Regional Trading Blocs in the Global Economy*, National Planning Association, Washington, D.C.

Bergsten, C.F. (1991) Rx for America: export-led growth, *Economic Insights*, Institute for International Economics, Washington, D.C., 2–6.

Bhagwati, J.N. (1988) *Protectionism, The 1987 Ohlin Lectures*, MIT Press, Cambridge, Mass.

Brown, S. (1993) Free trade brings high tax, *Christian Science Monitor*, 12 June.

External Affairs and International Trade Canada (1993) *NAFTA: What's It All About?*, Government of Canada, Ottawa.

Grinspun, R. and Cameron, M.A. (1993) *The Political Economy of North American Free Trade*, St Martin's Press, New York.

Holstein, W.J. and Borrus, A. (1990) Is free trade with Mexico good or bad for the US?, *Business Week*, 14 March.

Hufbauer, G.C. and Schott, J.J. (1993) *NAFTA: An Assessment*, Institute for International Economics, Washington, D.C.

IMF (1992) *International Financial Statistics Yearbook*, IMF, Washington, D.C.

IMF (1993) *International Financial Statistics, Vol, XLVI*, IMF, Washington, D.C.

International Trade Reporter (1990a) *Current Report*, International Trade Reporter, The

National Bureau of National Affairs, Inc., Washington, D.C., 19 September.
International Trade Reporter (1990b) *Current Report*, International Trade Reporter, The National Bureau of National Affairs, Inc., Washington, D.C., 26 September.
International Trade Reporter (1991a) *Current Report*, International Trade Reporter, The National Bureau of National Affairs, Inc., Washington, D.C., 9 January.
International Trade Reporter (1991b) *Current Report*, International Trade Reporter, The National Bureau of National Affairs, Inc., Washington, D.C., 6 February.
International Trade Reporter (1993) *Current Report*, International Trade Reporter, The National Bureau of National Affairs, Inc., Washington, D.C., 19 May.
Investment Canada (1991) *Investing in Canada*, Government of Canada, 5, 2.
MacPherson, A.D. and McConnell, J.E. (1992) *Canadian-owned establishments in Western New York*, Occasional Paper No.4, Canada-US Trade Center, University at Buffalo, Buffalo, N.Y.
Maggs, J. (1991) U.S. labor fights Mexico trade pact, *Journal of Commerce*, 6 February.
Mishra, U.N. (1990) Mexico outlines trade goals, *Mexico City News*, 20 September.
Nanto, D.K. (1990) Asian responses to the growth of trading blocs, in: R.S. Belous and R.S. Hartley (eds.) *The Growth of Regional Trading Blocs in the Global Economy*, National Planning Association, Washington, D.C., 85–115.
New York Times (1990) AFL-CIO opposed to trade agreement, *New York Times*, 12 June.
Porter, M.E. (1991) *Canada at the Crossroads: The Reality of a New Competitive Environment*, Business Council on National Issues, Ottawa.
Rugman, A.M. (1991) *Diamond in the rough: Poter and Canada's international competitiveness*, Research Programme Working Papers No.44, Ontario Centre for International Business, Toronto.
Russett, B.M. (1967) *International Regions and the International System: A Study in Political Ecology*, Rand McNally & Company, Chicago.
Smith, W.R. (1993) *The NAFTA Debate, Part 1: A Primer on Labor, Environmental, and Legal Issues*, The Heritage Foundation, Washington, D.C.
Testa, W.A. (1992) Trends and prospects for rural manufacturing, *Regional Economic Issues*, Federal Reserve Bank Working Paper No.12, Chicago.
United States General Accounting Office (1992) *North American Free Trade Agreement: U.S.-Mexican Trade and Investment Data*, United States General Accounting Office, Washington, D.C.
Weintraub, S. (1990) *Transforming the Mexican Economy: The Salinas Sexenio*, National Planning Association, Washington, D.C.
Wonnacott, R.J. (1990) *U.S. Hub-and-spoke bilaterals and the multilateral trading system: commentary*, C.D. Howe Institute, Paper No.23, Ottawa.

Figure 7.1 Latin America

7 Regional integration in Latin America: the revival of a concept?

Robert Gwynne
University of Birmingham

The discussion of Latin American integration immediately introduces the problem of definition. Latin America is essentially a cultural term that dates from the latter half of the nineteenth century (Blakemore and Smith, 1983). The term was intended to stress the region's historical relationship with the Latin nations of Europe, and also 'to differentiate the vast area impregnated with Iberian culture from the other, Anglo-Saxon America to the north' (Blakemore and Smith, 1983, 533). Thus the regional term 'Latin America' definitely includes the 16 mainland nations of Spanish origin to the south of the Rio Grande (Mexico, Guatemala, El Salvador, Honduras, Nicaragua, Costa Rica, Panama, Colombia, Venezuela, Ecuador, Peru, Bolivia, Chile, Paraguay, Uruguay and Argentina), the one mainland country of Portuguese origin (Brazil) and the two independent island republics of Spanish origin in the Caribbean (Cuba and the Dominican Republic). The term should not strictly include the mainland enclaves of the Guianas or Belize, nor the great majority of Caribbean islands colonised by North European nations (Britain, France and the Netherlands); more debatably the term should not include Puerto Rico, a Commonwealth territory of the USA. This paper will concentrate on schemes of regional integration in the 19 clearly defined countries of Latin America (Figure 7.1).

The concept of regional integration and its application throughout the world has both economic and political connotations (see Chapters 1 and 2). Furthermore, there can often be significant conflicts between the economic and political objectives of integration. This would appear to be very much the case in Latin America The political objectives of integration have been to develop a pan-Latin American unity in world fora; one objective here has been to develop a hemispheric counterbalance to the hegemony of the United States. The economic objectives, meanwhile, of regional integration have been to maximise economic growth. However, the record in Latin America of economic growth

Continental Trading Blocs: The Growth of Regionalism in the World Economy
Edited by R. Gibb and W. Michalak

shows that multilateralism has been a much more productive strategy than regionalism; and that if regionalism is to create a more successful national economy then links with the United States rather than with other Latin American countries are to be preferred.

Undoubtedly, regionalism and multilateralism, in both theory and practice, have significantly influenced the political and economic development of Latin America. Furthermore, it could be argued that each concept has been prominent at different stages in the historical evolution of the continent. The broad patterns of regionalism and multilateralism have changed with significant events in the political and economic evolution of Latin America. This paper will briefly examine the history of regionalism and multilateralism in Latin America before examining their recent evolution.

HISTORICAL BACKGROUND

In the colonial period, it could be argued that regionalism was enforced by the governing powers, Spain and Portugal, particularly by the former. By the end of the eighteenth century, Spain was organising its colonial territories through four regional Viceroyalties – New Spain (Mexico and Central America), New Granada (the northern Andes), New Castile (the southern Andes) and that of the River Plate (created in 1776). Regionalism was enforced through an administrative structure of subsidies to special regions, known as the *situados*. *Situados* were sums of money sent from colonial territories to others, either to cover more or less permanent gaps between government earnings (mostly from taxes) and expenses (mostly official salaries) or for temporary purposes (Morris, 1981). Considerable volumes of money were transferred in this way. For example, in 1771 2.3 million pesos were transferred from Lima to the River Plate area, a figure that was equivalent to the total value of imports going into the booming Potosi area of Bolivia, at that time the major silver-producing region of the world. *Situados* were fundamentally political and administrative payments, organised and directed from Spain in order to smooth over regional differences in wealth and taxation.

In Hispanic America, the colonial period had been characterised by political systems of extreme centralisation (based on Madrid) and what could be termed enforced regionalism, as evidenced by the *situados*. The push for independence from Spain brought political energies that favoured not only the breaking up of the centralised system but also a distinct antipathy to regionalism. Thus, the period in which Hispanic America achieved independence from Spain witnessed the disintegration of the continent around regional loci of power (based in cities). Between 1810 and 1828, the Viceroyalty of the River Plate divided into four – Argentina (Buenos Aires), Paraguay (Asuncion), Bolivia (La Paz) and Uruguay (Montevideo) – and that of New Castile into Peru (Lima) and Chile

(Santiago). The Viceroyalty of New Granada lasted a little longer owing to the political energies and leadership of Simon Bolivar, but by 1830 it had nevertheless divided into Ecuador (Quito), Colombia (Bogota) and Venezuela (Caracas). The most dramatic fragmentation of all took place in Central America. In 1823, New Spain was divided into Mexico and the United Provinces of Central America, with the capital being located in the city of Guatemala. However, such a union proved untenable due to the rivalries and strife that broke out between the constituent city states. Sixteen years later, Central America divided itself into some of the smallest countries in the world based around the cities of Guatemala, San Salvador, Tegucigalpa (Honduras), Managua (Nicaragua) and San Jose (Costa Rica).

Portuguese America, meanwhile, remained intact. Singer (1973) has argued that this was closely linked to the weak administration of the Portuguese authorities and the considerable autonomy that regional elites enjoyed. Rather than break away from union, it was in the commercial interests of the regional elites (based in three port-towns) to maintain political integration. The two north-eastern towns of Salvador and Recife constituted the entrepots for the north-east's sugar economy, while Rio was the commercial hinge for the interior mining state of Minas Gerais.

A federal structure of government developed in Brazil – as opposed to the system of independent and highly centralised new states that evolved in Hispanic America. These politically centralised states rapidly came to be characterised by multilateralism – attempting to forge strong links with as many European and North American governments as possible but relatively disinterested in forging close regional ties. Economic factors were paramount. The newly independent countries tried to boost trade (for so long denied them by the Spanish Crown) and greater complementarity in trade was found with European as opposed to neighbouring states. Free trade and multilateralism thus became intertwined. Mineral and agricultural products were exported from the diverse Latin American countries to Europe and European manufactured products formed the bulk of imports. In the latter half of the nineteenth century, multilateralism became more pronounced as in-migration was encouraged from a wide variety of countries (Germany, Switzerland, Croatia, Syria, Russia, Japan) and inward investment from such countries as Britain and the United States expanded.

In general, multilateralism survived the First World War, despite the disruptions to trade that it caused and the beginnings of protectionist policies towards industry in the larger Latin American countries. However, the combination of free trade and multilateralism did not survive the Depression of the 1930s. The collapse in world trade between 1929 and 1933 started a major restructuring in Latin American countries. Each country started to turn its back on free trade and began to increasingly protect its economy, particularly its industrial sector:

With the onset of the Depression, prices of industrial imports rose relative to

> export prices as the terms of trade deteriorated, and rose relative to prices of non-traded and import-competing goods as national currencies were devalued. These changes greatly encouraged domestic production of manufactured commodities ... Import substitution became the basic source of growth, with just a few sectors accounting for most of the increase in value added. (Chu, 1983, 124)

The Depression caused a dramatic shift in economic orientation in Latin America. A set of free trading, outward-oriented countries were transformed into a group of inward-oriented, import-substituting countries. However, there was no real concomitant shift back from multilateralism to regionalism. This had to await the perceived failure of policies of import substitution.

IMPORT SUBSTITUTION AND ECONOMIC INTEGRATION

Inward-oriented industrialisation in Latin America during the 1930s and 1940s came to be seen as a strategy for economic growth. The theoretical formalisation of the strategy for government policy became known as import substitution industrialisation (ISI). Adherents of the policy came to envisage four stages of industrial production (Hirschman, 1968). The first stage saw the production of basic non-durable consumer goods such as textiles, foodstuffs and pharmaceuticals. This was followed by the production of consumer durable products, such as cookers, radios, televisions and the strategic motor vehicle industry; assembly began with a considerable ratio of imported parts. The third stage was seen as critical in the industrialising process as it had to promote 'intermediate' industries producing the inputs for companies set up during the first and second stages. Such crucial feedstock industries as steel, chemicals (fibres, dyes, acids, plastics) and aluminium were to be promoted alongside component manufacturing (small motors, gearboxes) for the durable goods sector. The final stage of the process would promote the development of the capital goods industry which would manufacture machinery and plant installations. It was seen as the task of government to plan and synchronise each subsequent stage in the process.

However, as ISI progressed through the 1940s and 1950s, it became evident that despite its theoretical attraction, high-cost industry normally resulted, particularly in the smaller countries where market size was limited. A United Nations study that analysed eleven non-durable and nine durable consumer good sectors in 1964 concluded that 'the prevailing situation in the region is characterised by high relative prices of manufactured products and this phenomenon cannot but affect the size of Latin America's market for the type of good' (Gwynne, 1985, 25). High costs and high prices behind high protective tariffs were due to a wide variety of factors, but one key structural problem was the lack of economies of scale provided by small markets, particularly in the critical second and third stages of the import substitution process. It was here that regional economic integration was seen as a possible solution, with countries grouping together to form a much larger market for manufacturing products produced within them.

The first major scheme to promote regional economic integration within a group of import-substituting states was the Latin American Free Trade Area (LAFTA), established on 18 February 1960 with the signing of the Treaty of Montevideo by Argentina, Brazil, Chile, Mexico, Paraguay, Peru and Uruguay. Ecuador and Colombia joined up in 1961, Venezuela and Bolivia in 1967 (see Figure 7.1). The Treaty provided for the elimination of tariffs and other trade restrictions 'in respect of substantially all' of the reciprocal trade of the member countries. This was to be achieved over a 12-year period. However, LAFTA's record through the 1960s was very disappointing (Griffin, 1969). The amount of the region's international trade that was intrazonal remained at only 10 per cent, and most of this (70 per cent) was accounted for by only three countries – Argentina, Chile and Brazil. These were the most industrialised countries in the 1960s. Thus the benefits of economic integration were seen to be concentrated in the small number of countries that were more industrialised than the remainder. In general, the move to tariff elimination was constantly being obstructed by Latin American governments. Within an import-substituting framework, governments wished to protect industrial firms in their own countries as much as possible and therefore resisted major reductions in intrazonal tariffs. Such reductions were thought of as benefiting the larger countries, particularly Brazil and Argentina, by expanding their intrazonal exports of manufactures, but not the smaller countries, whose industrial products were less competitively priced than those of the larger countries.

More successful as a scheme of economic integration was the Central American Common Market (CACM) also created in 1960 (see Figure 7.1). The CACM consisted of five small countries at comparable levels of development – Guatemala, El Salvador, Nicaragua, Costa Rica and Honduras. This enlarged market gave manufacturers a market of 10 million, up to ten times larger than previous national markets. Furthermore, rapid advance towards integration took place with a customs union virtually being formed by 1966, when intraregional free trade existed on all but 8 per cent of items and common external tariffs had been agreed on the great majority of items.

When LAFTA had been seen to stagnate as a scheme of economic integration, five of its smaller members geographically attached to the Andes (Chile, Bolivia, Peru, Ecuador and Colombia) formed the Andean Pact in 1969. The main motivator in its formation was Chile whose industrial development by import substitution had been severely constrained by small market size. The prospect of an enlarged market of 70 million attracted all five countries and later Venezuela in 1973. However, the pace towards economic integration was notably slower than in the CACM. In 1976, Chile left the Andean Group to pursue a policy of free trade in a multilateral context.

Both the Andean Pact and CACM suffered from two problems that occur when economic integration takes place within a regional group of inward-oriented countries. One problem concerns the unequal distribution of benefits among the countries concerned – normally concentrated in the more developed and larger countries of the grouping. In terms of industrial benefits, the locational policies of multinational firms are important. Such firms tend to locate in the largest and most developed market of an inward-oriented grouping for two reasons. First, the firm should have access to the best conditions within the grouping in terms of skilled labour, industrial suppliers, local capital markets and services. Second, the firm would have the best 'fall-back' position should the scheme's progress falter or deteriorate; although access to an enlarged regional market may no longer occur, the firm would nevertheless be located in the largest market of the grouping.

The second problem is that of trade diversion in the smaller and poorer countries of the regional grouping. One of the theoretical advantages attributed to economic integration is trade creation, when a low-cost supply replaces a high-cost one. However, in groupings of import-substituting developing countries, trade diversion can often be the result – that is, a low-cost supply replaced by a high-cost one (see Chapter 1). The case of tyre supply in Honduras illustrates the point. In the early 1960s, Honduras purchased tyres from the cheapest international source available, such as from Goodyear and Firestone plants in the United States. However, in 1966 Goodyear completed its tyre plant in Guatemala, a plant that was built under the CACM regional allocation of industry scheme (Schmitter, 1971). This scheme was supposed to allocate large-scale basic industries to individual CACM countries ensuring no internal competition (from rival firms within the CACM) and protective tariffs reducing competition from sources external to the CACM. However, for Honduras this effectively meant that consumers had to buy tyres from a relatively high-cost supplier; costs were high because of modest economies of scale and the monopoly position that the plant held within the CACM. Thus, Honduras had effectively suffered from a case of trade diversion with a high-cost supply replacing a low-cost one. Honduras would have to extract some definite advantages from the process of economic integration in order to counter-balance the cases of trade diversion. However, Honduras was unable to attract a firm to set up as part of the CACM's regional allocation of industry scheme.

The regional allocation of industry scheme had been set up at the outset of the CACM in 1960. In order that all CACM countries should benefit, it was stated that each country should receive an 'allocated' industry and that no country would receive a second 'allocated' industry before all five countries had received one. However, in 1963, largely at the initiative of El Salvador, the CACM approved an arrangement called 'The Special System for Promotion of Productive Activities' (Schmitter, 1971). According to this arrangement, the five countries would agree on the regional desirability of a list of industrial products, without specifying the precise location of future production. The first plant in each product sector which could prove that its effective capacity could cover 50 per cent of regional demand would be granted additional tariff protection against extrazonal imports. However, the industrial growth promoted by this scheme was very much concentrated in the two countries with the larger industrial structures, Guatemala and El Salvador.

The failure of the CACM's harmonisation policies to promote industrial development in Honduras, and thereby to counterbalance that country's increasing disadvantages of membership, led to the withdrawal of Honduras from full membership of the Market at the end of 1970. The precise timing of the withdrawal was precipitated by the outbreak of war between Honduras and El Salvador, the neighbour that Honduras identified as being the major beneficiary of the Market. Subsequently, the CACM stopped being a full customs union with the regional programmes of industrial development being abandoned and a series of bilateral trade agreements being signed between member countries.

The CACM showed that within an import-substituting framework, schemes of economic integration become characterised by the spatial concentration of economic benefits in the more industrially developed countries of the regional grouping. This problem was recognised by the Andean Group's 1969 Charter, the Cartagena Agreement. Special measures were taken to protect the economic interests of the poorer countries. Thus Bolivia and Ecuador were granted special concessions in terms of both intraregional trade and the common external tariff (Morawetz, 1974). Furthermore, priority was to be given to Bolivia and Ecuador in the allocation of industries under the regional sectoral programmes. However, in order to attempt to equalise the benefits from regional industrial programmes among all countries, highly bureaucratic negotiating machinery and complex regulations were set up. Indeed, such complexity surrounded the implementation of regional sectoral programmes that little real advance was made in the creation of regionally integrated as opposed to nationally integrated industries.

The unsuccessful implementation of the Andean motor vehicle programme provided a good example of a convoluted bureaucratic framework stifling a regional industry (Gwynne, 1980). In its original formulation at the end of the 1970s, the Andean Group bureaucracy for industrial programmes developed three categories of automobile production for international vehicle corporations and member countries. These were:

1. *Allocation* – this signified that an international vehicle corporation could develop a production centre (assembly and parts) in one member country and sell vehicles from this centre in all countries of the Andean Group.
2. *Assembly Agreement* – this permitted a member country that had received the allocation of a vehicle category (and had selected the corporation and vehicle) to authorise the corporation to assemble the vehicle in another member country.

3. *Coproduction Agreement* – this permitted member countries to specialise in the production of certain automobile parts on the understanding that there would be interchange between the automobile industries of member countries.

The Andean Group automobile programme subdivided vehicle assembly into cars (4 subcategories), pick-ups (2 subcategories), trucks (4 subcategories) and jeeps; this constituted 11 categories in all. By the end of the 1970s, the awards were as follows:

Bolivia	3 allocations, 4 assembly agreements.
Colombia	4 allocations, 6 assembly agreements, 1 coproduction agreement.
Ecuador	2 allocations, 3 assembly agreements, 1 coproduction agreement.
Peru	5 allocations, 1 coproduction agreement.
Venezuela	4 allocations, 1 assembly agreement, 1 coproduction agreement.

Disregarding coproduction agreements (for parts production), this programme effectively meant that during the 1980s, 18 vehicle production centres (with parts production) should have been constructed in the Andean Group alongside a further 14 assembly plants. Ostensibly, these 32 plants were to provide vehicles for a market that at its peak in the late 1970s averaged only 170 000 cars and 75 000 trucks a year. In terms of efficient economies of scale, such a regional market would have been best served by one car and one truck plant. The rationale for such a large number of plants was to equalise benefits between member countries of the Andean Group. However, the debt crisis of the 1980s significantly reduced the regional market for vehicles and effectively meant that the Andean Group Automobile Programme was never implemented. International vehicle corporations were not prepared to make the required investments, intraregional trade in vehicles generally stagnated (apart from trade between Colombia and Venezuela) and the complex schemes of allocation designed to equalise benefits in regional automobile manufacture remained in manuscript form.

The record of LAFTA, the CACM and the Andean Group showed that economic integration within an import-substituting framework was not only difficult but also likely to fail. Economic integration would be likely to fail owing to one of two scenarios. First, if there are relatively few checks and balances in the scheme of economic integration, there will inevitably be a spatial concentration of economic benefits in the more advanced countries; in this scenario, the demise would occur owing to the exit of countries that had not shared in the economic benefits. In the second scenario, a more elaborate system of checks and balances is established but this permits serious procrastination in the execution of programmes of integration, such as those of regional industrial allocation; in this scenario, participating countries become disillusioned with schemes of regional economic integration and orientate their industrial policies back to the national framework.

By the early 1980s, after two decades of attempts at economic integration within an import-substituting framework, the process of regional integration was deemed a failure and regionalism was largely dropped from the political agenda. In 1980, LAFTA was renamed; in many ways this was a symbol of the lack of success of economic integration over a 20-year period. Reference to a free trade area was removed as the *Asociacion Latinoamericana de Integracion Economica* (ALADI) or Latin American Integration Association (LAIA) was constituted. As a result, ALADI was less ambitious than LAFTA, attempting to foster economic complementarity through the establishment of bilateral and subregional sectoral agreements (for which it acts as an umbrella organisation). ALADI still contemplates the eventual establishment of a single common Latin American market. In 1984, it introduced a mechanism for multilateral concessions under the name of PAR (*Preferencia Arancelaria Regional* – or Regional Tariff Preferences). It has been difficult to negotiate, agree to and apply these preferences, particularly for the region as a whole. However, at a subregional level, there were limited successes, particularly within the MERCOSUR (*Mercado Comun del Cono Sur*) grouping, a topic that will be returned to later.

THE DEBT CRISIS AND MULTILATERALISM

Some commentators argue that the Latin American debt crisis of the 1980s has been an event that has brought in its wake a substantial restructuring of Latin American economies (Corbridge, 1993; Krugman, 1989). Economic restructuring has largely been undertaken in order that governments could 'cope' with the aftermath of the debt crisis and attempt to generate a more sustainable form of economic growth. The debt crisis left most Latin American countries with serious problems in their external accounts. Large debts meant large annual interest repayments but capital inflows dried up in 1982–84. In many countries, balance of payments problems were exacerbated by trade deficits. In order to cope with the crisis, Latin American governments had to boost exports (normally after devaluation), increase domestic interest rates (to promote savings) and ease the entry for foreign investment. In a decade when most of Latin America was in the economic doldrums, individual countries tried to boost trade with developed countries – particularly the USA, the European Community and Japan. In order to increase inward investment, links with firms in developed countries were nurtured. In this way, most of the economies of Latin American countries became distinctly more multilateral, with firms promoting trading and investment links with a wide variety of developed and newly industrialising countries. Intraregional links became distinctly secondary in consideration.

The example of Chile's trading patterns during the 1980s illustrates this point

Table 7.1 Destination of Chilean exports by world regions, 1980–1990

Destination	*1980*	*1990*	*% Change*
Latin America	1137	1075	–5.5
USA	586	1469	+150.7
EC	1900	3203	+68.6
Japan	507	1388	+173.8
Taiwan, S. Korea and Hong Kong	113	579	+412.4
Others	428	866	+102.3
TOTAL	4671	8580	+83.7

Source: Banco Central de Chile, 1981, 1991.

clearly (see Table 7.1). Between 1980 and 1990, the value of Chile's exports nearly doubled in nominal terms from $4671 million in 1980 to $8580 million in 1990. However, exports to other Latin American countries took a dive in the early 1980s and in 1990 they were still 5.5 per cent lower than they had been in 1980. While exports to Latin America languished, Chilean exports to the United States and Japan nearly trebled and exports to the EC rose to over the $3 billion mark. The most dramatic increase in Chilean exports was to the three newly industrialising countries of East Asia (Taiwan, South Korea, Hong Kong); exports in 1990 were more than five times greater than those of 1980. Chile was perhaps the most successful country in Latin America to extricate itself from the constraints of the debt crisis during the 1980s. However, it did so by promoting multilateral links in trade and investment rather than regional links. Thus, whereas Chilean regional exports to Latin America constituted 24.3 per cent of the total in 1980, they were responsible for only 12.5 per cent of the total a decade later.

The Chilean pattern of reducing regional trade in the 1980s was common for most Latin American countries, as well as for most regional trade groupings in Latin America. In 1980, 24.1 per cent of exports from countries of the Central American Common Market were intraregional; by 1990 this proportion had gone down to only 14.8 per cent (de la Torre and Kelly, 1992). Similar data for LAFTA/LAIA shows that intraregional exports fell from 13.7 per cent of total exports in 1980 to only 10.6 per cent in 1990. Intraregional exports of the Andean Group, meanwhile, have stayed at a miserable level (less than 5 per cent of total exports) throughout the 1970s and 1980s (de la Torre and Kelly, 1992).

However, it is within this context of increasing multilateralism that the renewed interest in economic integration is taking place in Latin America. This poses an apparent contradiction. The 1980s have shown that for countries to extricate themselves from the external constraints of the debt crisis, they must forge stronger trading, investment and technology links with developed countries – as the Chilean case demonstrates (Gwynne, 1993). However, the political emphasis in the 1990s on Latin American countries integrating within regional trading groups indicates a different strategy and a different economic agenda. Nonetheless, the important question must still be that of commitment. How committed are governments in individual countries to advancing regional integration in trade and investment as opposed to promoting multilateralism in trade and investment?

At this point, it is perhaps worth dividing up the countries of mainland Latin America into two groups, according to government integration policy:

1. Those countries committed to promoting economic integration with a developed country rather than regional economic integration (USA) – Mexico: Chile.
2. Countries apparently committed to regional integration as part of one of three trade groupings:
   a) Central American Common Market – Guatemala, El Salvador, Honduras, Nicaragua, Costa Rica.
   b) Andean Group – Venezuela, Colombia, Ecuador, Peru, Bolivia.
   c) MERCOSUR – Argentina, Brazil, Paraguay, Uruguay.

In the introduction, it was stressed that regional integration is both an economic and a political phenomenon. In this context, one could state that the first group of countries are emphasising economic criteria in their international plans. Meanwhile, the second group are considering regional political as well as economic factors in their international relations.

LATIN AMERICAN ECONOMIC INTEGRATION WITH THE UNITED STATES

Given the fact that Latin America experienced a continent-wide depression in the 1980s, it was those Latin American countries that actively shifted from regionalism (based in Latin America) to multilateralism that managed to end the 1980s with relatively dynamic economies. The two best examples were Mexico and Chile. The Mexican case is studied in depth in Chapter 6, but two points are worth emphasising.

First, compared to the Chilean case, the issues of multilateralism and regionalism are not so clear-cut. In contrast to much of Latin America, Mexico has always been closely linked in terms of trade, investment and technology transfer to the United States. Its historical regional economic links have very much been with the United States, even though it has often resisted close political ties. Thus, in 1980, at the height of Mexico's oil-inspired economic boom (when closer political and economic integration with the United States was not on the agenda), 64.7 per cent of Mexico's exports went to the United States and 58.8 per cent of its imports were sourced from its northern neighbour (see Chapter 6).

Therefore Mexico has always been characterised by strong economic ties with the United States, particularly in terms of trade, rather than with other Latin American countries.

Second, Mexico's shift to free trading policies in the 1980s and its commitment to multilateralism have had the impact of strengthening its economic ties with the United States even more. By 1991, Mexico's exports to the United States had risen to 74.5 per cent of the total, whilst imports sourced from the US had risen at an even faster rate – to 70.7 per cent of Mexican imports. Thus Mexico's shift to free trading policies has substantially increased its trading links with the US. Furthermore, these increased economic links are now complemented by a desire for more political cooperation (see Chapter 6).

However, closer economic ties with the United States are now seen as a target for many Latin American countries. In 1992, Chile's Aylwin government invested considerable effort in preparing the basis for a Free Trade Agreement (FTA) with the United States. When the Chilean president visited President Bush in mid-1992, he was rewarded with a diplomatic coup. President Bush promised that 'after Mexico, Chile would be next'. In other words, once the North American Free Trade Agreement between the USA, Canada and Mexico had been fully agreed, the US government would next consider an FTA with Chile. The replacement of Bush by Clinton as American president has not apparently changed this commitment. Thus Chilean policy is to wait for this promised FTA with the United States and not actively to pursue any schemes of regional economic integration in Latin America. This does not rule out the signing of FTAs on a bilateral basis with other countries. Indeed, Chile has already signed FTA agreements with Mexico and Venezuela and an economic complementarity agreement with Argentina. It is further considering FTAs with Colombia, Bolivia, Ecuador and Brazil.

The policy of the Bush administration was to hold out the reward of an FTA for the governments of countries that had genuinely shifted to free trade policies; hence the promise to Chile. This could be seen as part of Bush's much-trumpeted 'Enterprise for the Americas Initiative' in 1990. There were three key pillars to the initiative – trade, investment and debt. However, the shift to genuine policies of free trade as well as geopolitical considerations have been the criteria used for negotiating framework agreements within this initiative – with Chile, Venezuela and the MERCOSUR group. These framework agreements identify areas of mutual interest and establish general negotiating principles. Topics include barriers to trade in goods and services, investment and intellectual property.

Apart from Chile's desire to negotiate a bilateral FTA with the USA, Costa Rica, Venezuela and Ecuador have expressed a similar desire and commitment. Could this be the basis for a future hemispheric scheme of economic integration? Two factors presently dampen prospects in this area. First, the Clinton administration in the USA appears less committed to the principles of free trade than the former Republican administration; in the short term, then, brakes may be put on the process. Secondly, if United States' governments in the future are to expand free trading arrangements with Latin American countries, they will do so at best only slowly and as representing 'rewards' for countries that have made genuine changes to free trading policies. According to Whalley:

> It is the smaller countries [of Latin America] who, in the main, are seeking safe-haven trade arrangements with the United States, driven in part, by fears of a collapse in the multilateral trade system, and of future increases in US trade barriers .... This places the US in an extremely strong negotiating position with potential bilateral partners. A price can be extracted sequentially from each partner for a safe-haven agreement. In turn, such safe-haven agreements seem likely to move increasingly towards exclusionary arrangements against third parties. (Whalley, 1992, 139)

The conclusion, then, is that there may be some limited movement towards hemispheric integration, but that it will be a long, drawn-out process and that movement will slow rather than quicken in the future.

REGIONAL INTEGRATION SCHEMES: SIGNS OF REVIVAL?

Although there is evidence of certain Latin American countries wishing to forge greater economic links with the United States, schemes of regional integration are still being pursued. However, there is perhaps a greater deal of realism about these schemes. According to a report by the European Commission (1993, 54), this greater degree of realism found in schemes of economic integration in developing countries is characterised by the need for good governance, the critical role of structural adjustment, the significance of offering more room to private initiative and the progressive integration of developing countries in the world economy.

Central American Common Market

Recently there have been fundamental improvements in the political and economic conditions of the region. The ending of hostilities in Nicaragua and El Salvador and the momentum provided by economic reform have boosted intraregional trade and the furthering of integration. Thus intraregional trade as a percentage of total trade increased from a low of 10.2 per cent in 1986 to 14.8 per cent in 1990 (during the 1970s, it was generally greater than 25 per cent). A Central American Parliament was inaugurated in October 1991 and a broad institutional reform of integration mechanisms was undertaken in December 1991 with the creation of SICA (*Sistema de la Integracion Centroamericana*) to watch over the three main integration pillars: the Central American Common

Market as the instrument for economic integration, the Central American Parliament as the instrument for political cooperation and the ODECA (*Organizacion de Estados Centroamericanos*) as the institutional instrument.

There are some signs that economic integration could be successfully revived in Central America. Intraregional trade is dominated by manufactures (responsible for nearly 75 per cent of the total); this is in direct contrast with extraregional exports of which more than two-thirds are agricultural. Thus the regional market is particularly important for manufactures; about 40 per cent of manufactured exports from El Salvador, Guatemala and Nicaragua are intraregional, and about 30 per cent of those from Costa Rica and Honduras.

However, there is also evidence that one should be sceptical about the revival of the CACM. Costa Rica has shown a commitment to closer trading ties with the United States and Nicaragua and Honduras participate weakly in terms of intraregional exports (both together account for only 10 per cent). In contrast one country, Guatemala, accounts for nearly half of intraregional exports. At present, the regional market maintains significance only for El Salvador and Guatemala, for whom the region accounts for about 20 per cent of their total exports. As in the early 1970s, the distribution of benefits from economic integration in Central America is distinctly asymmetrical.

Andean Group

As has been previously noted, the Andean Group's attempts in the 1970s and early 1980s at the regional programming of industry and the liberalisation of intraregional trade did not prove successful. Furthermore, the Andean Group showed itself to be distinctly antagonistic towards foreign investment, particularly in the 1970s. However, recent initiatives (notably the 1987 Quito Protocol and the 1991 Caracas Act) show a more outward-oriented and less interventionist approach; hostility towards foreign investment has been reversed. Can one forecast a revival of economic integration in the Andean Group?

In early 1992, a common external tariff covering the Andean Group was finally put into operation (BMI, 1992a). The common external tariff has four levels: 5, 10, 15 and 20 per cent, with the 20 per cent rate due to be eliminated in 1994; some special exemptions were made for the motor vehicle industry until 1994. Furthermore, in early 1992 a free trade zone was created covering Colombia, Venezuela and Bolivia. In February 1993, Ecuador joined the zone (BMI, 1993a). Owing to Peru's grave economic crisis, it has been voluntarily excluded from these recent moves towards integration and the creation of a free trade zone.

The substantial increase in the relative significance of manufactures within intraregional trade (from around 20 per cent in 1970 to 50 per cent in 1989, and an estimated 83 per cent in 1992) is probably the most important development in the region's trade. Nevertheless, the regional market accounted for only 13 per cent of the member countries' total manufactured exports in 1989 – down from 24.5 per cent in 1980, a boom year for Colombian manufacturing exports to Venezuela. Indeed, intraregional trade tends to be dominated by Venezuela and Colombia which together account for about 60 per cent of intraregional exports. Colombia's exports to its partners totalled $996 million in 1992, around 14 per cent of its total exports. Of the other countries, rapid growth in intraregional exports has only occurred in Ecuador (now accounting for about 20 per cent of the total).

Intraregional trade is likely to grow strongly over the next few years, although most countries in the region are also looking to widen free trade links to incorporate other Latin American countries and the United States. A free trade pact between Venezuela, Colombia and Central America (except Costa Rica) came into operation in July 1993. Venezuela and Colombia have agreed to lower their tariff barriers for some Central American products over a five-year period. In February 1993, a preliminary agreement was reached between Venezuela, Colombia and Mexico (the so-called G3) on establishing a free trade zone from 1994.

Thus the push to free trade within the Andean Group and between members of the Andean Group and other neighbouring countries has been moving forward rapidly during the early 1990s. There are still numerous problems that beset integration in the Andean Group, most notably poor, communications, of particular importance in a mountainous region with numerous physical barriers. However, a renewed commitment to free trade and a more realistic approach to competition should mean steady growth in intraregional trade during the 1990s.

Mercosur

The revival of regional integration among developing countries has manifested itself not only in a reformulation of existing schemes but also in the establishment of new ones. The *Mercado Comun del Cono Sur* or Southern Cone Common Market (MERCOSUR) was established in March 1991 by the Treaty of Asuncion. It is an extension, to a large extent, of a process of integration between Brazil and Argentina initiated in 1986 as one of ALADI's partial complementarity agreements. The Treaty started operating in November 1991 and its members are Argentina, Brazil, Paraguay and Uruguay. Chile has expressed an interest in joining, but only when a free trade zone is established. At the end of 1992, it was decided that January 1995 should be the deadline for a full common market, when all internal trade barriers should be removed and a single external tariff ranging from 0 to 20 per cent should be imposed (BMI, 1993b).

Up to now, MERCOSUR has had a reasonable record of meeting its deadlines. Furthermore, intraregional trade has been experiencing rapid growth. It grew by 40 per cent between 1991 ($5.1 billion) and 1992 ($7.1 billion). However, serious imbalances are already becoming apparent, particularly between the two major markets of Brazil and Argentina. The main problem is a lack of harmonisation of macroeconomic policies between the two countries, an objective that was stressed in the Treaty of Asuncion. For example, in 1992 and 1993 the Argentine economy was characterised by a fixed and somewhat overvalued exchange rate but had achieved relative monetary stability. In contrast, Brazil was experiencing daily devaluations and domestic hyperinflation (BMI, 1993c). The result was that Brazil's exports to MERCOSUR boomed (up 79 per cent in 1992) while Argentina's exports rose much more slowly (up 12.6 per cent in 1992). Thus Brazil has a large surplus in trade with Argentina (calculated at $1.5 billion in 1992). This led Argentina temporarily to increase import protection by raising the tariff on imported Brazilian goods from 3 to 10 per cent at the end of 1992 (BMI, 1992b).

CONCLUSIONS

Regional integration is once again on the political agenda in Latin America in the 1990s. However, the objectives and nature of regional integration are very different to those of the 1960s. In the 1960s, regional economic integration was intimately linked to state planning, albeit within the wider structure of a market economy. In terms of regional Latin American trade, governments supported the concept of managed trade rather than free trade. The decade of the 1980s changed Latin American perspectives on state planning and managed trade. The debt crisis and its concomitant economic problems brought a new realism to Latin American governments in terms of economic policy. Since the early 1980s, most governments have endeavoured to shift their national economies to freer trade, less government intervention and an encouragement of inward investment. In essence, this has signified that Latin American governments have been attempting to integrate their national economies more closely into that of the wider, international economy.

However, such closer integration of Latin American economies with the world economy produces risks as well as benefits. Since the 1930s, economic policies in Latin America have been strongly affected by the Depression and the lesson that the world economy can turn against less developed, primary producing countries with particular hostility. Although economic policy making in Latin America has been forced to become more outward-oriented since the mid-1980s, politicians still emphasise the problems and pitfalls of closer integration with the world economy. At least three issues are significant:

1. Low prices for primary product exports, particularly evident during the mid-1980s, but perhaps returning as an important issue in the mid-1990s.
2. The power of multinationals in terms of investment power, marketing and the generation of technology.

3. The perception that developed countries are closing off their markets to the exports of developing countries. The creation of a single European market in 1992 was certainly seen in this light by many Latin American politicians and businesspeople (see Chapter 3). In general, they see trade policy in developed countries shifting away from multilateralism to increasing regionalism.

Latin American governments have lobbied about low primary product prices in such fora as the United Nations Commission on Trade and Development (UNCTAD) for nearly half a century. Concrete improvements have rarely resulted. Latin American governments appear resigned to the permanence of low primary product prices and respond by encouraging growth in the manufacturing sector. Meanwhile, Latin American governments have adopted a more balanced approach to inward investment by multinationals since the widespread nationalisations of the 1970s. There is now a recognition of the important benefits that inward investment by multinationals can bring.

However, the perception of regional free trade areas expanding in developed countries and restricting the exports of developing countries has been met in the late 1980s and 1990s by a concerted movement to develop regional free trade areas in Latin America. As has been noted, old regional trade groupings have been revived and new ones forged. In addition, a whole series of bilateral free trade agreements have been signed under the umbrella of ALADI.

For Latin American governments and their peoples, one crucial question is the relationship with the United States as the one major developed country in the region (trade with Canada is generally small). Some countries, such as Mexico, Chile and Costa Rica, have strongly supported closer economic ties with the United States through regional free trade agreements. However, closer economic integration with the United States does not have such a high priority in all Latin American countries. There are perhaps two reasons for this. First, the shift to freer trade in most Latin American countries is still considerably behind that achieved by those countries pressing for closer integration with the United States (Mexico, Chile and Costa Rica). Secondly, there is the idea that increased economic integration with the United States will act to increase US hegemony in the Americas. Such countries as Brazil, Venezuela and Argentina, whilst recognising the need for good economic relationships with the United States, also like to maintain a certain distance from the regional hegemonic power; they prefer to be seen as non-aligned rather than pro-US.

At the same time, closer integration in the whole of the Americas also depends strongly on the attitude of the US government. The very long debates over the North American Free Trade Agreement in Congress show that the US government is unable to move fast on such issues, when questions of employment and environment are raised (see Chapter 6). Meanwhile, there are influential commentators who think that, after the eventual passing of NAFTA, the United States government should reassert its commitment to multilateralism and not develop any further regional trade areas (FTAs) in Latin America. Bhagwati argues that:

> If the United States pushes FTAs only southward, it will certainly invite a defensive, if not retaliatory, bloc in Asia. Divisions will be sharpened and the world economy fragmented into four blocs: an expanded EC, a NAFTA extended to the Americas, a Japan-centered Asian bloc, and a marginalised group of developing countries, many with low incomes and only just turning to export-oriented strategies. (Bhagwati, 1993, 161)

Thus it is argued that the United States government must stop what Bhagwati sees as a worldwide shift to regional trading blocs and reassert its commitment to multilateralism through the GATT. However, for Latin America this would mean an abrupt end to closer integration with its northern neighbour.

What are the most likely trends in Latin American economic integration during the 1990s? The arguments so far presented lead one to conclude that two, somewhat conflicting, processes will occur. First, the United States government, whilst maintaining its broad allegiance to multilateralism, will gradually expand regional free trade agreements in Latin America. The emphasis will be on *gradually*, as Congressional approval will be increasingly slow and painstaking. But the process will be justified as one that encourages the extension of free trade in Latin America; the signing of a bilateral free trade agreement with a Latin American country will be seen as a reward for the successful and wholehearted shift to free trade.

Secondly, there should be gradual progress in the evolution of Latin American regional free trade areas – CACM, the Andean Group, MERCOSUR. The more realistic attitude towards freer trade and inward investment in these integration schemes will mean that the mistakes of the 1960s will not be repeated. However, increasing economic integration will not be a harmonious process. Tensions will emerge between countries as benefits are seen to be asymmetrically distributed within the member countries of integration schemes. However, the political imperative of the need for closer regional integration within a world economy increasingly divided into trading blocs should mean that Latin American governments will be more willing to bury their differences and disputes than they were in the 1960s.

References

Banco Central de Chile (1981) *Precios y Cantidades Fisicas de Principales Productos de Exportacion e Importacion, Diciembre 1980*, Santiago.

Banco Central de Chile (1991) *Indicadores de Comercio Exterior Diciembre 1990*, Santiago.

Bhagwati, J.N. (1993) Beyond NAFTA: Clinton's Trading Choices, *Foreign Policy*, 91, 155–162.

Blakemore, H. and Smith, C.T. (1983) *Latin America: Geographical Perspectives*, Methuen, London.
BMI (Business Monitor International) (1992a) CET on free trade, *Andean Group*, 9, 975.
BMI (1992b) Uncertainty over MERCOSUR, *Southern Cone*, 9, 1083.
BMI (1993a) Andean Pact Trade Soars, *Andean Group*, 10, 1119.
BMI (1993b) MERCOSUR summit sets single tariff, *Southern Cone*, 10, 1108.
BMI (1993c) MERCOSUR trade up, *Southern Cone*, 10, 1144.
Chu, D.S.C. (1983) The great depression and industrialisation in Colombia, in: A. Berry (ed.) *Essays on Industrialisation in Colombia*, Centre for Latin American Studies, Arizona, 99–142.
Corbridge, S. (1993) *Debt and Development*, Basil Blackwell, Oxford.
de la Torre, A. and Kelly, M.R. (1992) *Regional Trade Arrangements*, International Monetary Fund, Occasional Paper No.93, Washington, D.C.
European Commission (1993) The European Community as a world trade partner, *European Economy No.52*, European Commission, Brussels.
Griffin, K. (1969) *Underdevelopment in Spanish America*, Allen & Unwin, London.
Gwynne, R.N. (1980) The Andean Group Automobile Programme, *Bank of London and South America Review*, 14, 160–170.
Gwynne, R.N. (1985) *Industrialisation and Urbanisation in Latin America*, Johns Hopkins University Press, Baltimore.
Gwynne, R.N. (1993) *Chile 1993: Report on Government, Economy, the Business Environment and Industry*, Business Monitor International, London.
Hirschman, A.O. (1968) The Political Economy of Import-substituting Industrialisation in Latin American Countries, *Quarterly Journal of Economics*, 82, 1–32.
Krugman, P. (1989) New trade theory and less developed countries, in: G. Calvo (ed.) *Debt stabilisation and development: Essays in memory of Carlos Diaz-Alejandro*, Basil Blackwell, Oxford.
Morawetz, D. (1974) *The Andean Group: A Case Study in Economic Integration Among Developing Countries*, MIT Press, Cambridge, Mass.
Morris, A. (1981) *Latin America: Economic Development and Regional Differentiation*, Hutchinson, London.
Schmitter, P.C. (1971) Central American integration: spill-over, spill-around or encapsulation?, *Journal of Common Market Studies*, 9, 18–30.
Singer, P. (1973) *Economia politica da urbanizaçao*, Ediçoes CEBRAP, São Paulo.
Whalley, J. (1992) CUSTA and NAFTA: Can WHFTA be far behind? *Journal of Common Market Studies*, 30, 125–141.

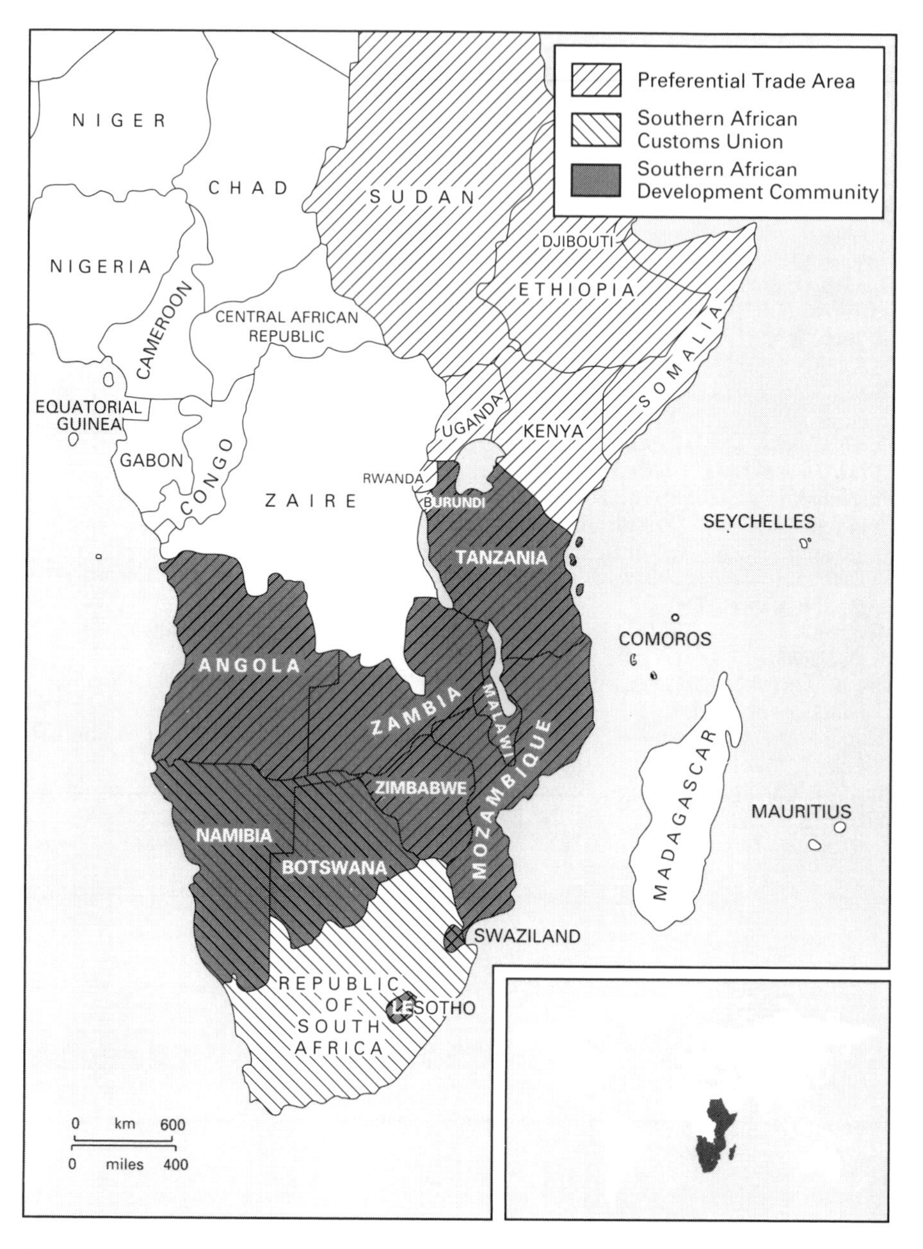

Figure 8.1 Southern Africa

8 Regional economic integration in post-apartheid southern Africa

Richard Gibb
University of Plymouth

The momentous changes and developments throughout southern Africa[1] since 1989 have helped remove many obstacles that previously hindered attempts to promote regional economic integration. Despite Africa's dismal track record in this field, most of the governments within the region support the idea of closer economic integration as a means of promoting economic development and fostering 'south-south' collective reliance. For Less Developed Countries (LDCs), regionalism is regarded as a means of reducing the exploitative dependency relationships arising from limited and unspecialised internal markets which lack economies of scale (Gibb, 1993). By enhancing scale economies and promoting a more rational use of the factors of production, regional economic integration is perceived to be an appropriate mechanism for the expansion of trade, income and bargaining power. This chapter therefore aims to examine the prospects and problems confronting regional economic integration within southern Africa (Figure 8.1). Focus is concentrated on South Africa's continuing regional dominance and appropriate mechanisms and institutional safeguards which may be adopted in order to restrain this dominance in the interests of the region as a whole.

The 1980 Lagos summit meeting of the Organisation of African Unity (OAU) endorsed the objective of establishing a unitary African common market by the year 2000. As a first step towards this ambitious goal, the OAU called for the creation of regional common markets such as southern or west Africa. Up to 1989, there was very little cause for optimism concerning the formation of a southern African common market that included South Africa and most of the majority-ruled states of the region. However, within the space of two years three major events, both within the region and internationally, created a climate favourable to regional economic integration. First, the white-minority controlled government presided over by F.W. de Klerk committed itself to abandoning apartheid and negotiating a new constitutional future, a decision reinforced by

Continental Trading Blocs: The Growth of Regionalism in the World Economy
Edited by R. Gibb and W. Michalak

an extraordinary 68 per cent vote in support of constitutional change in the 1992 South African whites-only referendum. In his speech of 2 February 1990 de Klerk, in addition to reversing the ban on political parties and releasing Nelson Mandela, stated that:

> southern Africa now has an historical opportunity to set aside its conflicts and ideological differences and draw up a joint programme of reconstruction .... Unless the countries of southern Africa achieve stability and a common approach to economic development rapidly, they will be faced with further decline and ruin. (House of Assembly, 1990)

Secondly, the collapse of the last European empire, the USSR, had a profound impact on many spheres of economic, political and social activity throughout the western and third worlds. Anglin (1990) argues that events in eastern Europe had a 'pervasive and profound' effect on the policies, opinions and perceptions emanating from southern Africa. There were two principal effects. The disintegration of the Soviet economy in 1989 wholly discredited the Soviet model for the command economy. The 'socialist' states of southern Africa could no longer depend on economic or political support from either eastern Europe or the Soviet Union. In addition, the end of the cold war enabled the west to impose 'political conditionality' (Hofmeier, 1990) on development aid, emphasising economic liberalisation and political democratisation. Simply put, the collapse of the Soviet Union removed the one-party Stalinist political and economic model as a realistic option available to third world countries. The third factor promoting some form of economic cooperation within southern Africa was the threat posed by the European Community's single market programme and its objective to create an internal market by 31 December 1992 (see Chapters 1 and 3). The threat of a 'fortress Europe' highlighted the benefits of enhanced bargaining powers in a world economy likely to be dominated by a few major trading blocs.

South Africa's commitment to political reform, the demise of the centrally planned economic model and the threat posed by the creation of world trading blocs, have collectively helped create the most encouraging climate for regional economic integration in southern Africa since 1948. However, as Davies (1990a) and Tjonneland (1989) observe, in the post-apartheid debate limited attention is being paid to the regional dimension and discussion on the future organisation of relations between southern African economies is almost non-existent. This lack of attention is compounded by an unfortunate polarisation in previous academic works. With a few notable exceptions (Lemon, 1991; Maasdorp, 1986; Vale, 1989a; Walters, 1989), most of the literature concerned with the political economy of southern Africa can be divided into the 'interdependence' and 'dependence' schools. The interdependence school has traditionally focused on an empirical investigation of economic transactions based on the premise that such exchange is mutually beneficial to all participants. This explanation of events is associated with the neoclassical framework based on liberal economic theory (see, for example, Leistner 1981, 1985; Malan, 1983a). Conversely, the dependency school, based on a neomarxist radical analysis, interprets such transactions as being to the disadvantage of the majority-ruled states of the region and to the benefit of South Africa. From the dependency viewpoint, destroying the pattern of economic transactions between South Africa and the neighbouring states was an inseparable part of the struggle to end apartheid. This extreme polarisation in the academic literature led to an unfortunate lack of analytical rigour. In evaluating the academic bias of regional studies, Sillem notes that:

> Hanlon, Johnson and Martin are the official rhetoricians of the Zimbabwe government and the Front Line states .... Although both works are essential to an understanding of the subject [destabilisation], each lacks interpretation. The authors have remained partial in their selection of material and have occasionally used suspect evidence. (Sillem, 1988, 9)

Similarly, Blumenfeld, in a detailed review of economic cooperation in southern Africa, observes that in the dependency school:

> any analysis which implied that there was anything beneficial, normal or even defensible in the structure and content of economic relations was to be denounced, not merely as analytically invalid, but morally reprehensible and politically unacceptable. (Blumenfeld, 1991, 5)

Against this background, the point is made that up to 1990/91, very little impartial academic analysis was orientated towards examining the opportunities and costs associated with regional economic integration. The current chapter therefore attempts to redress this imbalance. Any analysis of appropriate strategies for the post-apartheid situation must first begin with a study of the existing pattern of regional economic interaction.

COLONIALISM, CAPITALISM AND APARTHEID DESTABILISATION: CREATING A LEGACY OF 'DEPENDENCE'

The Southern African Development Coordination Conference[2] (SADCC, Figure 8.2) was a regional economic organisation established in 1980 by the leaders of the Frontline States.[3] In the Lusaka Declaration of 1980 the member states of SADCC proclaimed their commitment to 'the reduction of economic dependence, particularly, but not exclusively, on the Republic of South Africa'. SADCC maintained that the region's dependence on South Africa was 'not a

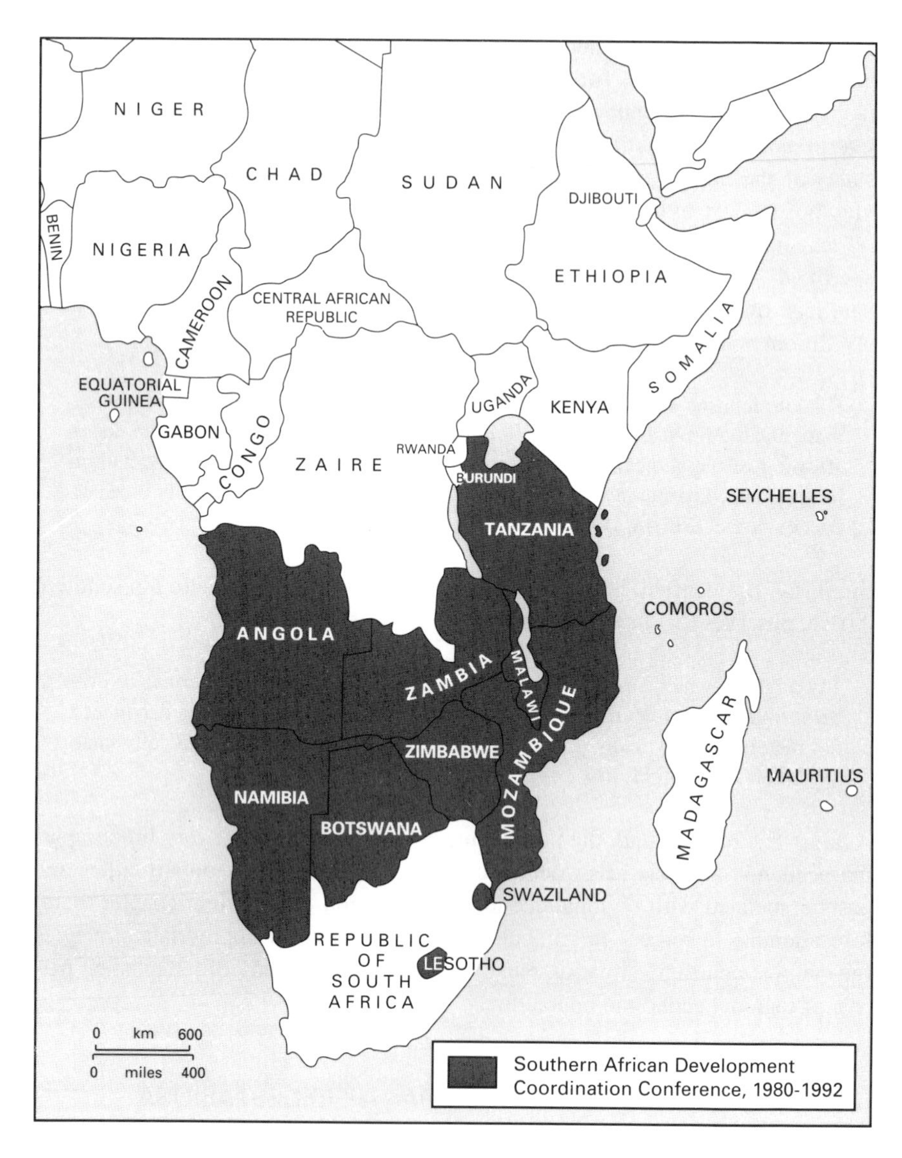

Figure 8.2 Member states of the Southern African Development Coordination Conference, 1980–1992

natural phenomenon' nor 'simply the result of a free market economy' (Hanlon, 1984, 4). Dependence was regarded as an inefficient and unnatural legacy inherited from the colonial past. Historically, the colonial era developed a form of capitalism in southern Africa that produced a high level of structural integration between South Africa and the regional economy (Davies, 1990b). However, the spatial configuration of development established by this particular blend of colonialism and capitalism was characterised by intense inequalities. The core areas of capital accumulation were concentrated in South Africa whilst the other states of the region were peripheralised and forced to function as service economies (du Pisani, 1991). The core areas then grew at the direct expense of the periphery. Labour, capital, services and raw materials were attracted to the poles of capital accumulation at the cost of the rest of the region.

The inequalities and structural disparities created by the policies of colonial administrations and capitalism have been compounded by the policies of successive South African governments. Since the early 1960s, South Africa has pursued a foreign policy which aims to promote further economic integration within southern Africa. The 'outward policy', 'détente' and 'CONSAS' (Constellation of Southern African States) all strove to establish diplomatic links and enhance economic ties with the African continent (Gibb, 1991). By establishing a 'co-prosperity sphere', a 'common market' or a 'constellation of independent states', Pretoria envisaged that the primacy of economic factors would in some way remove the political conflict and tensions throughout the region. Perhaps the most successful attempt to promote regional economic integration within the sub-continent, albeit centred on South Africa, is the Southern African Customs Union (SACU). A formal customs union between various parts of southern Africa can be traced back to the late nineteenth century, though the present SACU arrangement is based on an agreement signed in 1969. At present, the SACU area includes the BLS states (Botswana, Lesotho and Swaziland), Namibia and South Africa (Figure 8.3). Although SACU predates the development of customs union theory, it corresponds closely to traditional integration theory (Gibb, 1993). The SACU is clearly a customs union in the strict sense of the word: there are no classic trade barriers between member states, and a common external tariff is imposed on goods originating from outside the customs union (see Chapter 1).

However, for most of the majority-ruled states of southern Africa, Pretoria's regional strategy amounted to little more than 'apartheid as a foreign policy' (Green, 1981a). Even those states belonging to the SACU with a considerable level of dependency on South Africa, both economic and political, joined SADCC in an attempt to lessen their ties with the apartheid regime. Pretoria's belief that economic factors could outweigh political considerations proved to be unrealistic. The manifest failure of these policies led South Africa to adopt an aggressive military policy coupled with economic coercion. The operational complexities and success of 'apartheid destabilisation' have attracted a great deal of attention and have been adequately dealt with by many researchers (see, for example, Martin and Johnson, 1986; Msabaha and Shaw, 1987; and Chan,

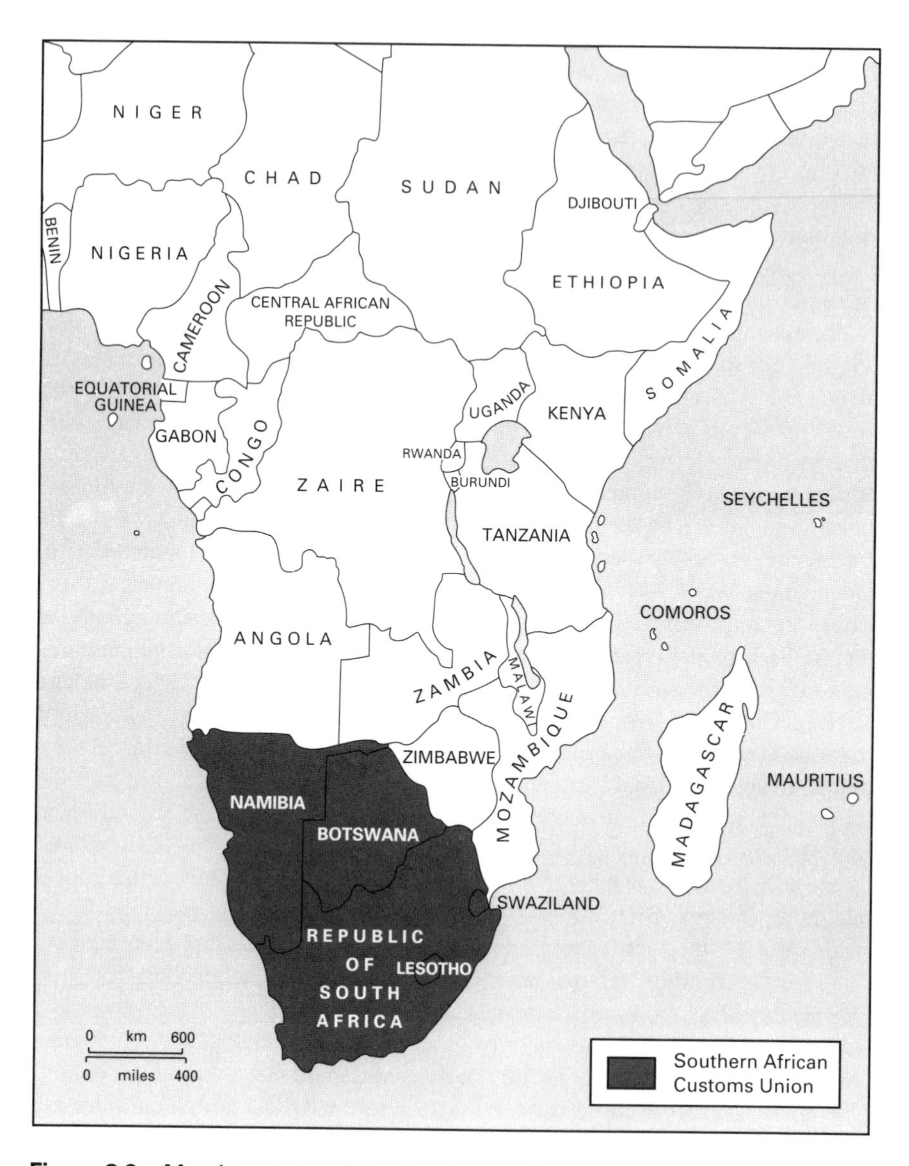

Figure 8.3 Member states of the Southern African Customs Union (SACU)

1990a). Focus here is on the contemporary nature and extent of structural integration that exists between the economies of South Africa and the region as a result of the combined impacts of colonialism, capitalism and apartheid destabilisation.

Any examination of international economic relations within southern Africa has to appreciate just how dominant the Republic of South Africa is. The basic realities of economic power are portrayed in Table 8.1. The acute imbalances and disparities are well illustrated in terms of GDP and GNP per capita. These figures indicate that in 1989, South Africa, which had a population less than half that of the SADCC member states, had a GDP nearly three times that of the SADCC group combined (SADCC, 1992). The unequal distribution of wealth was not, however, the only factor promoting uneven development. The structural impact of South Africa's campaign of aggression and destabilisation has been debilitating. The dismal record of destabilisation is estimated to have resulted in 1.5 million deaths in the original nine SADCC member countries between 1980 and 1988. Furthermore, the economies of the nine suffered considerable losses, both as a direct and indirect result of destabilisation, amounting to an estimated $62.46 billion (United Nations and SADCC estimates – SADCC, 1992).

Given South Africa's absolute and relative economic strength, it is not surprising that it should dominate intraregional trade. However, the nature and evolution of Pretoria's foreign trading relations with Africa are difficult to examine in detail due to the absence of published trade statistics. Although South Africa traded substantially with many African countries throughout the sanctions period, no statistics were released in order to safeguard this trade. It is, however, possible to acquire an accurate picture of the overall level of intraregional trade by using figures from a diverse number of secondary sources. The present analysis of intraregional trading transactions will first examine the importance of southern African trade to South Africa and then consider SADCC/SADC's relative level of 'dependence' on South Africa.

Table 8.1 SADC countries and South Africa – basic indicators 1989

	Population *millions*	*Area* *sq. km. ('000)*	*GNP per capita US$*	*GNP US$ billions*
Angola	9.7	1247	610	7.72
Botswana	1.2	582	1600	2.50
Lesotho	1.7	30	470	0.34
Malawi	8.2	118	180	1.41
Mozambique	15.3	802	80	1.10
Namibia	1.7	824	1030	1.65
Swaziland	0.7	17	936	1.66
Tanzania	23.8	945	130	2.54
Zambia	7.8	753	390	4.70
Zimbabwe	9.5	391	650	5.25
SADC	79.6	5708	363	28.87
S. Africa	35.0	1221	2470	80.37

Source: SADCC, 1992.

TRADING TRANSACTIONS

As a result of South Africa's relatively sophisticated manufacturing sector, it is in a good position to export a wide range of consumer and producer goods to a region that, with the exception of Zimbabwe, lacks a developed industrial economy. The African, and particularly southern African, market is an important one for South Africa. In 1980 each SADCC member, with the exception of Tanzania and Angola, had more trade with South Africa than with all the SADCC partners combined. According to the Economist Intelligence Unit (Hanlon, 1984, 67), in the early 1980s total SADCC trade with South Africa was approximately seven times higher than intra-SADCC trade, with South African exports to SADCC of approximately $2 billion a year and South African imports from SADCC of $300 million. However, as a result of political tensions, ideological differences and a deliberate policy of redirecting trade away from South Africa, coupled to a recession in the world economy, South Africa's exports to the rest of Africa fell between 1980 and 1985 (excluding trade within the SACU) from 1412 million to 710 million (Muirhead, 1988). Notwithstanding this reduction in trade, South Africa continued to record a healthy balance of payments surplus with African countries, which remained an important export market for manufactured goods (representing approximately 20 per cent of South Africa's non-gold exports). According to Maasdorp (1989), SADCC received 82 per cent of South Africa's non-gold African exports in 1985.

Following the reduction in trade between South Africa and Africa in the 1980 to 1985 period, South African exports have been increasing gradually, particularly since 1988. A 1988 SADCC economic survey estimated that the value of South Africa's exports to SADCC member states was more than five times the corresponding level of imports.

According to the first set of figures released since the 1986 embargo on trade statistics. South African exports to the region amounted to $1.6 billion in 1990 and imports totalled $278 million, leaving Pretoria with a $1.23 billion trade surplus. If these figures are added to South Africa's trade with the three countries then belonging to SACU (Namibia joined SACU on independence), South African exports to southern Africa in 1990 amounted to $4.1 billion (with approximately $2.5 billion destined for SACU). Therefore, the African market currently accounts for approximately 25 per cent of Pretoria's non-gold exports. Conversely, South Africa's imports from the rest of the African continent are minimal, accounting for a mere 1.6 per cent of total South African imports. Excluding SACU countries, Pretoria's principal African trading partner is Zimbabwe, which purchased 25 per cent of South Africa's African exports and supplied more than 60 per cent of its African imports (Table 8.2). Even under sanctions, South Africa managed to retain a healthy export market with SADCC member states.

Table 8.2 South Africa's Leading African trading partners, Excluding Sacu

Country	*Imports from South Africa (R million)*	*Export to South Africa (R million)*
Zimbabwe	1062	442.0
Zambia	494	7.0
Zaire	453	22.0
Mozamabique	432	30.0
Malawi	378	81.0
Mauritius	301	14.0
Réunion	128	0.4
Morocco	55	5.0
Madagascar	52	1.0
Angola	50	0.6
Côte d'Ivoire	49	44.0
Seychelles	45	0.3
Egypt	42	6.0

Source: African Markets Monitor, 1991.

SADCC therefore relied upon South Africa to supply a considerable proportion of its total imports. One of the principal objectives of SADCC, as outlined in the 1980 Lusaka Declaration, was to reduce economic dependence on the Republic of South Africa (Green, 1981a). SADCC at first managed to diminish the relative importance of South African trade. Between 1980 and 1986, SADCC's imports from South Africa fell from 30 to 24 per cent of total imports and exports dropped from 7 to 4.5 per cent (SADCC, 1989). However, in the early 1990s South Africa remained an important trading partner of all SADCC states with the exception of Angola and Tanzania. According to The Economist (1990) and the Development Bank of Southern Africa (DBSA, 1990), South African imports dominated the markets of Lesotho, Swaziland (both 90 per cent), Botswana (80 per cent) and Namibia (75 per cent), and were the single most important source for Malawi (40 per cent), Zambia and Zimbabwe (both 20 per cent). The equivalent figures for exports destined for South Africa were Lesotho at 55 per cent, Swaziland, Malawi, Namibia and Zimbabwe at 47, 37, 20 and 19 per cent respectively, and diamond-rich Botswana at 6 per cent.

The overall picture is therefore one of South African dominance, leaving Pretoria with a multi-billion dollar regional trading surplus. Whilst South Africa is dependent upon the markets of southern Africa for the sale of manufactured products, the African market (excluding SACU) represents a mere 9.6 per cent of non-gold exports and 1.6 per cent of imports. Furthermore, throughout the 1980s South Africa pursued a deliberate policy of reducing its dependence on southern Africa in two important relationships in which it had historically been a buyer: transport and migrant labour (SADCC, 1992). South Africa's use of the transport facilities in neighbouring states fell dramatically throughout this period. For example, South African traffic using the port of Maputo fell from 2.98 million tonnes in 1981 to 479 000 tonnes in 1987 (Gibb, 1991). In the field of migrant labourers, Davies (1990b) points out that foreign workers as a proportion of the mine labour force fell from 60 per cent in 1975 to 40 per cent in the mid-1980s.

Given this asymmetrical pattern of economic interdependence between South Africa and SADCC, the key question facing any scheme of regional economic integration is how to respond and manage these acute spatial inequalities in a way that will benefit the whole region. It is to this question that the present chapter now turns.

THE TRADITIONAL APPROACH TO REGIONAL INTEGRATION

Traditional integration theory or customs union theory is based upon concepts derived from classical and neoclassical international trade theory, such as comparative advantage, equalisation of costs and the free movement of the factors of production. In this framework, emphasis is placed upon trade liberalisation, competition, and free market forces. The theory and practice of traditional economic integration is rooted in the experience of western Europe, particularly the European Community (EC) and the European Free Trade Association (EFTA). Integration theory encourages the elimination of discriminatory tariff and quota barriers between participating member state economies. The overall benefits associated with a customs union depend on the balance between 'trade creation' and 'trade diversion' (see Chapter 1). In theory, a customs union will benefit members through the welfare-generating effects of trade creation and damage members if the opposite occurs.

Since 1989, the South African government under the presidency of F.W. de Klerk has abandoned the aggressive and coercive tactics of the destabilisation era and promoted regional economic cooperation according to the principles of traditional integration theory. This policy is rooted in the domestic policies being pursued within South Africa, based upon the unleashing of market forces. A renewed emphasis has been placed upon the needs of export-led growth replacing import substitution as a means of promoting economic development. The South African policy is to promote exports with the aid of a 'General Export Incentive Scheme' (GEIS). As du Pisani (1991) states:

> Export led growth as opposed to import substitution, provides the new economic rationality – and as such it forms an integrated part of South Africa's new diplomacy (du Pisani, 1991, 38)

The South African government therefore aims to mould regional economic cooperation around a market-orientated framework based upon trade liberalisation and the free movement of capital, services and goods. This form of post-apartheid regional development has been promoted in order to secure the principles of capitalism, private property and enterprise, and to protect the perceived economic, political and regional interests of South Africa with a majority-ruled government. In other words, the government of F.W. de Klerk is eager to secure the ground rules upon which future regional cooperation will be based so as to limit the policy options available to a post-apartheid government.

South Africa's new regional policy centred upon trade liberalisation and unfettered market forces is described by Davies (1990b), du Pisani (1991) and Tostensen (1990) as a revised 'Constellation of Southern African States' (CONSAS). 'CONSAS mark two' is considered to be a sanitised and deracialised form of an old CONSAS project promoted by P.W. Botha as part of the 'total strategy'. At the 1979 Carlton Conference in Johannesburg, P.W. Botha outlined the principles upon which the original CONSAS project was to be based. CONSAS was to be centred upon the free market economy and trade liberalisation, with the government and the private sector working in harmony towards the collective goal of regional cooperation. In an examination of CONSAS, Uys observes that:

> The government would provide the policy .... while the private sector provided the goods, services and employment that, hopefully, would lead to a more equitable regional distribution of economic development in southern Africa. (Uys, 1988, 244)

Notwithstanding this goal of equitable development, in 1981 SADCC described the CONSAS project as aiming to:

> turn the free states of southern Africa into little more than bantustans. Constellation is simply apartheid as a foreign policy. (SADCC, 1981)

By assuming that economic concerns would override political tensions, ideological differences and military conflicts, the original CONSAS project overestimated the integrative powers of trade. However, in the 1990s, once many of the political and ideological tensions have been removed, the CONSAS policy could well be reactivated. This would be based upon expanding the existing economic relations within the region and would secure for South Africa a pivotal economic and political role in post-apartheid southern Africa. The present policy of promoting minimal governmental interference, with a reliance on business operating within a free market to generate growth and welfare, will in all likelihood result in the existing regional disparities persisting well into the post-apartheid future. Regional economic integration pursued along the lines of traditional integration theory may even serve to enhance already severe inequalities by denuding the more peripheral areas of overseas investments. In a classic customs union-type agreement, the entire southern African market could be serviced from South Africa. For example, it was announced in 1992 that a number of South African auto manufacturers plan to expand their distribution agreements with international franchisers to allow them to export finished vehicles to African countries. BMW and Toyota of South Africa have already established agreements whereby service and distribution networks will be established throughout the region with production and assembly plants focused exclusively on South Africa (Business International, 1992).

The creation of a regional market based upon trade liberalisation and free market principles does not adequately address the problem of size disparities and the threat of deindustrialisation in countries like Zambia and Zimbabwe. It also threatens to promote trade diversion amongst SADC member states. According to Hanlon (1987, 1989), South Africa is a high-cost producer of manufactured goods which has been able to secure a profitable regional market through destabilisation, economic coercion and the dominance of key economic sectors by South African companies. In a customs union or common market, member states of the Southern African Development Community (SADC) would be under pressure to purchase high-cost South African goods at the expense of cheaper products from outside the union which would be prevented from gaining market access by a protective external tariff. Trade diversion would then take place to the detriment of SADC countries. However, proponents of traditional integration theory would dispute the fact that the elimination of discriminatory barriers to trade would ensure the strengthening of South Africa's hegemonic position in the region. Such a counterargument would be supported by neoclassical equilibrium theory, which suggests that spatial inequalities are but a temporary feature on the economic landscape which will, over time, be removed as resources flow to areas of better comparative advantage. In other words, free traders automatically assume that enhanced economic cross-border interaction will be to the benefit of all those countries concerned. This is certainly the philosophy underpinning South Africa's regional policy which has been encouraging business to expand into the once restricted African markets. Davies (1991) notes that:

> state officials – prominent among them Foreign Minister Pik Botha – have made a major effort to cast themselves in a new role as champions of African development and there have been reports that 'an ambitious development programme which would lead to a southern African Common Market' has been put to nine of the ten SADCC states by the [South African] Department of Foreign Affairs. (Davies, 1991, 2)

The threat of traditional economic integration perpetuating discrepancies in favour of the most developed country has led to the promotion of alternative strategies that explicitly reject trade liberalisation and the free market approach.

STRATEGIES FOR EQUILIBRIUM

Many SADC countries have experienced the polarising effects of regional integration. The collapses of the Portuguese Community, the Central African Federation and the East African Community were, in part, owing to industrial development being concentrated on the most economically developed member state; Portugal, Southern Rhodesia and Kenya respectively. The unhappy experience of these past attempts to promote regional economic integration persuaded SADCC to reject trade liberalisation as an organising principle for regional cooperation. Furthermore the African National Congress (ANC), supported by a majority of regional scholars (Hanlon, 1987; Vale, 1990; Davies 1991; du Pisani, 1991), also reject the indiscriminate opening-up of the southern African market to South African business. Unfettered market forces, it is argued, will be detrimental to the long-term development of the whole region, including South Africa. In the words of Vale:

> Long-term economic growth in southern Africa lies in developing the region as a whole and not focusing attention on one centre of economic power alone. (Vale, 1990, 9)

The sentiments expressed by Vale accurately reflect the policy stance of the ANC. In a discussion document on economic policy, the ANC (1990) recognises the significant opportunities that exist for a post-apartheid South Africa to expand economic relations with the rest of southern Africa. Increased trade, investment and the provision of services are identified as likely beneficial results. At the same time, however, the ANC notes:

> it is imperative that we take account of existing regional imbalances and the damage caused to the economies of southern Africa by apartheid destabilisation. A future democratic government should actively seek to promote greater regional cooperation along new lines which would not be exploitative and which will correct imbalances in current relationships. (ANC, 1990, 17)

Similar sentiments have been expressed by both SADCC and SADC which, whilst recognising the threat posed by the South African economy, made it clear that a South Africa free of apartheid and the dream of exerting economic and military hegemony over its neighbours will be welcome to join as the eleventh member state. The SADCC theme document of 1992 states that:

> the real question is not whether some form of integration will be attempted in southern Africa, but rather on what terms and principles it will be constructed .... Taking account of the well-known extreme disparities and relations of dependency which exist between the ten and South Africa, will also be indispensable to any regional integration strategy (SADCC, 1992, 19–20)

The pertinent question that therefore needs to be asked is how regional economic cooperation/integration will be organised in order to ensure a more equitable distribution of benefits and losses. No consensus has yet been established other than to reject the traditional economic approach. However, a number of regional scholars have identified possible ways to proceed.

Hanlon (1987), in one of the first studies of post-apartheid South Africa and its neighbours, argues that the nature of regional cooperation will depend less on the political colour of a democratic South African government and more on its perception of whether it sees itself as a regional power. The critical question is whether a democratic government accepts the present inequalities as history or promotes redistributive policies. In order for South Africa to treat its neighbours as equals, Hanlon encouraged SADC/C to negotiate on behalf of all its members so as to prevent the smaller and economically weaker states being played off one against another. However, in an extraordinarily bizarre conclusion, the institutional framework that Hanlon identifies as being the most appropriate for southern Africa is a South African one-party state:

> Finally, any government in South Africa will face heavy internal demands for jobs and increased spending on health, education and development. There will be little room to manoeuvre and limited scope for making concessions to the neighbouring states. Here the nature of the government could make a difference – a one-party state may feel more able to make initial concessions to build a better long-term relationship with the neighbouring states (Hanlon, 1987, 437)

Although writing 12 months before the collapse of east-central Europe, the evident failure of the awesomely corrupt one-party state system to generate meaningful economic growth should have prevented Hanlon from reaching such a conclusion (Kolakowski, 1978). Far from promoting an equitable distribution of wealth and development, the one-party state structure would in all probability have exactly the opposite impact.

In a more detailed study of South-southern African relations after apartheid, Davies (1990b) also rejects trade liberalisation and the indiscriminate opening-up of markets as a framework for regional cooperation. Instead of advancing some form of regional institutional framework, Davies argues for a 'qualitative transformation of existing relations' based upon the principle of reciprocity. Those areas in which South Africa now dominates the regional economy should be restructured so as to give more benefits to the ten SADC states. Davies identifies policies in trade, water, electricity, minerals, regional transport, agriculture and, particularly, migrant workers, that would increase South Africa's structural dependence on the rest of southern Africa and provide economic benefits to the region as a whole. This sectoral approach to promoting more equitable regional development is based upon a strategy of planned market intervention. For example, in the field of migrant labour Davies observes that the level of recruitment is determined by the demands of mining capital and not the interests of supplier states. For the supplier states, migrant workers represent a substantial proportion of the wage earning population (in 1985, Lesotho 86 per cent and Mozambique 20 per cent) and provide an important source of revenue through remittance payments. In order to prevent the detrimental economic impacts associated with fluctuating demand levels, Davies proposes two main policy initiatives: first, migrant labourers to be accorded the same rights as South African citizens, including the right to settle in the industrial areas of South Africa; and secondly, an end to the fluctuating demand levels for migrant labour, with binding agreements on the numbers to be employed.

In the sensitive area of transport, Davies proposes a policy of SADC states using only SADC ports and for a greater use of these ports by South Africa. As far as trade is concerned, intra-SADC trade should be prioritised over trade with South Africa which should be sensitive to the development of local industries. Overall, Davies proposes a substantial restructuring of the existing economic relations based upon planned interventionism in order to advance a 'qualitative transformation'. Whilst this approach could stimulate a redistributive trend in favour of SADC, it appears to be a rather random collection of ideas that fail to address adequately the central issue of restructuring the existing regional institutions, which will presumably provide the framework for future cooperation. Furthermore, the policies identified run the risk of being too inflexible. For instance, Zimbabwean and Swazi sugar exporters have both threatened to switch from Maputo to the South African port of Richards Bay as a result of pilfering (African Markets Monitor, 1991). In 1990, 17 000 tons of Zimbabwean and Swazi sugar, worth approximately $3.5 million, were stolen from Maputo port. As a result, Swaziland started to use the port at Durban as a temporary measure. Although Mozambican ports should in theory be the cheaper option, efficiency and security problems make them uneconomic. In such a situation, the policy of SADC trade using only SADC ports would be detrimental to the exporting state. Equally, a policy that prescribes a certain number of migrant workers, irrespective of supply and demand, may result in a further 'internalisation' of mine workers and a substantial reduction of migrant labourers to the detriment of the supplier states.

The original SADCC approach to regional economic cooperation provides another possible strategy for equitable development. A unique aspect of the SADCC policy was its deliberate avoidance of free market forces. It explicitly rejected the notion of free trade areas and common markets as an appropriate mechanism for regional integration. The SADCC project does not therefore fit anywhere into traditional economic integration theory. According to Lee (1989, 352) SADCC was:

> the first serious attempt by a group of third world nations to reject totally traditional customs union or integration theory, or modifications thereof, as solutions to their regional problems.

This rejection was outlined clearly by the SADCC's Executive Secretary in February 1988:

> Our approach to trade integration theory is not based on orthodox trade liberalisation strategies .... the reduction or even elimination of tariffs and other barriers to trade does not always yield increased trade, in the absence of tradable goods. (Hanlon, 1989, 65)

In the Lusaka Declaration, which officially launched SADCC on 1 April 1980, member states committed themselves to creating a 'fabric of regional cooperation and development' in order to create 'genuine and equitable regional integration' (Hanlon, 1984). This equitable development was to be achieved by reducing the level of dependence on the Republic of South Africa and, most importantly, project coordination and planning with the aim of enhancing intraregional trade. SADCC adopted a policy of 'balanced trade' as opposed to free market trade, within which industrial production was to be managed, planned and coordinated as opposed to being exposed to free market forces.

Although SADCC built up a reputation for success (Hanlon 1984, 1989; Weimer, 1991), it failed to achieve two of its principal objectives. First and foremost, SADCC was unsuccessful in promoting intraregional trade and that trade which did take place was to the benefit of one or two member states. Throughout the 1980s, intra-SADCC trade continued to account for a negligible proportion of the total trade of SADCC member states. As outlined in Table 8.3, this trade actually declined in both relative and absolute terms during the 1980s. Furthermore, Zimbabwe dominated intra-SADCC trade, being involved in 80 per cent of the value of such trade (Maasdorp, 1989). Far from promoting equitable development, SADCC actually contributed to industrial polarisation focused upon Zimbabwe's relatively developed and diversified economy. Secondly, SADCC manifestly failed to achieve what it set itself as a primary goal, and perhaps even its *raison d'être*, the reduction of dependence on South Africa. Despite SADCC being successful in reducing the dependence of most

Table 8.3 Intra-SADCC trade

	SDR million*	*% of Total Trade*
1981	548	4.7
1982	536	4.7
1983	495	4.5
1984	512	4.5
1985	417	3.8
1986	384	4.2

*SDR = Special Drawing Rights
Source: SADCC, 1992.

member states on South Africa's transport and communication sectors, South Africa–SADCC trade has increased, in both relative and absolute terms, since 1985. There is little doubt that the ten SADCC states, both individually and collectively, were more dependent on South Africa in 1992 than they were in 1980. Given SADCC's failure to promote intraregional trade and its inability to reduce member states' dependence on South Africa, it would appear to be a most inappropriate framework to be used to incorporate a democratic South Africa. The failure of SADCC is a result of many factors and not only, or even principally, South African-sponsored destabilisation. SADCC itself identified the primary problem to be:

> inappropriate policies and poor management affairs, exacerbated by inadequate public accountability, the erosion of basic individual and corporate freedoms and the rule of law. (SADCC, 1992, 2)

There are two important lessons to be learnt from the SADCC experiment that are of relevance to the present discussion. First, project coordination, along the lines suggested by Davies (1990b), has not been successful in promoting trade, reducing dependence or creating an equitable distribution of the benefits and losses associated with cooperation. Second, a more equitable form of regional integration will depend not only on South Africa's regional philosophy but also on southern Africa's ability to put its own house in order through a process which centres upon political democratisation and economic liberalisation.

The 1992 SADCC theme document recognised that 'project coordination' had a limited impact in promoting deeper or wider cooperation and integration. The limited progress made by SADCC in promoting regional integration resulted in the Development Community (SADC) replacing the Coordination Conference (SADCC) on 18 August 1992. The desire to promote deeper economic cooperation and integration arose from a renewed realisation of the benefits to be derived from restructuring member states' economies on a regional scale so as to create a more productive and cooperative basis. Whilst it is too early to examine in detail the integrative strategy of SADC, it is possible to identify a few basic principles and ideas. The new organisation has moved away from explicitly rejecting any notion associated with traditional integration theory and moved towards accepting, in part, the trade liberalisation and free market approach to integration. The SADC Treaty (SADC, 1992) established a framework of cooperation that prioritised deeper economic integration, common economic and political systems and a common foreign policy. In particular, it called for the free movement of capital, goods, labour and people as well as freedom of movement, residence and employment throughout the region.

IN CONCLUSION

A fully democratic South Africa will have far-reaching implications for the region as a whole and for regional economic integration in particular. The policy of apartheid destabilisation has been abandoned and the scene looks set for cooperation and coordination to replace conflict and confrontation. Furthermore, at the beginning of the 1990s, a unique set of both regional and international forces have combined to produce an environment suitable for regional integration. Great opportunities therefore exist for the freeing of resources from security and defence towards economic development, and for South Africa to use its dominant economy for the purposes of regional growth. However, as this chapter illustrates, an apartheid-free South Africa will at the same time present a serious economic challenge to the region. If the *status quo* prevails, South Africa, because of its territory, population, mineral resources, infrastructural development and industrial power, will continue to serve as the focus for regional economic growth and retain, and perhaps even intensify, its hegemonic position. Consequently, there is a very real danger that because of this marked asymmetrical interdependence, the present favourable environment for cooperation will not last. Many of the regional economic models previously adopted, whether based upon traditional liberal integration theory or strategies designed to promote equilibrium, do not appear appropriate to the task of promoting development whist at the same time addressing the legacy of dependence. Since none of the models is suitable for post-apartheid southern Africa, the region 'will have to develop its own model of cooperation' (Datta, 1989) sympathetic to the unique characteristics of the subcontinent. This may have already started with the creation of SADC. However, it would be premature to evaluate whether SADC represents a significant departure from the integrative strategy adopted by SADCC.

Alternatively, Maasdorp (1992) has promoted the idea of converting SACU into a Southern African Common Market (SACM). This would have the benefit of building regional cooperation on the basis of trade liberalisation, an approach supported by the World Bank, whilst at the same time incorporating a substantial element of redistribution. Recognising the disadvantageous impact a customs union may have when undeveloped countries join together with a more advanced partner, a built-in compensation factor was incorporated into the 1969 SACU revenue-sharing formula (Walters, 1989). However, South Africa could in no way afford to extend the current revenue-sharing formula to incorporate the whole subcontinent. Indeed a democratic South Africa may feel less inclined to support such a generous redistributive mechanism to existing SACU member states (The Economist, 1990). Furthermore, the ability of South Africa to finance development throughout the region, in terms of developing infrastructure, manufacturing and technological expertise, will be constrained substantially by its need to address the economic and political inequalities inherited from the apartheid era. Although SADC expects a post-apartheid South Africa to 'fill the role of present co-operating partners [i.e. aid agencies] through the provision of aid and investment capital to member states' (Kongwa, 1991), it is increasingly recognised that South Africa will have neither the economic nor political strength to assume such a role. The success of regional economic integration will therefore depend on the support and commitment to southern Africa exhibited by the international community.

Notes

1. For the purposes of the present study, southern Africa includes the 10 countries of the Southern African Development Community (Botswana, Lesotho, Swaziland, Mozambique, Tanzania, Zambia, Zimbabwe, Angola, Malawi and Namibia) and South Africa.
2. On 18 August 1992, the Southern African Development Community (SADC) replaced the Southern African Development Coordination Conference (SADCC). The new organisation has the same ten member states and is committed to regional economic integration as opposed to the cooperation approach advocated by SADCC.
3. The 'Frontline States' grouping includes Angola, Botswana, Mozambique, Tanzania, Zambia and Zimbabwe.

Acknowledgement

The author would like to acknowledge the support of an ESRC grant (No. R000233303) for this research into southern Africa.

References

Africa Markets Monitor (1991) Exporters may switch to South African ports, *Africa Markets Monitor*, 7, The Economist Group, London.

ANC (African National Congress) (1990) *Discussion Document on Economic Policy*, African National Congress, Johannesburg.

Anglin, D. (1990) Southern African responses to Eastern European developments, *Journal of Modern African Studies*, 28, 431–455.

Blumenfeld, J. (1991) *Economic Interdependence in Southern Africa: From Conflict to Cooperation*, Pinter, London.

Business International (1992) Economic cooperation in Southern Africa: critical issues for companies doing business in South Africa, *Business International*, 2, The Economist Group, London.

Chan, S. (1990a) *Exporting Apartheid: Foreign Policies in Southern Africa 1978–1988*, Macmillan, London.

Datta, A. (1989) Strategies for regional cooperation in post-apartheid Southern Africa – the role of Non-Governmental Organizations, in: B. Oden and H. Othmen (eds.) *Regional Cooperation in Southern Africa: A Post-Apartheid Perspective*, The Scandinavian Institute of African Studies, Seminar Proceedings, No. 22.

Davies, R. (1990a) Post-aparteid scenarios for the Southern African region, *Transformation*, 11, 12–39.

Davies, R. (1990b) *Key Issues in Reconstructing South-Southern African Economic Relations After Apartheid*, Centre for Southern Africa Studies, University of the Western Cape.

Davies, R. (1991) *Southern Africa Into the 1990s and Beyond*, paper presented to a conference on current and future prospects for the political economy of Southern Africa, 15–19 May, Broederstroom.

DBSA (Development Bank of Southern Africa) (1990) *Financial Resources and Capital Investment Prospects in the Common Monetary Area*, paper presented to the conference on the 'New post-apartheid South Africa and its neighbours', Maseru, 9–12 July.

du Pisani, A. (1991) *Ventures into the Interior: Continuity and Change in South Africa's Regional Policy 1948–1991*, paper presented to the conference 'Southern Africa into the 1990s and beyond', Magaliesburg, 15–19 April.

The Economist (1990) South Africa survey: after apartheid, *The Economist*, 3 November, 1–24.

Gibb, R.A. (1991) Imposing dependence: South Africa's manipulation of regional railways, *Transport Review*, 11, 19–39.

Gibb, R.A. (1993) A common market for post-apartheid Southern Africa: prospects and problems, *South African Geographical Journal*, 75, 28–35.

Green, R.H. (1981a) First steps towards economic liberation, in: A.J. Nsekela (ed.) *Southern Africa: Towards Economic Liberation*, Rex Collings, London.

Hanlon, J. (1984) *SADCC: Progress, Projects and Prospects*, Report No. 182, The Economist Intelligence Unit, London.

Hanlon, J. (1987) Post-apartheid South Africa and its neighbours, *Third World Quarterly*, 9, 437–449.

Hanlon, J. (1989) *SADCC in the 1990s: Development on the Front Line*, Special Report No. 1158, The Economist Intelligence Unit, London.

Hofmeier, R. (1990) Politische Konditionierung von Entwicklungshilfe in Afrika, *Afrika Spectrum*, 25, 167–179.

House of Assembly (1990) Opening address of State President F.W. de Klerk, Republic of South Africa, *Debates of Parliament (Hansard)*, Second Session, Ninth Parliament, 2 February.

Kolakowski, L. (1978) *Main Currents of Marxism: Volume 1. The Founders; Volume 2. The Golden Age; Volume 3. The Breakdown*, Oxford University Press, Oxford.

Kongwa, S. (1991) SADCC: Creating a new vision for the future, *AI Bulletin*, 10, 1.

Lee, C. (1989) *Options for Regional Co-operation and Development in Southern Africa*, unpublished Ph.D. thesis, University of Pittsburgh, Johnstown, Penn.

Leistner, G.M.E. (1981) Towards a regional development strategy for Southern Africa, *Southern African Journal of Economics*, 4, 349–364.

Leistner, G.M.E. (1985) *Southern Africa: the market of the future*?, paper presented at the 24th Annual Meeting of the Bureau of Market Research, 1984, University of Pretoria, Africa Insight, 15, 17–21.

Lemon, A. (1991) Apartheid as a foreign policy: dimensions of international conflict in Southern Africa, in: N. Kliot and W. Stanley (eds.) *The Political Geography of Conflict and Peace*, Belhaven, London.

Maasdorp, G. (1986) The Southern African nexus: dependence or interdependence?, *Indicator South Africa: Economic Monitor*, 4, 5–19.

Maasdorp, G. (1989) *Economic Relations in Southern Africa – Changes Ahead*?, paper presented to the conference on 'South and Southern Africa in the 21st century', Maputo, December.

Maasdorp, G. (1992) Economic prospects for South Africa in Southern Africa, *South Africa International*, January, 121–127.

Malan, T. (1983a) Regional economic cooperation in Southern Africa, *Africa Insight*, 1, 43–51.

Martin, R. and Johnson, P. (eds.) (1986) *Destructive Engagement*, Zimbabwe Publishing

House, Harare.

Msabaha, I. and Shaw, T. (1987) *Confrontation and Liberation in Southern Africa*, Westview Press, Boulder.

Muirhead, D. (1988) Trade and trade promotion, in E. Leistner and P. Esterhuysen (eds.) *South Africa in Southern Africa: Economic Interaction*, Africa Institute, Pretoria.

SADC (South African Development Community) (1992) *Towards a Southern African Development Community*, SADC, Gabarone.

SADCC (South African Development Coordination Conference) (1981) *SADCC2-Maputo*, SADCC Liaison Committee, SADCC, Gabarone.

SADCC (1989) *SADCC Annual Progress Report 1989–90*, SADCC, Gaborone.

SADCC (1992) *SADCC Theme Document*, Maputo conference, 29–31 January, SADCC, Gabarone.

Sillem, T. (1988) *South African Destabilization of the Front Line: The Case of Zimbabwe*, unpublished M.Phil. thesis, Trinity Hall, University of Cambridge, Cambridge.

Tjonneland, E.N. (1989) South Africa's regional policies in the late post-apartheid periods, in: B. Oden and H. Othman (eds.) *Regional Cooperation in Southern Africa: A Post Apartheid Perspective*, The Scandinavian Institute of African Studies, Seminar Proceedings, No. 22, Uppsala.

Tostensen, A. (1990) Les défis des années 90 pour SADCC, in: Conseil Canadien pour la coopération internationale, *SADCC vers la décennie 90: un défi pour le Canada*, Conseil Canadien pour la coopération internationale, Ottawa.

Uys, S. (1988) The short and unhappy life of CONSAS, *South Africa International*, 4, 243–248.

Vale, P. (1989a) Integration and disintegration in Southern Africa, *Reality: A Journal of Liberal and Radical Opinion*, May, 7–12.

Vale, P. (1990) Starting Over: Some Early Questions on a Post-Apartheid Foreign Policy, *Southern African Perspectives*, a Working Paper Series, No. 1, Centre for Southern African Studies, University of the Western Cape.

Walters, J. (1989) Renegotiating dependency: the case of the Southern African Customs Union, *Journal of Common Market Studies*, 38, 29–52.

Weimer, B. (1991) The Southern African Development Coordination Conference (SADCC): Past and future, *Africa Insight*, 2, 78–88.

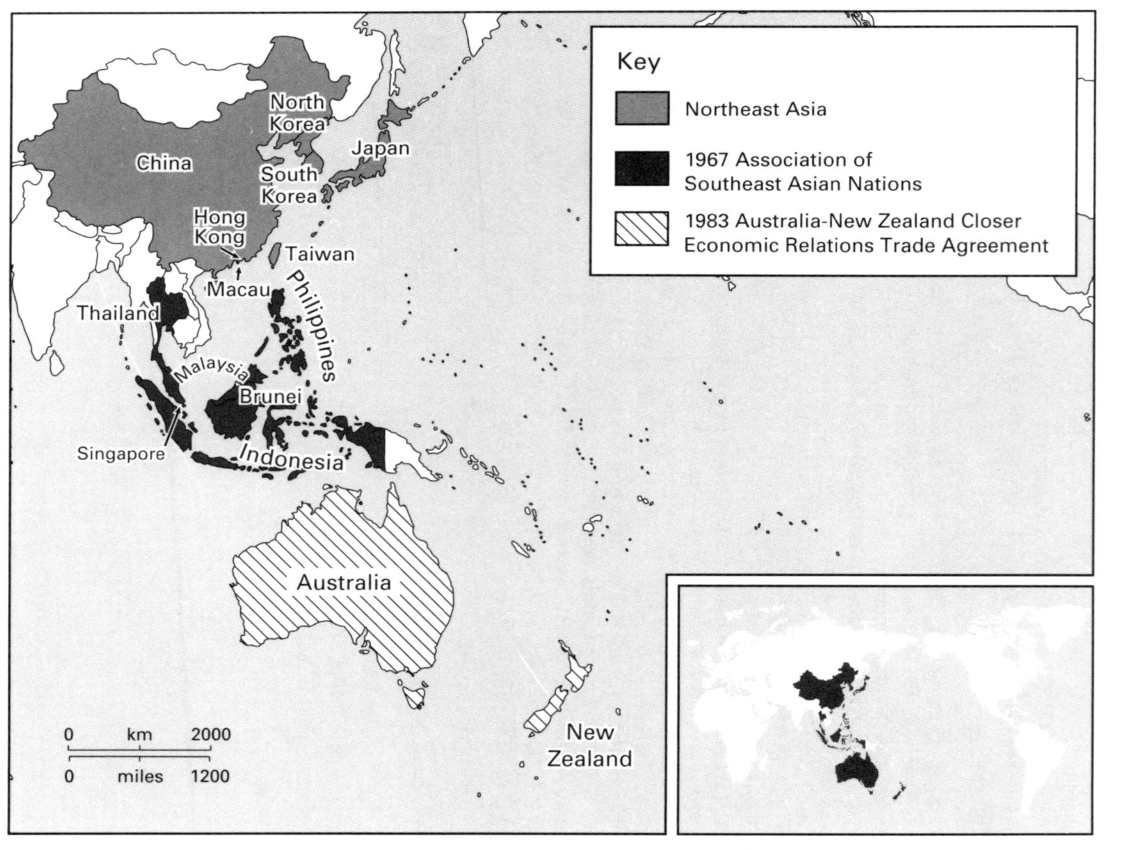
Key
Northeast Asia
1967 Association of Southeast Asian Nations
1983 Australia-New Zealand Closer Economic Relations Trade Agreement
China
North Korea
South Korea
Japan
Hong Kong
Macau
Taiwan
Philippines
Thailand
Malaysia
Brunei
Singapore
Indonesia
Australia
New Zealand
0 km 2000
0 miles 1200

Figure 9.1 The West Pacific Rim

9 The West Pacific Rim

Rupert Hodder
The Chinese University of Hong Kong

REGIONALISM IN THE WEST PACIFIC RIM

Any discussion of regionalism in relation to international trade in the countries of the West Pacific Rim (East Asia and Australasia) tends to be confused by the practice of regarding Japan – rather like the United States – as one of three great existing trading blocs of the world economy. In the western rim of the Pacific, however, it is only in the centre, in the six countries of the Association of South East Asian Nations (ASEAN), that anything approaching an integrated regional economic organisation can be said to exist (Figure 9.1). Otherwise – and this includes Australia and New Zealand, with their very limited Closer Economic Relations Trade Agreement (CERTA) – the priority of multilateralism over regionalism is marked and seemingly persistent.

This is not to suggest that attempts have not been made to set up various forms of integrating structures in the region. In part, these attempts have resulted from difficulties experienced in the Uruguay round of the General Agreement on Tariffs and Trade (GATT) from a perceived need to counter the rise of the European Community (EC) and North American trade blocs, and from an increasingly self-confident recognition that there is in the western rim of the Pacific an almost limitless potential for intraregional trade and economic cooperation. Certainly since the Second World War there have been several, though largely abortive, attempts to establish effective groupings of one kind or another. These include the Pacific Free Trade and Development (PAFTA) conference series, the Pacific Basin Economic Council (PBEC), the Pacific Economic Cooperation Conferences (PECC), and the Asia Pacific Economic Cooperation (APEC) process, although none of these can in any sense be regarded as a trading bloc. Moreover, many of these groupings are pan-Pacific in their geographical coverage, including countries along the eastern rim of the Pacific.

There are several reasons why, with the limited exception of ASEAN, no substantial or wideranging regional trading group has yet evolved in this dynamic and increasingly significant part of the global economy. First, there are many more powerful centrifugal forces at work here than in Europe or North America.

Continental Trading Blocs: The Growth of Regionalism in the World Economy
Edited by R. Gibb and W. Michalak

This is a region of exceptional heterogeneity in its environments, peoples, economies, levels of development, cultures and historical, including colonial, relations. It comprises a series of islands, island groups, peninsulas and littorals extending for over 7000 miles from north to south, and is controlled by 17 national governments of widely different political complexions. In no sense, then, can the West Pacific Rim be referred to as a 'continental' bloc, even though China and Australia are clearly continental in scale and location. In this largely insular and peninsular world, trade and outward-looking attitudes have long been far more characteristic of this region than elsewhere; and a complicated network of trading relations with other parts of the region, and with other parts of the world, has long existed.

A further reason for the low priority given to regionalism here relates to the dominant position in the region of Japan, which, together with the US and the EC, forms the focus of so much of the current discussion on international trade and the debate on multilateralism versus regionalism. In terms of any potential Asian Pacific trading bloc Japan's presence is vital. As one writer noted with some prescience in 1970, Japan's value:

> both as a supplier and as a customer, to all countries trading in the Pacific is likely to be enhanced during the next [1970–80] decade. Japan herself might be expected, *A PRIORI*, to welcome any movement towards the elimination of trading restrictions, for, as an aggressive competitor for world markets, her interests seem to lie in the greatest possible freedom for international transactions. (Corbett, 1970, 43)

But it is now clear that the Japanese do not regard regionalism as the only or the best way to achieve global mutilateralism; nor, as yet, does Japan or her neighbours believe that a Japan-centred regional bloc is necessary or even desirable in a bloc-infested world. The Japanese today seem inclined to reject formal schemes to tie their economy more closely to East Asia's. While Japan has been active in promoting cooperation globally and within Asia – Japan joined the GATT 1955, the OECD in 1957, and is an active member of UNECAFE (United Nations Commission for Australasia and the Far East) – the Japanese have followed and seem set to continue a policy of building bilateral and ad hoc trading and foreign direct investment links, not only within the region but throughout the world. As one report puts it, Japan's growing influence in Asia is based not upon explicit treaties of the type that westerners strive for (such as NAFTA or the EC), but on 'murkier understandings, like personal trust and favours' (The Economist, 1993b). In Japan, as in most of the region, the preference for multilateralism reflects the fact that the post-war history of rapid economic growth has been built up very largely through export-led industrialisation and trade liberalisation.

For all these reasons, regionalism is not as yet a powerful force along the western rim of the Pacific. Loose arrangements for discussion, consultation and mutual support are in place in the form of APEC, and ASEAN is already regarded as having the status of a trading bloc of some global significance. But regionalism seems unlikely to gain further ground unless the existence of powerful trading blocs elsewhere – notably in Europe and the Americas – and continued problems with the Uruguay round of the GATT pose such a threat to multilateralism that the region is forced to consider creating its own bloc structures.

In discussing regionalism in the West Pacific Rim, therefore, two points must be borne in mind. First, the major player, Japan, views regionalism as a means to an end which might best be achieved by a non-regional, bilateral and multilateral approach. Secondly, any regionalism that exists in this part of the world is expressed through subregional groupings, often of a very loose and ad hoc nature. To facilitate the discussion, reference will be made throughout to the following subregional groupings of economies: (i) *Northeast Asia* (consisting of Japan, the two Koreas, China, Hong Kong, Macao and Taiwan); (ii) *ASEAN* (comprising the six economies of Malaysia, Brunei, Indonesia, Singapore, Thailand and the Philippines), together with Vietnam; and (iii) *Australia and New Zealand (Figure 9.2).*

CHARACTERISTICS OF INTERNATIONAL TRADE

There are three main characteristics of international trade of central relevance to the present discussion. First, the West Pacific Rim is accounting for an increasing proportion of global trade in both exports and imports. Second the direction of trade is changing in that dependence on the US appears to be weakening and intraregional trade is increasing in relative importance. Third, the composition of trade is changing to the extent that primary products are decreasing as a proportion of all trade.

The growth of international trade in the region has certainly been remarkable:

> During the late 1970s and early 1980s, East Asian countries increased their share of European and North American industrial markets, albeit from a low base. Much of this trade was dominated by Japan, joined later by Hong Kong, South Korea, Taiwan and Singapore, then by several of the ASEAN countries – notably Malaysia, Indonesia and the Philippines – and finally, most recently, by China. (Hodder, 1992, 65)

While available statistics are of uneven quality and of little value for comparative purposes, the general trend is clear enough. From 1982 to 1988, the growth of export volume from east Asia (even excluding Japan) was just over 12 per cent per annum, a rate double that of South Asia, three times that of the Middle East, North Africa and Latin America, and six times higher than in sub-Saharan

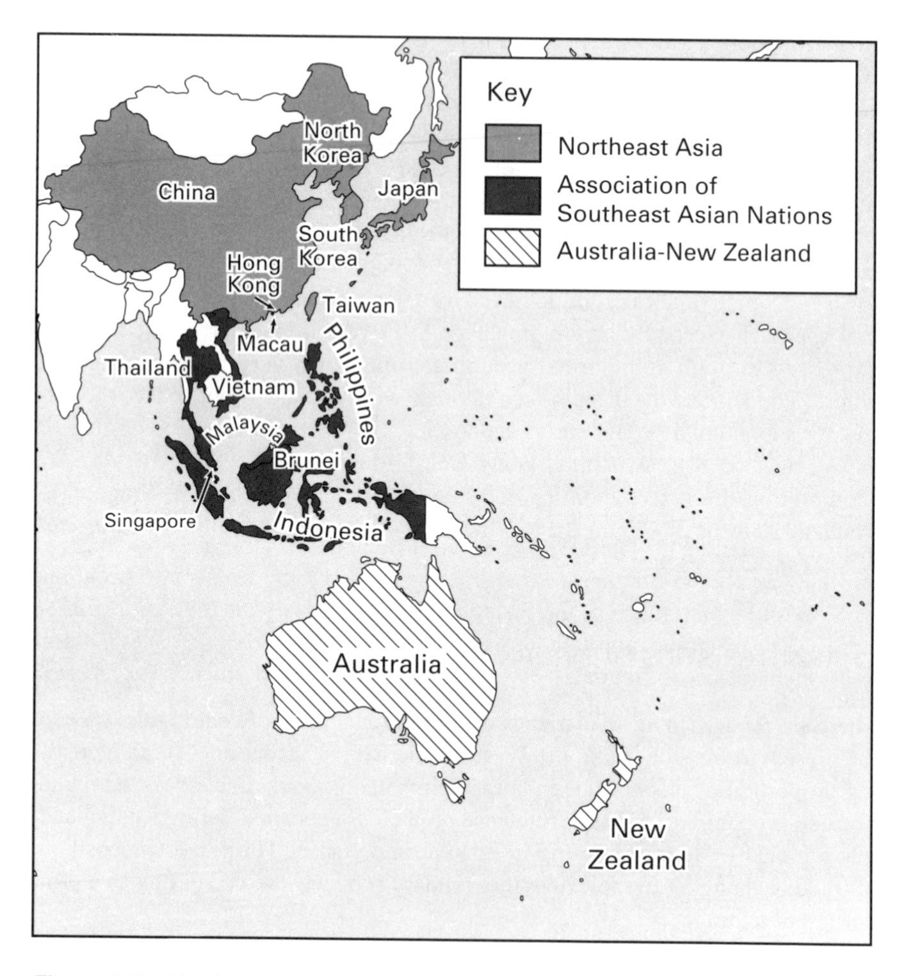

Figure 9.2 Northeast Asia, ASEAN and Australia–New Zealand

Africa. The region's share of total world trade increased from 14.3 per cent in 1971 to 15.3 per cent in 1976, 18.2 per cent in 1981, and 22.8 per cent in 1984. Over the period from 1971 to 1984 exports from the region grew at an average annual rate of 17.9 per cent compared with world exports, which grew at an average rate of 14.2 per cent over the same period (Hodder, 1992, 67 – 8). In 1990 total exports from the West Pacific region, including Australia and New Zealand but excluding North Korea and Vietnam, stood at $695 153 million, or 22 per cent of the world's total; and of this 43 per cent was accounted for by Japan. As for imports, the West Pacific Rim accounted for 21 per cent of world trade, of which Japan took 33 per cent (World Bank, 1992).

Some countries have done rather better than others. According to the 1990 figures China, Taiwan, South Korea and Hong Kong all experienced average annual growth rates in trade of 10 per cent or more during the period 1980 – 1990. In international trading terms, South Korea rose between 1965 and 1987 from obscurity to rank fifth among all countries in the Pacific region. During the same period Singapore rose from fifth to fourth place; Japan remained in second place behind the US; and Indonesia rose from thirteenth to eighth place. On the other hand, Malaysia and the Philippines lost ground, the Philippines actually dropping out of the top ten countries in ranking during the same period (Hodder, 1992, 68). According to 1990 figures, China, Taiwan, South Korea and Hong Kong all experienced average annual growth rates in trade of 10 per cent or more during the 1980s.

This is not the place to go into the causes of this generally dramatic rise in prosperity and trade. But it is important to emphasise that the most powerful initial stimulus and catalyst came from the US. Determined to protect its own strategic interests against communism and imperialism, the US provided massive aid and open markets to Japan, helping Japan to rise from the ruins of her economy after 1945. The US also set up a triangular trading network between the US, Japan and Southeast Asia, thereby diffusing the benefits of economic growth throughout much of the region. Other factors were the traditional trading skills of many groups throughout the region, the existence of overseas Chinese throughout much of Southeast Asia, and the geographical advantages for trade of this insular and peninsular world (Chan, 1990b).

As for the direction of trade, there are two main points to make. The first is concerning the changing role of the US in West Pacific trade. From 1970 to 1988 Japan's total trade with the US declined slightly, while its trade with the rest of the West Pacific Rim increased markedly; but the shares of Japanese, Korean and Taiwanese exports shifted sharply away from the US and towards East Asian markets. ASEAN countries also experienced a generally predominant import dependence on East Asian trading partners, while US trade correspondingly showed some decline in the import share from East Asia in the late 1980s, although her export share remained steady. The second point is the corollary of the first: that intraregional trade has become increasingly significant. Over the 1971–84 period intraregional exports grew at an average annual rate of 19.1 per cent, compared with an average annual growth rate in extraregional exports of 17.4 per cent. Indices compiled on inward and outward trade linkages suggest that Pacific Basin countries are becoming more interdependent, or integrated; the proportion of world trade (exports and imports) between the countries of the region is growing more quickly than trade with countries outside the region (Dixon and Drakakis-Smith, 1993)

The important exception is Japan which has expanded into a major world trading power and continues to play an important role in the changing triangular trade with east Asia and the US. The old triangular trade referred to earlier in this chapter has gradually given way to a new triangular pattern 'consisting of a strong flow of components, capital goods, and consumer goods from Japan to the NICs and other Pacific countries, and a strong flow of final manufactures from those countries to the US' (Petri, 1992, 42). Nevertheless, Japan plays a leading, indeed critical, role in intra regional trade and is increasingly reliant on trade in Asian markets, which now accounts for 41 per cent of Japan's total trade, while North America accounts for only 30 per cent. Japan seems well placed to tap the potentially almost limitless market in China, and is now beginning to dominate the trade and investment structures of most countries in the region. The four NICs are also beginning to compete increasingly successfully in this intraregional trade; and the Overseas Chinese in Southeast Asia have already established a close network of business links with the Chinese in Taiwan, Singapore, Hong Kong and elsewhere:

> If China's trade expands as expected, these Overseas Chinese in Southeast Asia will be well placed to take advantage of the business opportunities open to them. Working together, the Japanese and Chinese, including the Overseas Chinese, are likely to develop and dominate the already impressive existing trading networks within the countries of the West Pacific Rim. (Hodder, 1992, 75)

In the composition of trade, although no longer in the direction of trade, there is some evidence to support the suggestion that the colonial pattern of trade still persists in the region:

> Especially if we divide trade flows to distinguish between flows of crude materials (raw materials and primary products) and flows of manufactured goods, then the colonial pattern of trade is very much alive. (Gibson, 1990).

This is particularly true of the Southeast Asian sector. Despite the growth of the industrial sector in Southeast Asia, most of the imports into the leading trading countries are still manufactured goods, whereas a clear majority of the exports are crude raw materials. With the exception of Singapore, there is some evidence in most of Southeast Asia of the former colonial pattern of trade in which countries export raw materials and primary products and import most of their manufactured goods. This contrasts sharply with Japan, where the opposite is true: 55 per cent of its imports are crude materials and primary products and 99 per cent of its exports are manufactured goods. As for Hong Kong and South Korea, a clear majority of both their imports and exports are manufactured goods (Hodder, 1992). This division of the West Pacific Rim into resource-rich countries in the south and resource-poor countries in the north helps to explain the importance of the intraregional trade referred to earlier.

Changes are occurring, however. Regarding exports, it is clear that changes in the composition of exports from countries in the region are now more generally toward manufactured goods and away from foodstuffs and agricultural raw materials. The composition of imports is similarly changing. There is now a tendency for foodstuffs to constitute a smaller share of imports, reflecting the tendency for the proportion of income spent on food to fall as incomes rise and the trend towards higher agricultural productivity and food self-sufficiency in many of the countries. Similarly, the proportion of agricultural raw materials, ores and metals in total imports is declining – a trend consistent with increased emphasis on value-added manufacturing activity (United Nations, 1990b).

Until the late 1980s and early 1990s many of the countries of the West Pacific Rim developed as competitors to one another in an effort to be principal suppliers to the largest industrialised markets of North America and Europe, as well as Japan. But with the recent onset of global recession, a corresponding reduction in consumption in the industrialised world, and signs of protectionism in the EC and the US, there now appears to be a relative decline in such trade and an increase in horizontal intraregional trade:

> Although this perceived trend might portend a more formal integration in the future, it is more likely simply a reaction which has forced the economies of the region to look at local and regional markets as the means to maintain rapid industrialisation while the markets of the rest of the industrialised world are in recession. (Loesch, 1993, 1)

PROBLEMS AND OPPORTUNITIES IN THE SUBREGIONS

In the absence of any strong movement towards regionalism among the economies of the West Pacific Rim as a whole, it seems sensible to examine its problems and opportunities within the context of the three subregions identified earlier – Northeast Asia, ASEAN and Australia–New Zealand.

North east Asia

It is now widely believed that during the 1990s and beyond Northeast Asia might well become the centre of gravity of the world economy and will account for a growing share of world trade and financial flows. Certainly its volume of trade, both in size and as a proportion of world trade, has grown dramatically since the Second World War. The composition of exports and imports has also changed, driven by the quest for higher value-added exports to absorb rising costs. Significant increases in per capita incomes and greater openness to trade have also changed the composition of imports. These shifts have been accompanied by changes in the direction of trade, with a marked reduction in dependence on the US market for all countries except China. Throughout Northeast Asia there will be rising intraregional trade and a relative decline in the economic role of the US; heightened technological competition; and the growth in importance of manufacturing trade, together with increased intra-industry trade. At the same time it seems that growth in the most successful economies – Japan, Taiwan, South Korea and Hong Kong – will increasingly become domestic-led rather than export-led, reflecting an international environment less dependent than formerly on external economic conditions (East Asia Analytical Unit, 1992).

The main trading problems faced by Northeast Asia can perhaps best be illustrated by Japan, although much of the discussion on Japan's trade is bedevilled by a number of misconceptions, of which two are particularly relevant here. One is that Japan is too often regarded as merely an offshoot of the Western world. However, Japan is neither Asian nor Western. Her isolation from Asia has been exaggerated by China's self-imposed isolation since 1949, thereby cutting Japan off from her cultural roots. But with the recent opening-up of China there is now nothing to stop Japan operating as energetically in Asia as it already does globally. The other important misconception is the stereotypical image of Japan as a country unusually dependent on foreign trade. Yet its trade dependence is not at all remarkable if dependence is measured simply by the ratio of imports to national income. Its dependence on foreign trade is only substantial if measured in physical terms – reflecting the great need for food and raw materials inevitable in a resource-poor country like Japan. This is indeed a problem with which Japan has long had to contend, but it has provided a major incentive for Japan's remarkable export-led industrialisation in the recent past (Corbett, 1970).

Two problems concerning Japan which receive a good deal of attention relate to friction between Japan and the US. At the time of writing US measures are targeting the Japanese government's procurement practices in construction and superconductors, while at the same time complaining about abuses of intellectual property rights in Japan's neighbours, notably Taiwan. Lying at the root of this problem is of course the US trade deficit: in 1992, 58 per cent of the $84 billion trade deficit was accounted for by trade with Japan and a further significant amount by other countries in Northeast Asia. It is for this reason especially that US protectionist sentiments seem sometimes to be moving against the GATT and NAFTA (see Chapter 6). And in this connection it is important to recognise the recent change in balance between free trade and managed trade in the US. Until quite recently, the US was the acknowledged champion of free trade and multilateralism. Japan was regarded as pursuing managed or mercantilist policies (Nester, 1990), although Bhagwati (1991) challenges the stereotypes of Japan that fuel sentiments supporting managed trade and aggressive unilateralism. Now, however, it seems that the US has resorted to aggressive and unilateral, if selective, tactics in trade policy while Japan is posing as the bulwark of open trade and multilateralism. The current trade disputes are concerned particularly with 'fair' trade and managed trade; and regionalism seems to be gaining ground in North America and Europe.

The change in Japan during the late 1980s and early 1990s is that it is now becoming an important consumer. Thus its problems are no longer solely those of an exporter; Japan now builds new factories abroad every day, setting up truly global operations for the first time. The Japanese are now anxious to buy a wide range of imports and are looking for ways in which buyers can express their individuality rather than conformity. In this sense Japan now has a mature economy, and a number of other economies in the region are about to follow suit.

China is now also in some conflict with the US, but in this case the ostensible causes of friction include human rights violations, as well as intellectual property rights and the Chinese 1992 trade surplus of over $18 billion with the US. This trade deficit relates especially to textiles and the garment trade, but in dealing with China the US has to be careful because Chinese products go mainly through Hong Kong; Hong Kong's trade with China rose from $1 billion in 1978 to $28 billion in 1990. At the moment Beijing seems prepared to stand up to the US, apparently being willing even to wreck a £65m trade deal for telecommunications with a US company – the biggest ever Sino-American deal – if the US attach human rights and other conditions to Chinese trade (Mirsky, 1993). The US is also threatening to remove Most Favoured Nation (MFN) status from China. This is a serious matter for China, for without MFN status duty on Chinese exports would increase from 8 per cent to 40 per cent. The US takes up to 30 per cent of China's annual exports of $85 billion; and in 1992 exports to the US from China amounted to $26.5 billion in which China had a trade advantage of over $18 billion.

The opportunities for increased economic power and influence in Northeast Asia are immense, particularly for Japan, which is building links with the whole of the West Pacific Rim in trade and investment. Between 1984 and 1989, the annual flow of Japan's direct investment to East Asia leapt fivefold. Yet this did not reflect any deliberate targeting of Asia, for during this period Japan's direct investments in Europe and America multiplied even faster. The share now going to Asia is increasing, however. Since Japan's long-term growth rate at 3–4 per cent is around half that of East Asia, Japanese companies are interested in Asians as consumers as well as cheap workers. Thus in 1992 Toyota captured nearly 25 per cent of the car market in the ASEAN countries, and hopes to have captured 30 per cent in 1993. Japan is also moving the location of some of its industries abroad into many parts of Northeast and Southeast Asia, partly because of rising wages and labour shortages in Japan and because of the appreciation of the yen, which has eroded competitiveness (Ariff, 1991a). Japan's production is now on a global scale, based on the international division of labour and specialisation; the same process is spreading to other parts of the West Pacific Rim (Economist, 1993b).

While there is no recognisable trading group, let alone a trading bloc, in

Northeast Asia, there has been some growth of what are called 'concept zones' in this subregion. Four examples of this trend are: (i) Integrated Southern China (South China, Hong Kong and Taiwan); (ii) The Japan Sea Rim (Japan, China, North and South Korea, and the Pacific coast of Russia); (iii) The Yellow Sea Rim (China, North and South Korea and part of Japan); and (iv) The Tumen River Project (China, North Korea, the Pacific coast of Russia and South Korea). However, these 'concept zones' are in no sense trading groups; they are simply loose, ad hoc arrangements designed to achieve specific aims and mutual advantage among the participating members. In most cases they exist more as concepts than as fact (East Asia Analytical Unit, 1992).

It is clear that Northeast Asia has enormous potential for continued successful economic growth and power, not just within the West Pacific Rim but also globally. If, as seems likely, China opens up more fully to trade and investment opportunities and energetically pursues its policy of export-led industrialisation, then Japan, China and the NICs in Northeast Asia could easily dominate the world's trade and markets. But this is most likely to occur not through the route of regionalism, but through the kind of route to multilateralism now being followed quite deliberately by the economies of Northeast Asia.

The Association of Southeast Asian Nations – ASEAN

ASEAN is the nearest approximation to a trading bloc in the West Pacific Rim. It was established in Bangkok in 1967 'to accelerate economic progress and to increase the stability of the Southeast Asian Region .... to collaborate .... for the expansion of their trade, including the study of the problems of international commodity trade' (Europa, 1992). It began with four members, but now includes a total of six – Malaysia, Singapore, Thailand, Indonesia, Brunei and the Philippines.

While ASEAN as a whole is regarded as a bloc, it is an extremely open and fluid one and has been anxious to develop links and arrangements with many other groupings, including the EC. ASEAN has taken the first step to regional economic cooperation, especially in communications and transport and, to a more limited extent, in tariff reductions. It is trying to set up a free trade area among its members (ASEAN Free Trade Area, or AFTA). This operates a two-stage tariff reduction: the first stage involves cutting tariffs to 20 per cent within 5–8 years; the second stage will see tariffs lowered to 0.5 per cent in the following 7–10 years, although different plans relate to different product categories. AFTA could prove valuable if it creates incentives for Japan and other capital-surplus countries in Asia to locate their plants in ASEAN to take advantage of lowered tariffs for goods produced within the bloc.

ASEAN's intraregional trade is still only 12 per cent of its total trade (which is only 3 per cent of all world trade). In world terms, then, it is small. Yet a number of observers suggest that ASEAN is likely to be the next big capitalist bloc to join the US, EC and Japan among the developed regions of the world. Not only does it have the institutional framework, but it also has the great advantage of having at its core the successful NIC of Singapore. Although in terms of natural resources it is by far the poorest of the ASEAN countries, Singapore has easily the highest levels of prosperity. Yet Singapore is the most trade dependent state in ASEAN, and 25 per cent of its exports go to the US. It is also vulnerable in that while it enjoys heavy inward investment, it has few of its own multinational corporations (MNCs). On the other hand, Singapore is less vulnerable to protectionist retaliation than Taiwan and South Korea because of its liberal trade policy and its limited restrictions on quotas or on export and import licences. What Singapore needs above all is not to withdraw into any closed trading bloc, but to secure a larger market share in Japan, Southeast Asia, the countries of the former European communist bloc, China and the EC. All this will have been helped by Singapore shifting its entrepot functions from exporting Southeast Asian goods to being a gateway for goods entering the region. It will also be helped if Singapore realises its aim of replacing Hong Kong as Southeast Asia's financial centre after 1997.

As for Vietnam, which is establishing closer economic relations with ASEAN, it is possible that it will eventually become a member – and an important member – of ASEAN. As a result of *Doi moi* (renovation) it is already attracting foreign investment of $4 billion. It is resource-rich, with oil, mineral and agricultural resources, and is now the world's third largest exporter of rice. Its exports hit a new peak in 1992 and it has a trade surplus. But above all it has excellent human resources. Its population of 73 million provides potentially the second largest market after Indonesia in Southeast Asia. The population is also well educated, with an 88 per cent literacy rate, and it has current wage rates half those in Malaysia and Thailand. It is certainly an important potential market for the economies of Northeast Asia, as well as for the ASEAN countries. It also offers great opportunities for US trade and investment, although at the moment this is restricted by the US embargo on executing contracts with Vietnam. Its problems are poor infrastructure (especially transport), power shortages, corruption and a heavy bureaucracy.

It may well be that Vietnam will become a member of ASEAN before long. But whether it does or not, the ASEAN countries as a whole seem to have an excellent future with real opportunities for further development, internally, in their contacts with other parts of the West Pacific Rim, and with the EC and the US. While ASEAN is described as a trading bloc in international trade statistics, it does not think or act in any narrow, regional terms; it considers its present organisation not as an example of regionalism, but as a pragmatic and open means of participating more fully in the world economy and of playing its full role in the development of the West Pacific Rim.

Australia and New Zealand

Australia and New Zealand have their own arrangement known as CERTA, but these two countries are located so far away from the European and American blocs, and are so fearful of the implications of losing their markets in these traditional areas, that they are moving irresistibly towards closer relations with the rest of the West Pacific Rim, and especially with north east Asia. Their international trade is not as yet very great. Over 40 per cent of Australia's merchandise exports go to the north east Asian region (30 per cent to Japan); and almost 30 per cent of Australia's merchandise imports come from north east Asia (20 per cent from Japan). These figures compare with exports to the US and EC of 11 per cent and 12 per cent respectively, and imports from the US and EC of 24 per cent and 22 per cent respectively.

Australia and New Zealand, like all other economies in the West Pacific Rim, are worried by the spectre of the major trade blocs (especially in North America and the EC) becoming increasingly inward-looking and protectionist. Australia and New Zealand are also worried by any bilateral trade deals between the major players and about being marginalised, not only globally, but also within the West Pacific Rim.

According to official Australian sources:

> Australia's future economic growth will depend significantly on its success in increasing its economic interaction with Northeast Asia .... Successful economic performance in Australia will also be a prerequisite for Australia to maintain and extend its political and strategic relevance in the northeast Asian region and in the wider Pacific area. (East Asia Analytical Unit, 1992, 27–8)

There is believed to be a great need to internationalise the Australian and New Zealand economies. The problem is not just how to hold on to their market share in north east Asia, but also how to win back the market share lost in the 1980s. In addition there is a need to broaden the basis of their exports. Primary products accounted for 65 per cent of Australia's exports to Northeast Asia in 1990. Certainly Australia has a comparative advantge in agriculture, some minerals and energy resources, but the country needs to develop its manufacturing industries further, including value-added industries like food, fibre and minerals. A good example of this is in the development of elaborately transformed manufacturing (ETM), such as motor vehicle parts, office machines and telecommunications equipment (East Asia Analytical Unit, 1992).

As well as changing the composition of its exports to Northeast Asia, Australia needs to invest more heavily in the region. Its investment there is still small: representing in 1990 only one-twelfth of all Australian investment globally. There is much more investment by north east Asia in Australia, 17.5 per cent of which is from Japan. North east Asia, therefore, continues to present great opportunities for Australia's economic advancement. The rapid pace of growth and structural and technological changes in the region both encourage and require continued internationalisation of the Australian economy.

Protectionism

Whatever the specific problems and opportunities expressed in the different countries, it is the fear of protectionism in the western industrialised world that lies behind so much of the thinking and attitudes of the economies of the West Pacific Rim, so many of which are continuing to follow export-led industrialisation policies. It will be remembered that it was the openness of US markets that enabled Japan to recover after the Second World War. But the huge trade deficit in the US is encouraging protectionist attitudes. Over half the US trade deficit is with countries of the region. Protectionism in the US, Canada and the EC could lead the countries of the West Pacific Rim into retaliatory protectionism, resurgent nationalism and economic stagnation. Just as open international trade has been the key to the region's prosperity, so protectionism could be the cause of its decline (Hodder, 1992).

Most of the economies of the western rim of the Pacific fear that protectionism, the raising of tariff barriers, the continuation of subsidies, and the failure of the GATT to keep open the channels of world trade could well lead within a decade or so to the creation of separate trading blocs in the Americas and Europe. But with well over half the world's population, and including Japan, which many believe will by then have an economy as large as and probably wealthier than that of the US, the power of the West Pacific Rim to influence what happens to the world economic order might well become critical (Ariff, 1991b). Japan is already strengthening its trading and investment relationships with Australasia, the former Soviet Union and eastern Europe, the Americas and the countries of the EC.

Asia-Pacific Economic Cooperation – APEC

Since its launch in 1989, the APEC process has been regarded as a useful framework for supporting and strengthening the multilateral trading system throughout the western and eastern rims of the Pacific. It is designed to reflect the need for closer economic cooperation among its members through an open, non-formal arrangement, focusing on the practical development of economic cooperation, including trade promotion. It has little central bureaucracy – it is simply a body for international dialogue and consultation on regional economic issues. Nor is APEC designed to compete with other bodies in the region; indeed it is specifically designed to complement existing bodies, including ASEAN, the Pacific Economic Cooperation Conference (PECC) and the Pacific Basin Economic Council (PBEC).

In the context of the present discussion, two points about APEC need to be emphasised. First, it is in no sense a regional bloc, nor does it exist to encourage regionalism. It is firmly and unequivocally committed to multilateralism. Yet as far as trade is concerned, the issue of regional trade liberalisation has been an important focus of APEC's efforts to explore the basis for action to further gains from interdependence following the Uruguay Round of the GATT. One of APEC's underlying principles is to exert a strong positive influence on the future evolution of the global trading system. Its purpose is to reduce barriers to trade in goods, services and investment, but in a manner fully consistent with the GATT principles.

Secondly, APEC is pan-Pacific in its geographical coverage and includes all governments around the rim that are irrevocably committed to market-oriented, open trade economic strategies. It therefore includes Canada and the United States, as well as Australia and New Zealand, as well as much of Northeast Asia and ASEAN. From the point of view of the West Pacific Rim, therefore, the geographical scope of APEC is so broad that it seems to some governments to threaten their national interests and independence. Partly for this reason the Malaysian government proposed the setting up of the East Asia Economic Grouping (EAEG), now the East Asia Economic Council (EAEC), which has aims very similar to those of APEC, but which pointedly excludes the English-speaking countries of Canada, the United States, Australia and New Zealand.

Although the EC has expressed concern about APEC's size and potential power, this seems to reflect a misunderstanding of the purpose and style of APEC. It is so far simply a body for international dialogue and consultation on economic matters of mutual concern to its members around the rim of the Pacific Ocean. APEC is still in its infancy, yet it certainly poses no threat either to the EC or to any other emerging trade blocs in other parts of the world. Indeed, it appears to be hindered by the magnitude and vagueness of the agenda it hopes to undertake. It is also likely to be constrained by the national self-interest of at least some of its members. The Joint Policy Statement by the CSIS/APA Working Group (Okwuizumi, et al., 1992) recognised the current vagueness of APEC and argued that it should be developed in parallel with a strengthened US – Japan bilateral relationship, and that its programme should be complementary to multilateral commitments, and especially to the GATT. It could be especially useful in policy areas not currently covered by more broadly based international organisations. Thus the Joint Policy Statement argues for new arrangements, implemented through APEC, to integrate trade, investment, technology and aid relationships so as to maximise the development impact consistent with environmental concerns and the need to minimise trade-distorting effects.

APEC could also be useful in other ways. If present global multilateral efforts continue to be thwarted, the emphasis in trade policy may become sectoral or industry-based rather than regional or bilateral. While these policies are not necessarily mutually exclusive, it is clear that 'APEC is well placed to provide a focus for sectoral or regional initiatives while keeping them consistent with multilateral principles' (East Asia Analytical Unit, 1992, 23).

APEC is certainly ideally suited to provide a framework for the interests of globally oriented economies throughout the West Pacific Rim. Japan has supported Australia's recent suggestion that the 15-country APEC should be strengthened to encourage liberal trading in the region as a whole. Australia has called for solid progress at APEC's next ministerial meeting in Seattle in November 1993, and is urging a framework for harmonising trade and investment rules and resolving trade disputes among its members.

REGIONALISM AND SECURITY

Since the end of the Second World War the issue of security has played a significant role in the growth and nature of trade in the West Pacific Rim. Central to this issue is the changing relationship between Japan and the United States. It has already been pointed out that US strategic interests determined to a large extent its attitudes towards the rebuilding of post-war Japan and in particular towards the provision of open market opportunities for Japanese goods in the US. At least until the end of the Cold War, strategic interests were always at the centre of American trade policy towards the West Pacific Rim. Since then, however, security matters have begun to play a much greater role in the thinking and interests of countries in Northeast Asia and ASEAN.

Keeping its distance from the US in security matters has not been on the agenda for Japan in the past 30 years, except for the first opposition party which regarded the US–Japan Treaty and the existence of the Self Defence Force as unconstitutional. Although the Japanese public fiercely opposed the continuation of the US–Japan Security Treaty before its renewal in 1960, the Japanese are well aware of the continued need to guard against any potential communist threat. Today it appears that most Asian governments in the region feel that a continued strong US presence and a close US understanding of the region's problems are essential for the region's future security, without which any progress in trade and economic prosperity will have little meaning. It is true that many Asian countries still nurse animosities and mistrust among themselves, and especially against Japan, because of its wartime military occupations (Kusakabe, 1993). But the fear now is that the end of the Cold War and of any clear-cut superpower divide will create a vacuum in which further tensions could grow. After all China remains communist, the situation in Russia is extremely fluid and potentially dangerous, and there are still communist governments in North Korea and Vietnam.

The nations of Southeast Asia hope that new arrangements can be made to help the US retain a diminished but still substantial military presence to discourage potential regional powers, particularly Japan and China, from extending their strategic interests and power in the region. Some authorities cite the ASEAN principle which does not approve of the existence of non-ASEAN military bases in the region; they argue that reliance on US forces undermines the spirit of Asia based on mutual trust and cooperation, and they do not see a realistic possibility of any military threat from Japan (Kusakabe, 1993). The majority view in Southeast Asia, however, seems to be that East Asia must try to avoid a pre-emptive move by any power to secure spheres of influence or areas of dominance, before a new balance is gradually reached, and is concerned that 'a rapid change in the regional balance of power would be triggered by an untimely US withdrawal (Kusakabe, 1993, 3). The US military presence in East Asia has been underpinned by a network of essentially bilateral alliances, the linchpin of which has been the US–Japan security pact. Japan, so it is now generally accepted, does not possess territorial ambitions and it is, in any case, not in a state of military independence. The continued presence of the US in the Asian region must give relief to the countries throughout the west Pacific. This policy of retaining and even enhancing security ties with the US is now accepted by most countries in the region.

This stance is especially logical in view of the possibility of Korean unification, which is already causing much concern in Japan, and of specific worries in the US over North Korea's nuclear programme. Faced with the prospect that the US might have less interest in the region in the absence of a Soviet threat, almost every Asian state in the region is stepping up its military expenditure and nervously eyeing that of its neighbours. Enhanced security ties between Japan and the US seem to be in the interests not only of the US and Japan, but also of all countries in the West Pacific Rim. For without security and political stability there can be no sustained progress in the multilateral, open trading system upon which the prosperity of the West Pacific Rim countries must depend.

FUTURE COOPERATION AND INTEGRATION

There are many possible scenarios representing policy alternatives among the nations of the West Pacific Rim, and most of them revolve around Japan, in particular around Japan's relations with the United States (Kusakabe, 1993). One scenario suggests the increased integration of the economies of Japan and the United States; this implies closer ties between the two major global economies, with the result that Japan would appear to turn its back on any wider Pacific Asian community. A second, 'tripolar' scenario represents Japan's supposed inclination to isolate itself from the United States in security affairs but to enhance the basic trilateral economic relations between Japan, the United States and the EC. This scenario would confirm and perpetuate the notion of three great trading blocs referred to in much of the literature. A third scenario represents Japan as rejecting altogether the Japan–United States alliance. In these circumstances Japan would focus on developing its economic ties, whether formal or informal, with the nations of the West Pacific Rim; and a likely outcome of this would be the emergence of a West Pacific trading bloc to counter the aggressive and protectionist tendencies of other trading blocs. Yet another, more drastic, scenario predicts the development of two large trading blocs: (i) Japan and China, together with East Asia/Australasia; and (ii) India and her neighbours in South Asia. There would also be the two smaller blocs of the Americas and Europe.

More likely than any of these scenarios, however, is one in which the security ties of the west Pacific countries with the US are retained, or even enhanced, while in economic matters the economies of the region continue to develop their intraregional trade. The economies of the West Pacific Rim are likely to continue on their present course, expanding into and embracing not only the great potential of China, but also possibly of the Indian subcontinent and much of the former Soviet Union. At the same time the West Pacific economies will continue to look outwards rather than inwards, participating fully and enthusiastically in what they hope will emerge through the GATT as an increasingly multilateral global economy. Certainly there appears to be no compelling reason why the economies of the West Pacific Rim should move towards regionalism, and every reason why they should encourage a more aggressive and less tentative support for multilateralism in the rest of the world. It is possible that the erection of trade barriers elsewhere – for instance in a 'fortress Europe' – might force the West Pacific Rim economies reluctantly towards some form of de facto regionalism. But Bhagwati (1991) is probably correct in predicting that this is not likely to happen, and arguing that the GATT must declare that interregional trade between its main players remains of central importance to the world economy.

CONCLUSION

The case of the West Pacific Rim discussed in this chapter is revealing in many respects. In the first place, it emphasises a point made earlier in this book, that multilateralism and regionalism should not be viewed as discrete entities, representing entirely contrary and antithetical approaches to international trade (See Chapters 1 and 2). Certainly in the economies of the West Pacific multilateral, bilateral and regional approaches are all in evidence. As the CSIS-APA Joint Policy Statement put it

> the approach to Asia-Pacific economic cooperation presented here is eclectic, a combination of multilateral, regional, and bilateral relationships implemented in a mutually reinforcing way. The policy issues are not posed in terms of either ... or

> between multilateral, regional and bilateral arrangements, but of how all three can be pursued jointly toward agreed objectives (Okwuizumi et al., 1992, 22).

On the other hand there is little doubt of the preference for multilateral and bilateral approaches in the West Pacific Rim. There seems to be no compelling reason for embracing regionalism, especially in the wealthiest and most dynamic economies in north east Asia. It may be that this wealth and dynamism prove to be unsustainable in the medium or long term; and, as already emphasised, it is at least possible that any further growth of protectionism in a 'fortress Europe' and in America (NAFTA) might well force the economies of the West Pacific to consider the retaliatory measure of creating their own regional integrated trading bloc. Certainly there is the fear in East Asia that protectionism in the EC and NAFTA (see chapters 3 and 6) might well divert trade and investment away from the West Pacific. But even so, it seems most unlikely that regionalism will ever achieve priority over multilateralism and bilateralism in this part of the world. Cooperation is acceptable, but integration is not. This is so, not simply because of the essentially pragmatic nature of most governments in the region, nor because of the other imperatives already noted, but also because communist governments still exist – in China, North Korea and Vietnam. The economies of the western rim of the Pacific are therefore faced with adjustment problems internally and within the region as whole. The communist states face grave contradictions – between economic capitalism and communist dogma – and the pressures for greater political freedoms that rapid economic advance inevitably breeds.

More generally in these largely export-led economies, it is recognised that only in an open, liberalised world economy can countries expect to sustain their growth and prosperity; and most countries in the region support the view that while cooperation between economies is always beneficial, integration is not. It is also more appreciated here than in most other parts of the world that the distinction commonly made between international and domestic trade is to some extent misleading and arbitrary: that the hundreds of thousands of firms of all sizes, MNCs, Foreign Direct Investment (FDI) flows and offshore operations all require a trading environment – at home as well as abroad – unfettered as far as possible by restrictions imposed from some central regional authority.

Looking to the immediate future, it seems likely that even if the EC and North America retreat into the strictest forms of protectionism and regionalism – even to the extent that there seems to be no prospect of furthering the aims of the GATT – this need not have such serious and damaging effects in the West Pacific Rim as it would certainly have in Europe and North America. Mention has already been made of the enormous potential, particularly in north east Asia, for economic growth and for huge market expansion. The potential for increased intraregional trade appears almost limitless. Thus it could well be that the West Pacific Rim could withstand the development elsewhere of powerful trade blocs and the possible effects of a trade war far better than anywhere else in the world. This is one added reason why the EC and the United States need to consider very carefully their attitude to regionalism and to its role in furthering the critical aims of the GATT. While it may seem reasonable to argue that regionalism and multilateralism are not necessarily antithetical, and that regionalism is just one route to eventual multilateralism, this may not prove to be so in a world dominated in due course by the pragmatic, outward-looking, non-integrated, liberalised and essentially multilateral economies of the western rim of the Pacific.

References

Ariff, M. (1991a) *The Malaysian Economy: Pacific Connections*, Oxford University Press, Oxford.

Ariff, M. (ed.) (1991b) *The Pacific Economy: Growth and External Stability*, Allen & Unwin, Sydney.

Bhagwati, J.N. (1991) *The World Trading System at Risk*, Harvester-Wheatsheaf, Hemel Hempstead.

Chan, S. (1990b) *East Asian Dynamism: Growth, Order and Security in the Pacific Region*, Westview, Boulder, Col.

Corbett, H. (ed.) (1970) *Trade Strategy and the Asian-Pacific Region*, Allen & Unwin, London.

Dixon, C. and Drakakis-Smith, D. (eds.) (1993) *Economic and Social Development in Pacific Asia*, Routledge, London.

East Asia Analytical Unit (1992) *Australia and North-East Asia in the 1990s: Accelerating Change*, Government of Australia, Canberra.

The Economist (1993b) Japan ties up the Asian market, *The Economist*, 24 April, 79–80.

Europa (1992) *The Far East and Australasia*, Europa Publications, London.

Gibson, L.J. (1990) The Pacific Rim: Region or Regions, *Papers of the Regional Science Association*, 68, 1–8.

Hodder, R. (1992) *The West Pacific Rim*, Belhaven, London.

Kusakabe, E. (1993) *Japan's Role in the West Pacific Rim in the Next 50 Years*, unpublished M.Sc. thesis, London School of Economics and Political Science, London.

Loesch, K.R. (1993) *The Likelihood and Possible Implications of the Formation of a Free-Trade Bloc in the West Pacific Rim*, unpublished M.Sc. thesis, London School of Economics and Political Science, London.

Mirsky, J. (1993) China shows sting in dragon's tail with threat to trade deal, *The Times*, 18 May.

Nester, W.R. (1990) *Japan's Growing Power Over East Asia and the World Economy*, Macmillan, London.

Okwuizumi, K., Calder, K.E. and Gong, G.W. (eds.) (1992) *The US-Japan Economic Relationship in East and Southeast Asia*, Centre for Strategic and International Studies, Washington, D.C.

Petri, P.A. (1992) One bloc, two blocs, or more? Political economic factors in Pacific trade policy, in: K. Okwuizumi, K.E. Calder and G.W. Gong (eds.) *The US-Japan Economic Relationship in East and Southeast Asia*, Centre for Strategic and International Studies, Washington, D.C., 39–70.

United Nations (1990b) *Restructuring the Developing Economies of Asia and the Pacific in the 1990s*, United Nations, New York.

World Bank (1992) *World Development Report 1992*, Oxford University Press, New York.

10 Conclusion: international regionalism in perspective

Richard Gibb and
Wieslaw Michalak
*University of Plymouth and
Ryerson Polytechnic University*

At the beginning of this book, a number of key questions were posed in order to establish the scope of study and provide a focus for theoretical and empirical analysis. The most important question asked was whether regionalism constitutes a threat to multilateralism and undermines the principles and obligations of the GATT. This question arose from the observation that the globalisation of business appears, somewhat paradoxically, to have promoted regionalism as states try to control at the regional level what they have increasingly failed to manage at the national and multilateral levels. The sheer complexity of the issues involved leaves room for widely differing interpretations of the costs and benefits of multilateralism and regionalism to regional and national economies. The implications of regionalism for the global trading system remain a highly contentious issue charged with emotional political argument. Many argue that the multilateral and regional approaches are complementary. Others insist that regionalism contradicts the very principle of free trade.

From a neoliberal perspective, the analytical difficulty posed by regionalism is that it can be simultaneously trade diverting and trade creating for both participatory and non-participatory states. In other words, regionalism has the potential both to support and to erode the multilateral free trade system. As Chapter 1 demonstrates, theoretical analysis based on liberal economic principles is ambivalent and somewhat inconclusive in evaluating the impact and significance of trading blocs on economic performance, living standards, income and welfare. Clearly, there is no straightforward answer to the question of whether regionalism is merely a defensive response to the detrimental impacts of free trade or a strategy used to promote multilateral trade liberalisation. In part, this ambiguity reflects both the theoretical and pragmatic difficulties associated with the implementation and evaluation of the practical outcomes of the multilateral and regional approaches to free trade.

However, when the relationship between regionalism and multilateralism is

Continental Trading Blocs: The Growth of Regionalism in the World Economy
Edited by R. Gibb and W. Michalak

examined in relation to different systems of production, organisation and consumption, it is possible to identify a fundamental change in the nature and organisation of the capitalist economy. This change is based, in part, on the fact that multilateralism and regionalism are associated with different modes of regulation and a transformation is currently taking place away from Fordism to a more flexible system of production. The change in the mode of regulation from Fordism, supporting multilateralism and the GATT, to flexible specialisation has resulted in a new mode of international economic regulation based on regionalism and trading blocs. The rationale underpinning regionalism is regulation on an international scale attempting to preserve the requirements of flexibility at the same time as protecting domestic markets from outside competition. Regionalism is therefore a powerful force fundamentally transforming the world economy and the relationships between the world's principal trading players.

The empirical evidence and detailed case studies examined in this book both support and reject the neo-liberal and transformation theories. The epistemological gap evident between the theoretical frameworks and reality results from the complexity of interests, aspirations and ideas, sometimes contradictory, that have motivated regionalism. Regionalism is a phenomenon which, for a number of reasons, is not easily understood by recourse to the empirical and theoretical formulations discussed in the introductory chapters to this book. It is possible to identify at least four problem areas. First, as outlined in Chapter 1, regional trading arrangements take several forms and cover a multitude of different schemes, ranging from sectoral agreements to political unions. Although all forms of regionalism are geographically discriminatory, in that preferential terms of trade are granted exclusively to participants, the level and degree of preference varies enormously from region to region. To complicate matters further, there is no fixed relationship between the level and degree of integration and the extent of discriminatory behaviour.

Second, trading blocs, regardless of their level of integration, can be used as a vehicle to promote multilateralism or regionalism. For example, there is little doubt of the preference for the multilateral approach in the West Pacific Rim. As Chapter 9 makes explicitly clear, it is most unlikely that regionalism will ever achieve priority over multilateralism in this area of the world where cooperation is acceptable, but integration is not. Regionalism in the West Pacific Rim therefore supports the multilateral approach as an organising principle of international trade. On the other hand, the EC is often regarded as promoting discriminatory regionalism at the direct expense of multilateralism. To most non-EC countries, the Community's '1992' programme represented a threat to free trade. Indeed, there is little doubt that the internal market programme helped to accentuate calls for a more protectionist EC trade policy (Wise and Gibb, 1993). Anxiety over the internal economic consequences of a more competitive domestic market led some European business leaders and some governments to demand higher levels of protectionism for those industries adversely affected. A concern to maintain the generally high standards of social welfare and working conditions in the EC also led to a demand for more vigorous import controls. Although, as Chapter 3 concludes, the pressures and actions to protect certain European sectors for specific social, economic and strategic ends may be exaggerated, the Community is regarded, correctly or incorrectly, as promoting regionalism based in part on discriminatory protectionism. Different forms of regionalism in different parts of the world can therefore support or erode the GATT-sponsored process of multilateralism.

A third factor explaining the inability of empirical models and theoretical formulations to account for the policies, practices and trends evident in existing trading blocs is the presence of diverse and often contradictory motivations evident not only between different regional trading arrangements, but within individual trading blocs. Within the EC for example, the so-called 'Euro-pragmatists' want open transnational trade of a competitive nature without recourse to regulation or protectionism. The 'Euro-pragmatists' are keen to impose a liberal market philosophy on the process of European integration and to use the Community as a vehicle to promote multilateralism. On the other hand, some member states of the Community, keen to extend the EC beyond a simple common market, are more ready to accept a degree of regional protectionism in order to safeguard key economic sectors and the European way of life. Policies adopted by trading blocs are therefore the result of a diverse number of sometimes contradictory motivations and, as such, more often than not change over time.

Finally, not only is there a marked heterogeneity between the different forms of regional arrangement and amongst the member states comprising such arrangements, but there exists also a noticeable heterogeneity within individual member states. For example, in the USA attitudes towards NAFTA vary considerably. On the one hand, the anti-NAFTA campaign emphasises the detrimental environmental, social and human impacts that could arise from a free trade area linking developed countries with a developing state. The potential for job losses, falling living standards and increased pollution has generated a well-organised opposition to NAFTA. On the other hand, supporters of NAFTA point to the beneficial economic consequences of enhanced competition, improved productivity and expanded markets. As Chapter 6 illustrates, supporters and opponents of NAFTA have strong backing in Canada and Mexico as well as the USA.

Overall, regionalism is an extraordinarily diverse and complex phenomenon with a market level of heterogeneity evident between and within different trading arrangements. The following summary of the key themes and issues emerging from the case studies serves to illustrate the level of diversity existing amongst trading bloc arrangements.

TRADING BLOCS IN THE 1990S AND BEYOND: KEY THEMES AND AGENDAS

The success of regional integration in western Europe is one of the most important reasons for the present resurgence of regionalism in other parts of the world. However, the origins and practice of western European integration owe far more to the desire to prevent another military conflict in Europe than a belief in market forces and free trade (Chapter 3). For many Europeans, economic integration – although very important in itself – is only one step towards a deeper political unity of purpose and social harmony on the continent. The emergence of trading blocs cannot be fully understood outside this broader political context.

The different meanings and priorities attached to multilateralism and regionalism by the various member states of the EC are a source of internal conflict and confusion. Although the future direction and shape of the Community, and European integration generally, are far from certain, some trends are already visible. Arguably the most important development is the emergence of a 'multi-speed' or 'variable-geometry' Community which breaks with the tradition of building integration at the same speed in all member countries. A paramount factor influencing this change is the apparent inevitability of expansion, first to include EFTA countries, and then the newly democratised countries in eastern Europe.

Although intraregional 'small' integration in eastern Europe is important, the newly emerged democracies in this part of the European continent are reluctant to develop a new regional trading bloc (Chapter 4). East European countries are far from being each other's most important trading partners, and there are fears that regional integration, if imposed from above, would reintroduce the inefficiencies and absurdities of the centrally planned model of trade evident during the existence of the 'socialist trading bloc' – the CMEA. Moreover, eastern Europeans are reluctant to go along this path because of their fear of being locked outside the gates of the largest existing trading bloc inclined toward protectionism. This is why the best future for integration in eastern Europe appears to lie with a 'variable geometry' EC which would be better able to accommodate the vast disparities in levels of economic and social development between these two parts of Europe.

Initially the plans for wider economic integration in Europe, symbolised by the idea of the 'common European home', included also the republics of the ex-Soviet Union. However, such schemes and long-term objectives assumed that the Soviet Union would continue to exist as a single political and economic entity evolving towards a democratic political system. The disintegration of that country, coupled with continuing uncertainty about the future course of political and economic reforms, made such plans irrelevant. Indeed, a much more pressing issue for the ex-Soviet republics continues to be the management of their own internal economic and fiscal relations rather than integration with Europe (Chapter 5). In the face of a looming energy crisis and the collapse of the 'Ruble Zone', the republics are seeking to reconstruct economic integration between themselves in an attempt to lessen the economic and social costs of market transition. At present, the republics lack the political will and economic means to create multilateral or supranational institutions. The various regional schemes, including the CIS, which have emerged from the ashes of the former Soviet Union are unlikely to provide a solid basis for a regional trading bloc. It is entirely possible that no trading bloc resembling in its geographical scope the ex-Soviet Union will emerge at all.

North American integration is, in many ways, a response to the success of EC integration and the relentless competition from the Pacific Rim economies (Chapter 6). The principle of free trade finds its most ardent supporters in North America. Multilateralism is seen by many as critical to the past success of the continental economy. Paradoxically, it is also here that the origins of the newly found enthusiasm for protectionism can be found. There is no doubt that the idea of a 'pan-American' free market from Alaska to Tierra del Fuego has been conceived in order to strengthen and secure the position of the United States in the world economy. Although the relative decline of the United States as a world superpower may prove temporary, regionalism serves as one of the many policy options available to the American administration to arrest the downward trend. Similarly, the interests of Canada and Mexico in regionalism are purely economic and motivated by a strategic goal of securing markets for domestic industries. There are clear signs that neither Mexico nor, even less, Canada is prepared to relinquish in any significant manner its national sovereignty. The North American Free Trade Association, if ratified, will be first and foremost a free trade area with strong provisions against any form of further political union.

The desire of the USA to promote free trade on a more limited regional scale for a chosen and restricted number of members may eventually embrace certain parts of Latin America (Chapter 7). This process will in all likelihood be rather slow and gradual because of the ambiguous political role played by the United States in this region in the past. Moreover, the numerous failed attempts to promote continental integration in Latin America provide for a considerable degree of scepticism within the region towards any protectionist trading bloc scheme. The more realistic attitude toward the costs and benefits of regionalism and inward investment in new integration schemes such as CACM, the Andean Group and MERCOSUR may yield leverage to an argument for a multilateral rather than a protectionist type of regional bloc. Inevitably for a continent as diverse as Latin America, tensions will emerge between countries which do not benefit equally from the integration schemes. Despite these doubts, the increasing pressure of regionalism in other parts of the world should push Latin America towards greater regional integration.

Similar processes will have far-reaching implications for Africa as a whole and the southern part of it in particular (Chapter 8). The momentous changes taking place in South Africa will, for the first time since the introduction of the policy of apartheid, generate an environment suitable for regional integration and cooperation. South Africa, because of its economic authority in the region, will continue to serve as a focal point of any regional scheme. There is a real possibility, however, that a truly democratic South Africa will not necessarily result in lasting economic cooperation and integration in southern Africa. South Africa's enormous post-apartheid economic and social needs will probably take precedence over those of the rest of the region. It would be unrealistic to expect that, under present conditions, South Africa alone would be capable of providing the finance, infrastructure and expertise needed in the region.

The last region examined in this book is the Pacific Rim, in many ways the most important economic area to emerge after the Second World War. The relentless growth rates of the many economies of the region, including Japan, South Korea, Hong Kong, Singapore and Taiwan, have created what some consider the 'trading bloc of the twenty-first century'. Many analysts predict the Pacific Rim will be the single most dominant economic power to emerge in the world economy. Moreover, the resurgence of regionalism and the emergence of trading blocs elsewhere owe a great deal to the formidable ability of the Pacific Rim economies to compete with other traditionally dominant economies. Indeed, one of the most powerful impulses behind the creation of the Single Market in the EC and NAFTA negotiations in North America is Asian, or more specifically, Japanese competition. The leading economies of the Pacific Rim are almost without exception some of the most ardent supporters of multilateralism and the principle of free trade. In fact, there is a good deal of evidence that it is multilateralism and free trade which account for the formidable economic performance of the region. It is for that reason that Japan and the rest of the 'small tigers' are determined to resist not only integrative pressures at home but also attempts elsewhere to promote regional schemes based on protectionism. It is almost a unanimous opinion among leading analysts of the world economy that a protective trading bloc in the Pacific Rim would be disastrous not only for the less-developed economies in the region itself but also for most other regions including NAFTA and the EC. However, the sheer diversity of the region both in terms of its political and economic make-up make integration difficult. That said, a significant deterioration in the terms of market access to other regions of the world could, at short notice, produce a very powerful trading bloc centred on Japan and unified in its determination to fight 'fair' trade practices and reciprocal subsidies.

In conclusion, regionalism is a strikingly diverse and complex phenomenon. The ambivalent and often inconclusive evidence based on the neo-liberal analysis of regionalism would therefore appear to be mirrored in the policies and practices of existing regional arrangements. Trading blocs can promote multilateral free trade, support preferential regionalism and protectionism or promote simultaneously elements of both. This then raises the question of whether the emergence of trading blocs, with their collection of different policy objectives and ideals, really does constitute a fundamental change in the mode of regulation as well as the nature of contemporary economy and society. This thesis concerning a transformation in the mode of regulation, in sharp contrast to classical liberal economic theory, helps to explain the resurgence of international regionalism, with trading blocs representing the spatial expression of a new mode of international regulation. However, contrary to the extrapolations made by most of the social theories which envisage that a transformation to a more flexible mode of capital accumulation will be accompanied by a reduction in the role of the nation-state, this book provides evidence to suggest that nation-states allied into regional trading blocs are in fact becoming the principal actors reconstructing the international economic environment. In addition, the internal conflicts within trading blocs over the precise meaning and consequences of free trade or protectionism make social theory building even more problematic. For example, if the new post-Fordist mode of production and regulation is meant to signify a *gradual* departure from the well-established industrial methods of Fordism into a more flexible organisation of production at the local and regional levels, then many of the countries participating in the EC and NAFTA regional schemes fit this theory. These countries and regions are indeed undergoing the kind of transformation implied by social theories. If, however, these theories are meant to describe a *radical* departure away from Fordism and multilateralism in all parts of the world, then there is an abundance of evidence in this book to the contrary. There is evidence to suggest that the kind of regionalism taking place in eastern Europe, southern Africa, Latin America and the Pacific Rim is designed ultimately to strengthen the multilateral trading system and the type of international regulation associated with Fordism.

Clearly, many of the social theories designed to explain changes in the world economy and the changing relations of its principal actors focus almost exclusively on the highly developed Western world. Although these theories are important in that they attempt a genuine interpretation of the complex processes taking place in western Europe and North America, their relevance to other regions of the world is at best limited. The uniqueness of geographical space defies generalisations on the scale demanded by holistic social and economic theories. Once again, geography ruthlessly exposes the weaknesses and shortcomings of theories designed by social scientists to embrace too much without recourse to an intimate and thorough geographical knowledge. That does not mean that such theorising is without any value. However, any theory building which attempts to explain the phenomena of regionalism and trading blocs has to take into account the inherent diversity of existing geographic space and the ambiguities it injects into analyses. The empirical evidence of contemporary regionalism presented in this book clearly fails to provide *conclusive* evidence to support or reject the contention that a fundamental remodelling of the mode of regulation in international economic affairs is taking place.

In the end, the re-emergence of regionalism can perhaps be explained more simply by the thesis first advanced by Kindleberger (1973) that the 1930s recession was caused by the inability and unwillingness of countries to assume responsibility for stabilising the world economy. Regionalism in the 1990s may therefore be a result of the inability of the United States and the unwillingness of Japan to assume responsibility for the world economy. Only time and geography will tell.

References

Kindleberger, C.P. (1973) *The World in Depression 1929–1939*, Allen Lane, The Penguin Press, London.

Wise, M. and Gibb, R.A. (1993) *From Single Market to Social Europe: The European Community in the 1990s*, Longman, London.

Bibliography

Abescat, B.(1993) Le désaccord Bruxelles-Tokyo, *L'Express*, 22–28 juillet, 29.

Africa Markets Monitor (1991) Exporters may switch to South African ports, *Africa Markets Monitor*, 7, The Economist Group, London.

Aglietta, M. (1979) *A Theory of Capitalist Regulation*, New Left Books, London.

Alwin, J.A. (1992) North American geographers and the Pacific Rim: leaders and laggards, *Professional Geographer*, 44, 369–376.

Amin, A. (1989) Flexible specialisation and small firms in Italy: myths and realities, *Antipode*, 21, 13–34.

Amin, A. and Robins, K. (1990) The re-emergence of regional economies? The mythical geography of flexible accumulation, *Environment and Planning D: Society and Space*, 8, 7–34.

ANC (African National Congress) (1990) *Discussion Document on Economic Policy*, African National Congress, Johannesburg.

Anderson, K. (1991) Europe 1992 and the western Pacific economies, *The Economic Journal*, 101, 1538–1552.

Anglin, D. (1990) Southern African responses to Eastern European developments, *Journal of Modern African Studies*, 28, 431–455.

Ariff, M. (1991a) *The Malaysian Economy: Pacific Connections*, Oxford University Press, Oxford.

Ariff, M. (ed.) (1991b) *The Pacific Economy: Growth and External Stability*, Allen & Unwin, Sydney.

Bailey, J. (ed.) (1992) *Social Europe*, Longman, London.

Baker, S. (1990) Along the border, free trade is becoming a fact of life, *Business Week*, 12 June.

Balassa, B. (1962) *The Theory of Economic Integration*, Allen & Unwin, London.

Balassa, B. (1991) *Economic Integration in Eastern Europe*, Office of the Vice-President, The World Bank, Washington, D.C.

Balassa, B. and Bauwens, L. (1988) The determinants of intra-European trade in manufactured goods, *European Economic Review*, 32, 1421–1437.

Baldwin, R.E. (1992) Assessing the fair trade and safeguards laws in terms of modern trade and political economy analysis, *The World Economy*, 15, 185–202.

Barker, T.I.J. (1977), International trade and economic growth: an alternative to the neoclassical approach, *Cambridge Journal of Economics*, 1, 153–172.

Baum, J. (1993) The stumbling bloc, *Far Eastern Economic Review*, 6 May, 11–12.

Belkindas, M.V. and Sagers, M.J. (1990) A preliminary analysis of economic relations among the Union Republics of the USSR: 1970–1988, *Soviet Geography*, 31, 629–656.

Belous, R.S. and Hartley, R.S. (eds.) (1990) *The Growth of Regional Trading Blocs in the Global Economy*, National Planning Association, Washington, D.C.

Bergsten, C.F. (1990) The world economy after the cold war, *Foreign Affairs*, 69, 96–112.

Bergsten, C.F. (1991) Rx for America: export-led growth, *Economic Insights*, Institute for International Economics, Washington, D.C., 2–6.

Bhagwati, J.N. (1988) *Protectionism, The 1987 Ohlin Lectures*, MIT Press, Cambridge, Mass.

Bhagwati, J.N. (1990a) *Multilateralism at Risk: The Seventh Harry G. Johnson Lecture*, Princeton University Press, Princeton, N.J.

Bhagwati, J.N. (1990b) Departures from multilateralism: regionalism and aggressive multilateralism, *The Economic Journal*, 100, 1304–1317.

Bhagwati, J.N. (1991) *The World Trading System at Risk*, Harvester-Wheatsheaf, Hemel Hempstend.

Bhagwati, J.N. (1992a) *Regionalism and Multilateralism: an Overview*, World Bank and CEPR conference on new dimensions in regional integration, April 2–3 1992, Session 1, paper No. 1, Washington DC.

Bhagwati, J.N. (1992b) The threats to the world trading system, *The World Economy*, 15, 443–456.

Bhagwati, J.N. (1993) Beyond NAFTA: Clinton's Trading Choices, *Foreign Policy*, 91, 155–162.

Bhagwati, J.N. (1992c) Regionalism versus multilateralism, *The World Economy*, 15, 535–555.

Bhagwati, J.N. and Patrick, (eds.) (1990) *Aggressive Unilateralism: America's 301 Trade Policy and the World Trading system*, Harvester-Wheatsheaf, Worcester.

Biessen, G. (1991) Is the impact of central planning on the level of foreign trade really negative?, *Journal of Comparative Economics*, 15, 22–44.

Blakemore, H. and Smith, C.T. (1983) *Latin America: Geographical Perspectives*, Methuen, London.

Blumenfield, J. (1991) *Economic Interdependence in Southern Africa: From Conflict to Cooperation*, Pinter, London.

BMI (Business Monitor International) (1992a) CET on free trade, *Andean Group*, 9, 975.

BMI (1992b) Uncertainty over MERCOSUR, *Southern Cone*, 9, 1083.

BMI (1993a) Andean Pact trade soars, *Andean Group*, 10, 1119.

BMI (1993b) MERCOSUR summit sets single tariff, *Southern Cone*, 10, 1108.

BMI (1993c) MERCOSUR trade up, *Southern Cone*, 10, 1144.

Boissonnat, J. (1992) Au-delà du GATT, *L'Expansion*, 19 novembre-2 décembre.

Boyer, R. (1988) *In Search of Labour Market Flexibility: European Economies in Transition*, Clarendon Press, Oxford.

Boyer, R. (1990a) *The Theory of Regulation: A Critical Analysis*, Columbia University Press, New York.

Boyer, R. (1990b) *The Regulation School: A Critical Introduction*, Columbia University Press, New York.

Brada, J.F. (1992) *Regional Integration in Eastern Europe: Prospects for Integration Within the Region and With the European Community*, paper presented at World Bank and CEPR Conference on New Dimensions in Regional Integration, April 2–3, Washington, D.C.

Bradshaw, M.J. (ed.) (1991) *The Soviet Union: A New Regional Geography*?, Belhaven, London.

Bradshaw, M.J. (1993) *The Economic Effects of Soviet Dissolution*, Royal Institute of International Affairs, London.

Brooks, F.E.J. (1983) *The EEC and a southern African common market in legal perspective*, Unpublished Ph.D. thesis, University of Exeter.

Brown, S. (1993) Free trade brings high tax, *Christian Science Monitor*, 12 June.

Brusco, S. (1982) The Emilian model: productive decentralization and social integration, *Cambridge Journal of Economics*, 6, 167–184.

Brzeziński, Z. (1990) For Eastern Europe: A $25 Billion Aid Package, *The New York Times*, March 7.

Brzeziński, Z. (1991) To Strasbourg or Sarajevo? *European Affairs*, 5, 20–24.

Bulgakov, S. (1982) *Filosofiia khoziaistva*, (first published 1912) Chalidze Publications, New York.
Business Central Europe (1993) Open it up, *Business Central Europe*, May, 1, 7–9.
Business Eastern Europe (1993) The business outlook: Russia, *Business Eastern Europe*, 15 March, 4–5.
Business International (1992) Economic cooperation in Southern Africa: critical issues for companies doing business in South Africa, *Business International*, 2, The Economist Group, London.
Casanova, J.-C. (1993) La spéculation et l'Europe, *L'Express* 19 août, 16.
Cecchini, P. (1988) *The European Challenge: 1992 – The Benefits of a Single Market*, Wildwood House, Aldershot.
CEPR (Centre for Economic Policy Research) (1992) *Is Bigger Better? The Economics of EC Enlargement, Monitoring European Integration*, CEPR Annual Report, London.
Chan, S. (1990a) *Exporting Apartheid: Foreign Policies in Southern Africa 1978–1988*, Macmillan, London.
Chan, S. (1990b) *East Asian Dynamism: Growth, Order and Security in the Pacific Region*, Westview, Boulder, Col.
Chu, D.S.C. (1983) The great depression and industrialisation in Colombia, in: A. Berry (ed.) *Essays on Industrialisation in Colombia*, Centre for Latin American Studies, Arizona, 99–142.
Cooper, C. and Massell, B. (1965) A new look at customs unions theory, *Economic Journal*, 75, 742–747.
Cooper, J. (1993) *The Conversion of the Former Soviet Defence Industry*, Royal Institute of International Affairs, London.
Corbett, H. (ed.) (1970) *Trade Strategy and the Asian-Pacific Region*, Allen & Unwin, London.
Corbridge, S. (1993) *Debt and Development*, Basil Blackwell, Oxford.
Coriat, B. (1979) *L'atelier et le chronomètre*, Bourgois, Paris.
Coriat, B. (1990) *L'atelier et le robot*, Bourgois, Paris.
Coriat, B. (1991) *Penser à l'envers*, Bourgois, Paris.
Culbert, J. (1987) War-time Anglo-American talks and the making of the GATT, *The World Economy*, 10, 381–408.
Dam, K. (1970), *The GATT: Law and international economic organisation*, University of Chicago Press, Chicago.
Datta, A. (1989) Strategies for regional cooperation in post-apartheid Southern Africa–the role of Non-Governmental Organizations, in: B. Oden and H. Othmen (eds.) *Regional Cooperation in Southern Africa: A Post-Apartheid Perspective*. The Scandinavian Institute of African Studies, Seminar Proceedings, No. 22.
Dauvergne, A. (1992) Le champ des manoeuvres, *Le Point*, 28 novembre-4 décembre, 22–25.
Davies, R. (1990a) Post-apartheid scenarios for the Southern African region, *Transformation*, 11, 12–39.
Davies, R. (1990b) *Key Issues in Reconstructing South-Southern African Economic Relations After Apartheid*, Centre for Southern African Studies, University of the Western Cape.
Davies, R. (1991) *Southern Africa Into the 1990s and Beyond*, paper presented to a conference on current and future prospects for the political economy of Southern Africa, 15–19 May, Broederstroom.
Davila-Villers, D.R. (1992) Competition and cooperation in the River Plate: the democratic transition and MERCOSUR, *Bulletin of Latin American Research*, 11, 261–278.

Dawson, A.H. (1993) *A Geography of European Integration*, Belhaven, London.
DBSA (Development Bank of Southern Africa) (1990) *Financial Resources and Capital Investment Prospects in the Common Monetary Area*, paper presented to the conference on the New post-apartheid South Africa and its neighbours, Maseru, 9–12 July.
de la Torre, A. and Kelly, M.R. (1992) *Regional Trade Arrangements*, International Monetary Fund, Occasional Paper No. 93, Washington, D.C.
Delavennat, C. (1993) Protectionnisme: la fièvre monte, *Le Point*, 27 février-5 mars, 61.
Dellenbrant, J.A. (1986) *The Soviet Regional Dilemma*, M.E. Sharpe, New York.
de Melo, J. and Panagariya, A. (1992) *The New Regionalism in Trade Policy*, The World Bank, Washington, D.C.
De Michelis, G. (1990) Reaching out to the East, *Foreign Policy*, 79, 44–55.
Deyo, F.C. (1987) *The Political Economy of New Asian Industrialism*, Cornell University Press, Ithaca, NY.
Dicken, P. (1992a) *Global Shift: The Internationalization of Economic Activity, 2nd Edition*, Chapman, London.
Dicken, P. (1992b) International production in a volatile regulatory environment: the influence of national regulatory policies on the spatial strategies of transnational corporation, *Geoforum*, 23, 303–316.
Dienes, L. (1993) Prospects for Russian oil in the 1990s: reserves and costs, *Post-Soviet Geography*, 34, 79–110.
Dixit, A. and Norman, V. (1980) *Theory of International Trade*, Cambridge University Press, Cambridge.
Dixon, C. and Drakakis-Smith, D. (eds.) (1993) *Economic and Social Development in Pacific Asia*, Routledge, London.
Drache, D. and Gertler, M.S. (eds.) (1991) *The New Era of Global Competition: State Policy and Market Power*, McGill-Queen's University Press, Montreal.
Dunning, J.H. (1993) *Multinational Enterprises and the Global Economy*, Addison-Wesley, Wokingham.
du Pisani, A. (1991) *Ventures into the Interior: Continuity and Change in South Africa's Regional Policy 1948–1991*, paper presented to the conference 'Southern Africa into the 1990s and beyond', Magaliesburg, 15–19 April.
Dyker, D.A. (1983) *The Process of Investment in the Soviet Union*, Cambridge University Press, Cambridge.
Dyker, D.A. (1992) *Restructuring the Soviet Economy*, Routledge, London.
East Asia Analytical Unit (1992) *Australia and North-East Asia in the 1990s: Accelerating Change*, Government of Australia, Canberra.
The Economist (1990) South Africa survey: after apartheid, *The Economist*, 3 November, 1–24.
The Economist (1992), Free trade with luck, *The Economist*. October 17, 15–16.
The Economist (1993a) Multinationals: back in fashion, *The Economist*, March 27, 1–28.
The Economist (1993b) Japan ties up the Asian market, *The Economist*, 24 April, 79–80.
The Economist (1993c) A survey of the European Community: a rude awakening, *The Economist*, 3 July, 1–24.
The Economist (1993d) Norway not hooked, *The Economist*, 22 May, 41.
Economist Intelligence Unit (1991a) *USSR: Country Report No. 4*. The Economist Intelligence Unit, London.
Economist Intelligence Unit (1991b) *The Economist Country Profiles*, for South Africa and Namibia; Botswana; Lesotho; Swaziland and Zimbabwe; Malawi and Tanzania; Mozambique, Zambia and Angola, The Economist Intelligence Unit, London.
Ethier, W. (1982) National and international returns to scale in the modern theory of international scale, *American Economic Review*, 72, 389–405.

Europa (1992) *The Far East and Australasia*, Europa Publications, London.
Europe 2000 (1989) Foreign direct investment in the UK, *Europe 2000*, 4, 35–42.
European Commission (1987) *Treaties Establishing the European Communities and Treaties Amending These Treaties (Single European Act)*, Official Publications of the EC, Luxembourg.
European Commission (1991) *European Economy No. 45*, European Commission, Brussels.
European Commission (1992) *Treaty on European Union (Maastricht Treaty)*, Official Publications of the EC, Luxembourg.
European Commission (1993) The European Community as a world trade partner, *European Economy No. 52*, European Commission, Brussels.
Eurostat (1992) *Basic Statistics of the Community, 29th Edition*, Official Publications of the EC, Luxembourg.
L'Expansion (1992a) Guerre des tranchées au GATT, *L'Expansion*, 6–19 février.
L'Expansion (1992b) Au GATT, ce sont les pays les plus riches en céréales qui s'affrontent, *L'Expansion*. 5–18 mars.
L'Expansion (1992c) GATT: La France en difficulté, *L'Expansion*, 19 novembre-2 décembre.
L'Expansion (1992d) GATT: ne pas casser l'Europe, *L'Expansion*, 3–16 décembre.
External Affairs and International Trade Canada (1993) *NAFTA: What's It All About?*, Government of Canada, Ottawa.
Featherstone, K. (1989) The Mediterranean challenge: cohesion and external preferences, in: J. Lodge (ed.) *The European Community and the Challenge of the Future*, Pinter, London, 186–201
Fieleke, N.S. (1992) One trading world, or many: the issue of regional trading blocs, *New England Economic Review*, May/June, 3–20.
Finger, J.M. (1992) *GATT's Influence on Regional Arrangements*, The World Bank and CEPR conference on new dimensions in regional integration, April 2–3, 1992, Session V, Paper No.12, Washington, D.C.
Gaile, G.L. and Grant, R. (1989) Trade, power, and location: the spatial dynamics of the relationship between exchange and political economic strength, *Economic Geography*, 65, 329–337.
GATT (1986) *The Text of the General Agreement on Tariffs and Trade*, GATT, Geneva.
GATT (1992) *GATT: What It Is, What It Does*, GATT, Geneva.
Geldenhuys, D.J. and Venter, D. (1979) Regional cooperation in Southern Africa: A constellation of states?, *International Affairs Bulletin*, 3, 36–72.
Gertler, M.S. (1988) The limits to flexibility: comments on the post-Fordist vision of production and its geography, *Transactions of the Institute of British Geographers*, 13, 419–432.
Gertler, M.S. (1989) Resurrecting flexibility? A reply to Schoenberger, *Transactions of the Institute of British Geographers*, 14, 109–112.
Gertler, M.S. (1992) Flexibility revisited: districts, nation-states, and the forces of production, *Transactions of the Institute of British Geographers*, 17, 259–278.
Gertler, M.S. and Schoenberger, E. (1992) Commentary. Industrial restructuring and continental trade blocs: the European Community and North America, *Environment and Planning A*, 24, 2–10.
Gibb, R.A. (1991) Imposing dependence: South Africa's manipulation of regional railways, *Transport Reviews*, 11, 19–39.
Gibb, R.A. (1992) Regional integration must evolve one step at a time, *Business Day*, 13 August.
Gibb, R.A. (1993) A common market for post-apartheid Southern Africa: prospects and

problems, *South African Geographical Journal*, 75, 28–35.
Gibb, R.A. and Michalak, W.Z. (1993a) The European Community and Central Europe: prospects for political and economic integration, *Geography*, 78, 16–30.
Gibb, R.A. and Michalak, W.Z. (1993b) Foreign debt in the new East-Central Europe: a threat to European integration? *Environment and Planning C*, 11, 69–85.
Gibson, L.J. (1990) The Pacific Rim: region or regions, *Papers of the Regional Science Association*, 68, 1–8.
Gilpin, R. (1987) *The Political Economy of International Relations*, Princeton University Press, Princeton.
Glasmeier, A., Thompson, J.W. and Kays, A.J. (1993) The geography of trade policy: trade regimes and location decisions in the textile and apparel complex, *Transactions of the Institute of British Geographers*, 18, 19–35.
Globerman, S. (ed.) (1991) *Continental Accord: North American Economic Integration*, Fraser Institute, Vancouver.
Goskomstat SSSR (1988) *Promyshlennost' SSSR* (Industry USSR), Finansy i Statistika, Moscow.
Goskomstat SSSR (1991a) *Narodnoye khozyaystvo v 1990g* (The National Economy of the USSR in 1990), Finansy i Statistika, Moscow.
Goskomstat SSSR (1991b) *Soyuznyye respubliki: osnovnyye ekonomicheskiye i sotsial'nyyee pokazateli* (Soviet Republics: Basic Economic and Social Indicators), Goskomstat SSSR, Moscow.
Grabska, W. (1992) Realizacja Unii Europejskiej a problem terytorialnego rozszerzenia Europy (The European Union and a territorial expansion of the Community), *Sprawy Międzynarodowe*, 45, 35–50.
Granberg, A.G. (1990) Ekonomicheskiy mekanizm mezhrespublikanskikh i mezhregional'nykh otosheniy (Economic mechanism of interrepublican and interregional relations), in: V.I. Kuptsova (ed.) *Radikal'naya ekonomicheskaya reform*, Vysshaya Shkola, Moscow, 310–326.
Granberg, A.G. (1991) Inter-republican integration: Russia's position, *Business in the USSR*, December, 48–49.
Granberg, A.G. (1992) Mezhrespublikansiye ekonomicheskiy svyazi (Interrepublican economic relations), *Rossiyskoye Akademii Nauk*, 2, 3–14.
Granberg, A.G. (1993) The economic interdependence of the former Soviet Republics, in: J. Williamson (ed.) *Economic Consequences of Soviet Disintegration*, Institute for International Economics, Washington, D.C., 47–77.
Grant, R. (1993) Trading blocs or trading blows? The macroeconomic geography of US and Japanese trade policies, *Environment and Planning A*, 25, 273–291.
Green, R.H. (1981a) First steps towards economic liberation, in: A.J. Nsekela (ed.) *Southern Africa: Toward Economic Liberation*, Rex Collings, London.
Green, R.H. (1981b) Economic coordination, liberation and development: Botswana and Namibia perspectives, in: C. Harvey (ed.) *Papers on the Economy of Botswana*, Heinemann, London.
Greenaway, D., Hyclak, T. and Thornton, R.J. (eds.) (1989) *Economic Aspects of Regional Trading Arrangements*, New York University Press, New York.
Gregory, P.R. and Stuart, R.C. (1990) *Soviet Economic Structure and Performance*, 4th Edition, Harper & Row, London.
Griffin, K. (1969) *Underdevelopment in Spanish America*, Allen & Unwin, London.
Grimwade, N. (1989) *International Trade: New Patterns of Trade, Production and Investment*, Routledge, London.
Grinspun, R. and Cameron, M.A. (1993) *The Political Economy of North American Free Trade*, St Martin's Press, New York.

Gros, D. (1991) Regional disintegration in the Soviet Union: economic costs and benefits, *Intereconomics*, September/October, 207–213.
Gros, D. and Thygesen, N. (1992) *European Monetary Integration: From the European Monetary System to the European Monetary Union*, Longman, London.
Gustafson, T. (1989) *Crisis Amid Plenty: The Politics of Soviet Energy Under Brezhnev and Gorbachev*, Princeton University Pres, Princeton, NJ.
Gwynne, R.N. (1980) The Andean Group Automobile Programme, *Bank of London and South America Review*, 14, 160–170.
Gwynne, R.N. (1985) *Industrialisation and Urbanisation in Latin America*, John Hopkins University Press, Baltimore.
Gwynne, R.N. (1990) *New Horizons? Third World Industrialisation in an International Framework*, Longman, London.
Gwynne, R.N. (1993) *Chile 1993: Report on Government, Economy, the Business Environment and Industry*, Business Monitor International, London.
Haas, E.B. (1968) *The Uniting of Europe*, 2nd Edition, Stanford Press, Stanford.
Hamilton, F.E.I. (1990) COMECON: dinosaur in a dynamic world?, *Geography*, 75, 244–246.
Hanink, D.M. (1989) Introduction: trade theories, scale, and structure, *Economic Geography*, 65, 267–270.
Hanlon, J. (1984) *SADCC: Progress, Projects and Prospects*, Report No. 182, The Economist Intelligence Unit, London.
Hanlon, J. (1987) Post-apartheid South Africa and its neighbours, *Third World Quarterly*, 9, 437–449.
Hamon, J. (1989) *SADCC in the 1990s: Development on the Front Line*, Special Report No. 1158, The Economist Intelligence Unit, London.
Hansen, J.D., Heinrich, M. and Nielson, J.V. (1992) *An Economic Analysis of the EC*, McGraw-Hill, London.
Hansson, A.H. (1993) The trouble with the Ruble: monetary reform in the former Soviet Union, in: A Aslund and R. Layard (eds.) *Changing the Economic System in Russia*, Pinter, London,163–182.
Harrod, R.F. (1951) *The Life of Maynard Keynes*, Macmillan, London.
Harvey, D. (1989) *The Condition of Postmodernity: An Enquiry into the Origins of Cultural Change*, Blackwell, Oxford.
Harvey, D. and Scott, A. (1988) The practice of human geography: theory and empirical specificity in the transition from Fordism to flexible accumulation, in: W.D. MacMillan (ed.) *Remodelling Geography*, Blackwell, Oxford, 217–229.
Haus, L.A. (1992) *Globalizing the GATT: the Soviet Union's Successor States, Eastern Europe and the International Trading System*, The Brookings Institution, Washington, D.C.
Havrylyshyn, O. and Pritchett, L. (1991) *Trade Patterns After the Transition*, Working Paper, The World Bank, Washington, D.C.
Havrylyshyn, O. and Williamson, J. (1991) *Open for Business: Russia's Return to the Global Economy*, The Brookings Institution, Washington, D.C.
Hewett, E.A. (1976) A gravity model of CMEA trade, in: J.F. Brada (ed.) *Quantitative and Analytical Studies in East-West Economic Relations*, Indiana University Press, Bloomington, Ind., 35–46.
Hirst, P. and Thompson, G. (1992) The problem of 'globalization': international economic relations, national economic management and the formation of trading blocs, *Economy and Society*, 21, 357–396.
Hirst, P. and Zeitlin, J. (1991) Flexible specialization versus post-Fordism: theory, evidence and policy implications, *Economy and Society*, 20, 1–56.

Hodder, R. (1992) *The West Pacific Rim*, Belhaven, London.
Hofmeier, R. (1990) Politische Konditionierung von Entwicklungshilfe in Afrika, *Afrika Spectrum*, 25, 167–179.
Hojman, D. (1981) The Andean Pact: Failure of a Model of Economic Integration?, *Journal of Common Market Studies*, 20, 139–160.
Holstein, W.J. and Borrus, A. (1990) Is free trade with Mexico good or bad for the US?, *Business Week*, 14 March.
Hopkinson, N. (ed.) (1992), *Completing the Gatt Uruguay Round. Renewed Multilateralism or a World of Regional Trading Blocs?*, Wilton Park Paper No. 61, HMSO, London.
Horowitz, D. (1991) The impending 'second generation' agreements between the European Community and Eastern Europe: some practical considerations, *Journal of World Trade*, 25, 55–80.
Horrigan, B. (1992) How many people worked in the Soviet defense industry, *RFE/RL (Radio Free Europe/Radio Liberty) Research Report*, 1, 33–39.
House of Assembly (1990) Opening address of State President F.W. de Klerk, Republic of South Africa, *Debates of Parliament (Hansard)*, Second Session, Ninth Parliament, 2 February.
Hufbauer, G.C. and Schott, J.J. (1993) *NAFTA: An Assessment*, Institute for International Economics, Washington, D.C.
Hughes, H. (ed.) (1988) *Achieving Industrialisation in East Asia*, Cambridge University Press, Cambridge.
IMF (International Monetary Fund) (1992a) *Common Issues and Interrepublic Relations in the Former USSR*, International Monetary Fund, Washington, D.C.
IMF (1992b) *International Financial Statistics*, Vol. XLV, IMF, Washington, D.C.
IMF (1992c) *International Financial Statistics Yearbook*, IMF, Washington, D.C.
IMF (1993) *International Financial Statistics*. Vol. XLVI. IMF. Washington, D.C.
IMF, WB, OECD, EBRD (International Monetary Fund, The World Bank, Organisation for Economic Cooperation and Development, European Bank for Reconstruction and Development, 1991) *A Study of the Soviet Economy*, Vol.1 OECD, Paris.
International Trade Reporter (1990a) Current Report, *International Trade Reporter*, The National Bureau of National Affairs, Inc., Washington, D.C., 19 September.
International Trade Reporter (1990b) Current Report, *International Trade Reporter*, The National Bureau of National Affairs, Inc., Washington, D.C., 26 September.
International Trade Reporter (1991a) Current Report, *International Trade Reporter*, The National Bureau of National Affairs, Inc., Washington, D.C., 9 January.
International Trade Reporter (1991b) Current Report, *International Trade Reporter*, The National Bureau of National Affairs, Inc., Washington, D.C., 6 February.
International Trade Reporter (1991c) Current Report, *International Trade Reporter*, The National Bureau of National Affairs, Inc., Washington, D.C., 9 September.
International Trade Reporter (1993) Current Report, *International Trade Reporter*, The National Bureau of National Affairs, Inc., Washington, D.C., 19 May.
Investment Canada (1991) *Investing in Canada*, Government of Canada, 5, 2.
Irwin, D.A. (1992) *Multilateral and Bilateral Trade Policies in the World Trading System: An Historical Perspective*, The World Bank and CEPR conference on new dimensions in regional integration, April 2–3, 1992, Session IV, Paper No.9, Washington, D.C.
Ivanter, A., Kirichenko, N. and Khoroshavina, N. (1993) First-quarter CIS statistics indicate severe drop-off in GDP, *Commersant*, 2 June, 17–19.
Jacques, M. and Hall, S. (eds.) (1989) *New Times*, Lawrence & Wishart, London.
Jessop, R. (1990) Regulation theories in retrospect and prospect, *Economy and Society*,

19, 153–216.

Jessop, R., Kastendick, H., Nielsen, K. and Pedersen, O.K. (eds.) (1991) *The Politics of Flexibility: Restructuring State and Industry in Britain, Germany and Scandinavia*, Elgar, Aldershot.

Johnson, H.G. (1965) *The World Economy at the Crossroads: A Survey of Current Problems of Money Trade and Economic Development*, Clarendon Press, Oxford.

Johnston, R.J. (1989) Extending the research agenda, *Economic Geography*, 65, 338–347.

Judt, T. (1992) 'Ex Oriente Lux'? Post-celebratory speculations on the 'Lessons' of '89, in: C. Crouch and D. Marquand (eds.) *Towards Greater Europe? A Continent Without an Iron Curtain*, Blackwell, Oxford, 91–104.

Kalecki, M. (1990) *Collected Works. Vol.1 – Capitalism: Business Cycle and Full Employment*, Clarendon, Oxford.

Kalecki, M. (1991) *Collected Works. Vol.2 – Capitalism: Economic Dynamics*, Clarendon, Oxford.

Kaser, M.C. and Radice, E.A. (eds.) (1986) *The Economic History of Eastern Europe, 1919–1975*, Vol.2, Clarendon Press, Oxford.

Keenwood, A.G. and Loughead, A.L. (1992) *The Growth of the International Economy 1820–1990*, Routledge, London.

Kemp, M. and Wan, H. (1976) An elementary proposition concerning the formation of customs unions, *Journal of International Economics*, 61, 95–97.

Kennedy, P. (1993) *Preparing for the Twenty-First Century*, Harper Collins, London.

Kindleberger, C.P. (1973) *The World in Depression 1929–1939*, Allen Lane, The Penguin Press, London.

Kolakowski, L. (1978) *Main Currents of Marxism: Volume 1. The Founders; Volume 2. The Golden Age; Volume 3. The Breakdown*, Oxford University Press, Oxford.

Kongwa, S. (1991) SADCC: Creating a new vision for the future, *Al Bulletin*, 10, 1.

Korboński, A. (1964) COMECON, *International Conciliation*, September, 12–28.

Korboński, A. (1990) CMEA, economic integration, and perestroika, 1949–1989, *Studies in Comparative Communism*, 23, 47–72.

Kowalik, T. (1971) *Róża Luxemburg. Teoria akumulacji i imperializmu* (Rosa Luxemburg. Theory of Accumulation and Imperialism), Państwowy Instytut Wydawniczy, Warsaw

Köves, A. (1992) *Central and East European Economies in Transition: The International Dimension* Westview Press, Oxford.

Krause, L.B. (1991) European economic integration and the United States, in: J. Pinder *European Community: the Building of a Union*, Oxford University Press, Oxford, 173-4.

Krueger, A.O. (1992) Global trade prospects for the developing countries, *The World Economy*, 15, 457–474.

Krugman, P. (1980) Scale economies, product differentiation and the pattern of trade, *The American Economic Review*, 70, 950–959.

Krugman, P. (1983) New theories of trade among industrial economies, *The American Economic Review*, 73, 343–347.

Krugman, P. (1986) *Strategic Policy and the New International Economics*, Cambridge University Press, Boston.

Krugman, P. (1989) New trade theory and less developed countries, in: G. Calvo (ed.) *Debt Stabilisation and development: Essays in memory of Carlos Diaz-Alejandro*, Basil, Blackwell, Oxford.

Krugman, P. (1990a) *Increasing Returns and Economic Geography*, National Bureau of Economic Research, Working Paper Series No. 3245, Cambridge, Mass.

Krugman, P. (1990b) *Rethinking International Trade*, MIT Press, Cambridge, Mass.
Krugman, P. (1991) *Geography and Trade*, MIT Press, Cambridge, Mass.
Krugman, P. (1992) Does the new trade theory require a new trade policy?, *The World Economy*, 15, 423–441.
Kusakabe, E. (1993) *Japan's Role in the West Pacific Rim in the Next 50 Years*, unpublished M.Sc. thesis, London School of Economics and Political Science, London.
Lancaster, K. (1980) Intra-industry trade under perfect monopolistic competition, *Journal of International Economics*, 10, 151–175.
Langhammer, R.J. (1992) The developing countries and regionalism, *Journal of Common Market Studies*, 30, 211–232.
Langhammer, R.J., Sagers, M.J. and Lücke, M. (1992) Regional distribution of the Russian Federation's earnings outside the former Soviet Union and its implications for regional economic autonomy, *Post-Soviet Geography*, 33, 617–634.
Lash, S. and Urry, J. (1987) *The End of Organized Capitalism*, Polity, Cambridge.
Leborgne, D. and Lipietz, A. (1988) New technologies, new modes of regulation: some spatial implications, *Environment and Planning D: Society and Space*, 6, 263–280.
Leborgne, D. and Lipietz, A. (1989) *Pour éviter l'Europe à deux vitesse*, CEPREMAP, Paris.
Lee, C. (1989) *Options for Regional Co-operation and Development in Southern Africa*, unpublished Ph.D. thesis, University of Pittsburgh, Johnstown, Penn.
Leistner, G.M.E. (1981) Towards a regional development strategy for Southern Africa, *Southern African Journal of Economics*, 4, 349–364.
Leistner, G.M.E. (1985) Southern Africa: the market of the future?, paper presented at the 24th Annual Meeting of the Bureau of Market Research, 1984, University of Pretoria, *Africa Insight*, 15, 17–21.
Lemon, A. (1991) Apartheid as a foreign policy: dimensions of international conflict in Southern Africa, in: N. Kliot and W. Stanley (eds.) *The Political Geography of Conflict and Peace*, Belhaven, London.
Leyshon, A. (1992) The transformation of regulatory order: regulating the global economy and environment, *Geoforum*, 23, 249–267.
Lindberg, L. (1963) *The Political Dynamics of European Integration*, Oxford University Press, Oxford.
Lindberg, L. and Scheingold, S. (1970) *Europe's Would-Be Polity*, Prentice-Hall Englewood Cliffs, NJ.
Lintner, V. and Mazey, S. (eds.) (1991) *The European Community: Economic and Political Aspects*, McGraw-Hill, London.
Lipietz, A. (1977) *Le capital et son espace*, Maspero, Paris.
Lipietz, A. (1986) New tendencies in the international division of labour: regimes of accumulation and modes of regulation, in: A.J. Scott and M. Storper (eds.) *Production, Work, Territory: The Geographical Anatomy of Industrial Capitalism*, Unwin & Hyman, Boston, 16–40.
Lipietz, A. (1987) *Mirages and Miracles: The Crisis of Global Fordism*, Verso, London.
Lipietz, A. (1992) *Towards a New Economic Order: Postfordism, Ecology and Democracy*, Oxford University Press, Oxford.
Lipietz, A. (1993) The local and the global: regional individuality or regionalism?, *Transactions of the Institute of British Geographers*, 18, 8–18.
Lipsey, R. (1957) The theory of customs unions: trade diversion and welfare, *Economica*, 24, 40–46.
Lipsey, R. (1960) The theory of customs unions: a general survey, *The Economic Journal*, 70, 496–513.
Loesch, K.R. (1993) *The Likelihood and Possible Implications of the Formation of a*

Free-Trade Bloc in the West Pacific Rim, unpublished M.Sc. thesis, London School of Economics and Political Science, London.

Lu Fu-Chen and Salih, K. (eds.) (1987) *The Challenge of Asia-Pacific Cooperation*, Association of Development Research and Training Institutes of Asia and the Pacific, Kuala Lumpur.

Luxemburg, R. (1951) *The Accumulation of Capital*, (first published 1915) Routledge & Kegan, London.

Maasdorp, G. (1982) The Southern African Customs Union: an assessment, *Journal of Contemporary African Studies*, 1, 81–112.

Maasdorp, G. (1985) Squaring up to economic dominance, in: I. Rotberg (ed.) *South Africa and Its Neighbours: Regional Security and Selfinterest*, Lexington Books, Lexington.

Maasdorp, G. (1986) The Southern African nexus: dependence or interdependence?, *Indicator South Africa: Economic Monitor*, 4, 5–19.

Maasdorp, G. (1989) *Economic Relations in Southern Africa – Changes Ahead?*, paper presented to the conference on 'South and Southern Africa in the 21st Century', Maputo, December.

Maasdorp, G. (1992) Economic prospects for South Africa in Southern Africa, *South Africa International*, January, 121–127.

Machlup, F. (1977) *A History of Thought on Economic Integration*, Columbia University Press, New York.

MacPherson, A.D. and McConnell, J.E. (1992) *Canadian-owned establishments in Western New York, Occasional Paper No.4*, Canada-US Trade Center, University at Buffalo, Buffalo, NY.

Maggs, J. (1991) US labor fights Mexico trade pact, *Journal of Commerce*, 6 February.

Malan, T. (1983a) Regional economic cooperation in Southern Africa, *Africa Insight*, 1, 43–51.

Malan, T. (1983b) New dimensions in Southern African relations, *ISSUP Strategic Review*, March, 2–20, Institute for Strategic Studies, University of Pretoria.

Marer, P. (1991) Foreign economic liberalization in Hungary and Poland, *AEA Papers and Proceedings*, 81, 329–333.

Martin, R. and Johnson, P. (eds.) (1986) *Destructive Engagement*, Zimbabwe Publishing House, Harare.

Marx, K. (1967) *Capital: A Critique of Political Economy. Vol 2: The Process of Circulation of Capital*, (first published 1893) Progress Publishers, Moscow.

Matthews, J. (1983) Economic integration and cooperation in Southern Africa, in: J. Matthews (ed.) *Southern Africa in the World Economy*, McGraw-Hill, Johannesburg.

Matthews, J. 1984 Economic integration in Southern Africa: progress or decline? *South African Journal of Economics*, 3, 256–265.

McAuley, A. (1991a) Costs and benefits of de-integration in the USSR, *Moct-Most*, 2, 51–65.

McAuley, A. (1991b) The economic consequences of Soviet disintegration, *Soviet Economy*, 7, 189–214.

McConnell, J.E. (1986) Geography of international trade, *Progress in Human Geography*, 10, 471–483.

McDonald, F. and Dearden, S. (eds.) (1992) *European Economic Integration*, Longman, London.

Meade, J. (1955) *The theory of customs unions*, North-Holland, Amsterdam.

Miall, H. (1993) *Shaping the New Europe*, Royal Institute of International Affairs, Pinter, London.

Michalak, W.Z. (1993) Foreign direct investment and joint ventures in East-Central

Europe: a geographical perspective, *Environment and Planning A*, 25, 1573–1591.
Michalak, W.Z. and Gibb, R.A. (1992a) Political geography and eastern Europe, *Area*, 24, 341–349.
Michalak, W.Z. and Gibb, R.A. (1992b) The debt to the West: recent developments in the international financial situation of East-Central Europe, *Professional Geographer*, 44, 260–271.
Michalak, W.Z. and Gibb, R.A. (1993) Eastern Europe and the World System: an anti-systemic reply to Taylor and Johnston, *Area*, 25, 305–309.
Michalopoulos, C. and Tarr, D. (1992) *Trade and Payments Arrangement for States of the Former USSR*, The World Bank, Washington, D.C.
Michalopoulos, C. (1993) *Trade Issues in the New Independent States*, World Bank, Washington, D.C.
Michalski, A. and Wallace, H. (1992) *The European Community: The Challenge of Enlargement*, Royal Institute of International Affairs, London.
Milner, H. (1988) *Resisting Protectionism: Global Industries and the Politics of International Trade*, Princeton University Press, Princeton, NJ.
Mirsky, J. (1993) China shows sting in dragon's tail with threat to trade deal, *The Times*, 18 May.
Mishra, U.N. (1990) Mexico outlines trade goals, *Mexico City News*, 20 September.
Mitrany, D. (1933) *The Progress of International Government*, Allen & Unwin, London.
Mitrany, D. (1966) *A Working Peace System* (first published 1943), Quadrangle Books, Chicago.
Mitrany, D. (1975) *The Functional Theory of Politics*, Robertson, London.
Le Monde (1993) Deux visions incompatibles, editorial, *Le Monde*, 1–2 août.
Montagnon, P. (1989) EC could not become a trade fortress, *Financial Times*, 24 July.
Morawetz, D. (1974) *The Andean Group: A Case Study in Economic Integration Among Developing Countries*, MIT Press, Cambridge, Mass.
Morgan, R. (1992) Germany in the New Europe, in: C. Crouch and D. Marquand (eds.) *Towards Greater Europe? A Continent Without an Iron Curtain*, Blackwell, Oxford, 105–117.
Morris, A. (1981) *Latin America: Economic Development and Regional Differentiation*, Hutchinson, London.
Moulaert, F. and Swyngedouw, E.A. (1989) Survey 15: A regulation approach to the geography of flexible production systems, *Environment and Planning D: Society and Space*, 7, 327–345.
Msabaha, I. and Shaw, T. (1987) *Confrontation and Liberation in Southern Africa*, Westview Press, Boulder.
Muirhead, D. (1988) Trade and trade promotion, in: E. Leistner and P. Esterhuysen (eds.) *South Africa in Southern Africa: Economic Interaction*, Africa Institute, Pretoria.
Murphy, A.B. (1992) Western investment in East-Central Europe: emerging patterns and implications for state stability, *Professional Geographer*, 44, 249–259.
Mutharika, B.W. (1991) An interview: 'Eighteen into one does go'. *Africa South*, May, 29.
Mwase, N. (1986) Regional cooperation and socialist transformation: prospects and problems, *Journal of African Studies*, 1, 17–24.
Nanto, D.K. (1990) Asian responses to the growth of trading blocs, in: R.S. Belous and R.S. Hartley (eds.) *The Growth of Regional Trading Blocs in the Global Economy*, National Planning Association, Washington, D.C., 85–115.
Nester, W.R. (1990) *Japan's Growing Power Over East Asia and the World Economy*, Macmillan, London.
Ncube, P.D. (1991) *Economic Integration in Eastern and Southern Africa – An*

Evaluation of the Preferential Trade Area, Southern African Labour and Development Research Unit, Cape Town.
Newman, M. (1991) Britain and the European Community: the impact of membership, in: V.
Lintner and S. Mazey (eds.) *The European Community: Economic and Political Aspects*, McGraw-Hill, London, 146–164.
New York Times (1990) AFL-CIO opposed to trade agreement. *New York Times*, 12 June.
Noren, J.H. and Watson, R. (1992) Interrepublican economic relations after the disintegration of the USSR, *Soviet Economy*, 8, 89–129.
Nove, A. (1977) *The Soviet Economic System*, Allen & Unwin, London.
Nove, A. (1981) An overview, in: I.S. Koropeckyj and G.E. Shroeder (eds.) *Economics of Soviet Regions*, Praeger, New York, 1–8.
Nugent, N. (1989) *The Government and Politics of the European Community*, Macmillan, London.
O'Brien, R. (1992) *Global Financial Integration: The End of Geography*, Pinter, London.
OECD (1990) *Recent Developments in Regional Trading Arrangements Among OECD Countries: Main Implications for Third Countries and for the Multilateral Trading System*, Trade Committee, OECD, Paris.
Offe, K. (1985) *Disorganized Capitalism*, MIT Press, Cambridge, Mass.
Okwuizumi, K., Calder, K.E. and Gong, G.W. (eds.) (1992) *The US-Japan Economic Relationship in East and South East Asia*, Centre for Strategic and International Studies, Washington, D.C.
Ostry, S. (1992) The domestic domain: the new international policy area, *Transnational Corporations* 1, 7–26.
Oxley, A. (1990) *The Challenge of Free Trade*, Harvester-Wheatsheaf, Worcester.
Palankai, T. (1991) *The European Community and Central European Integration: The Hungarian Case*, Institute for East-West Security Studies, Occasional Paper Series, Westview Press, Boulder, Col.
Pallot, J. and Shaw, D.J.B. (1981) *Planning in the Soviet Union*, Croom Helm, London.
Patterson, G. (1989) Implications for the GATT and the world trading system, in: J. Schott (ed.) *Free Trade Areas and US Policy*, Institute for International Economics, Washington D.C., 353–366.
Pearson, C. and Riedal, J. (eds.) (1990) *The Direction of Trade Policy*, Blackwell, Oxford.
Penketh, K. (1992) External trade policy, in: F. McDonald and S. Dearden (eds.) *European Economic Integration*, Longman, London, 146–158.
Perlmutter, H. (1969) The tortuous evolution of the multinational enterprise, *Columbia Journal of World Business*, 4, 9–18.
Petit, P. (1984) *Slow Growth and the Service Economy*, Pinter, London.
Petri, P.A. (1992) One bloc, two blocs, or more? Political economic factors in Pacific trade policy, in: K. Okwuizumi, K.E. Calder and G.W. Gong (eds.) *The US-Japan Economic Relationship in East and South East Asia*, Centre for Strategic and International Studies, Washington, D.C., 39–70.
Pinder, J. (1991) *European Community: The Building of a Union*, Oxford University Press, Oxford.
Piore, M.J. and Sabel, C. (1984) *The Second Industrial Divide*, Basic Books, New York.
Pipes, R. (1970) *Struve, Liberal on the Left, 1870–1905*, Harvard University Press, Cambridge, Mass.
Pipes, R. (1980) *Struve, Liberal on the Right, 1905–1944*, Harvard University Press,

Cambridge, Mass.
Pollert, A. (1988) Dismantling flexibility, *Capital and Class*, 34, 42–75.
Pollert, A. (ed.) (1991) *Farewell to Flexibility?*, Blackwell, Oxford.
Porter, M.E. (1991) *Canada at the Crossroads: The Reality of a New Competitive Environment*, Business Council on National Issues, Ottawa.
Preeg, E.M. (1989) The GATT trading system in transition: an analytic survey of recent literature, *The Washington Quarterly* 12, 201–213.
Preeg, E.M. (1992) The US leadership role in world trade: past, present and future, *The Washington Quarterly*, 15, 81–91.
Prestowitz, C.V. (1988) *Trading Places. How We Allowed Japan to Take the Lead*, Basic Books, New York.
Prestowitz, C.V., Tonelson, A. and Jerome, R. (1991) The last gap of GATTism, *Harvard Business Review*, March-April, 130–138.
Reynard, P. (1950) The unifying force for Europe, *Foreign Affairs*, 28, 39–55.
Richardson, J.D. (1990) The political economy of strategic trade policy, *International Organisation*, 44, 107–135.
Robson, P. (1980) *The Economics of International Integration*, Allen & Unwin, London.
Rossi, E. (1944) *L'Europe de demain*, La Baconnière, Neuchatel.
Rugman, A.M. (1991) *Diamond in the rough: Poter and Canada's international competitiveness, Research Programme Working Papers No. 44*, Ontario Centre for International Business, Toronto.
Rugman, A.M. and Verbeke, A. (1990) *Global Corporate Strategy and Trade Policy*, Routledge, London.
Russett, B.M. (1967) *International Regions and the International System: A Study in Political Ecology*, Rand McNally & Company, Chicago.
SADC (South African Development Community) (1992) *Towards a Southern African Development Community*, SADC, Gabarone.
SADCC (South African Development Coordination Conference) (1981) *SADCC2-Maputo*, SADCC Liaison Committee, SADCC, Gabarone.
SADCC (1989) *SADCC Annual Progress Report 1989–90*, SADCC, Gaborone.
SADCC (1992) *SADCC Theme Document*, Maputo conference, 29–31 January, SADCC, Gaborone.
Sagers, M.J. (1991) Regional aspects of the Soviet economy, *PlanEcon Report*, 7, No. 1–2.
Sagers, M.J. (1992a) Regional industrial structure and economic prospects in the former USSR, *Post-Soviet Geography*, 33, 238–268.
Sagers, M.J. (1992b) Review of the energy industries of the former USSR in 1991, *Post-Soviet Geography*, 33, 487–515.
Sagers, M.J. (1993) Russian crude oil exports in 1992: who exported Russian oil?, *Post-Soviet Geography*, 34, 207–211.
Sayer, A. (1989) Postfordism in question, *International Journal of Urban and Regional Research*, 13, 666–695.
Sayer, A. and Walker, R. (1992) *The New Social Economy: Reworking the Division of Labour*, Blackwell, Oxford.
Schmitter, P.C. (1971) Central American integration: spill-over, spill-around or encapsulation?, *Journal of Common Market Studies*, 9, 18–30.
Schoenberger, E. (1988) Multinational corporations and the new international division of labour: a critical appraisal, *International Regional Science Review*, 11, 105–119.
Schott, J. (1989), More free trade areas, in: J. Schott (ed.) *Free Trade Areas and US Trade Policy*, Institute for International Economics, Washington, D.C., 1–59.
Schott, J.J. (ed.) (1990) *Completing the Uruguay Round: A Results-orientated Approach*

to the GATT Trade Negotiations, Institute for International Economics, Washington, D.C.

Schott, J.J. (1991) Trading blocs and the world trading system, *The World Economy*, 14, 1–17.

Schuman, R. (1950) The Schuman Declaration, in: European Commission, *Jean Monnet: a Grand Design for Europe*, European Documentation 5/1988, Official Publications of the EC, Luxembourg, 43–45.

Scott, A.J. (1986) Industrial organisation and location: division of labour, the firm, and spatial process, *Economic Geography*, 62, 215–231.

Scott, A.J. (1988) *New Industrial Spaces*, Pion, London.

Scott, A.J. and Storper, M. (eds.) (1986) *Production, Work, Territory: The Geographical Anatomy of Industrial Capitalism*, Unwin & Hyman, Boston.

Seliverstov, V. (1991) Inter-republican economic interactions in the Soviet Union, in: A. McAuley (ed.) *Soviet Federalism: Nationalism and Economic Decentralisation*, Leicester University Press, London, 108–127.

Senik-Leygonie, C. and Hughes, G. (1992) Industrial profitability and trade among the former Soviet Republics, *Economic Policy*, October, 354–386.

Shatalin, S.S. (1990) *Perekhod k rynku* (*Transition to the Market*), Ministry of Publishing and Mass Information Moscow.

Shaw, D.J.B. (1985) Spatial dimensions in Soviet central planning, *Transactions of the Institute of British Geographers*, 10, 401–412.

Sheepers, C.F. (1979), The possible role of a customs-union type model in promoting closer economic ties in Southern Africa, *Finance and Trade Review*, 13, 82–99.

Sked, A. (1992) The case against the Treaty, in: The European, *Maastricht Made Simple*, Special guide no. 1, The European, 27.

Sillem, T. (1988) *South African Destabilization of the Front Line: The Case of Zimbabwe*, unpublished M.Phil thesis, Trinity Hall, University of Cambridge, Cambridge.

Singer, P. (1973) *Economia politica da urbanização*, Edoçodes CEBRAP, São Paulo.

Slay, B. (1991) On the economics of interrepublican trade, *RFE/RL (Radio Free Europe/Radio Liberty) Research Institute: Report on the USSR*, 3, 1–8.

Smith, A. (1993) *Russia and the World Economy: Problems of Integration*, Routledge, London.

Smith, W.R. (1993) *The NAFTA Debate, Part I: A Primer on Labor, Environmental, and Legal Issues*, The Heritage Foundation, Washington, D.C.

Snyder, T. (1993) Soviet Monopoly in Williamson, J. (ed) *Disintegration*, Institute for International Economics, Washinton DC, 175–243.

Soviet Geography (1991) News notes, *Soviet Geography*, 32, 190–193.

Storper, M. (1992) The limits to globalization: technology districts and international trade, *Economic Geography*, 68, 60–93.

Storper, M. and Christopherson, S. (1987) Flexible specialization and regional industrial agglomeration: the case of the US motion picture industry, *Annals of the Association of American Geographers*, 77, 104–117.

Storper, M. and Scott, A.J. (1989) The geographical foundations and social regulation of flexible production complexes, in: J. Wolch and M. Dear (eds.) *The Power of Geography: How Territory Shapes Social Life*, Unwin & Hyman, Boston, 21–40.

Storper, M. and Walker, R. (1989) *The Capitalist Imperative: Territory, Technology, and Industrial Growth*, Blackwell, Oxford.

Taylor, F.W. (1947) *The Principles of Scientific Management* (first published 1911), Greenwood Press, Westport, Conn.

Taylor, M.J. and Thrift, N.J. (eds.) (1986) *Multinationals and the Restructuring of the*

World Economy, Croom Helm, Beckenham.

Teague, P. (1989) *The European Community: The Social Dimension*, Longman, London.

Testa, W.A. (1992) Trends and prospects for rural manufacturing, *Regional Economic Issues*, Federal Reserve Bank Working Paper No. 12, Chicago.

Thatcher, M. (1988) *Britain and Europe, Text of the Prime Minister's Speech at Bruges on 20 September, 1988*, Conservative Political Centre, London.

Thrift, N.J. and Leyshon, A. (1988) 'The gambling propensity': banks, developing country debt exposures and the new international financial system, *Geoforum*, 19, 55–69.

Thrift, N.J. (1990) The perils of the international financial system, *Environment and Planning A*, 22, 1135–1136.

Thurow, L. (1993) *Head to Head: The Coming Economic Battle Among Japan, Europe and America*, Nicholas Brealey, London.

Tickell, A. and Peck, J.A. (1992) Accumulation, regulation and the geographies of post-Fordism: missing links in regulationist research, *Progress in Human Geography*, 16, 190–218.

Tjonneland, E.N. (1989) South Africa's regional policies in the late post-apartheid periods, in: B. Oden and H. Othman (eds.) *Regional Cooperation in Southern Africa: A Post Apartheid Perspective*, The Scandinavian Institute of African Studies, Seminar Proceedings, No. 22, Uppsala.

Tomlinson, J. (1988) Can governments manage the economy?, *Fabian Tracts*, January, 524.

Tostensen, A. (1990) Les défis des années 90 pour SADCC, in: Conseil Canadien pour la coopération internationale, *SADCC vers la décennie 90: un défi pour le Canada*, Conseil Canadien pour la coopération internationale, Ottawa.

Tovias, A. (1991) A survey of the theory of economic integration, *Journal of European Integration*, 15, 5–23

Tugan-Baranovsky, M.I. (1913a) *Les crises industrielles en Angleterre*, Giard et Briere, Paris.

Tugan-Baranovsky, M.I. (1913b) *Sociale Theorie der Verteilung*, Scholz, Berlin.

Tugendhat, C. (1987) *Making Sense of Europe*, Pelican Books, London.

Tyson, L. (1987) *Creating Advantage: Strategic Policy for National Competitiveness*, BRIE Working Paper, Berkeley.

Tyson, L. (1992) *Who's Bashing Whom? Trade Conflicts in High-technology Industries*, Institute for International Economics, Washington, D.C.

Tyson, L., Dickens, W.T. and Zysman, J. (1988) *The Dynamics of Trade and Employment*, Ballinger, Cambridge, Mass.

United Nations (1990a) *Regional Trading Blocs: A Threat to the Multilateral Trading System: Views and Recommendations of the Committee for Development Planning*, Department of International Economic and Social Affairs, United Nations, New York.

United Nations (1990b) *Restructuring the Developing Economies of Asia and the Pacific in the 1990s*, United Nations, New York.

United States General Accounting Office (1992) *North American Free Trade Agreement: U.S.-Mexican Trade and Investment Data*, United States General Accounting Office, Washington, D.C.

Urwin, D.W. (1991) *The Community of Europe*, Longman, London.

Uys, S. (1988) The short and unhappy life of CONSAS, *South Africa International*, 4, 243–248.

Valance, G. (1993) France-Allemagne: le jour oÙ tout a craqué, *L'Express*, 5 août, 16–21.

Vale, P. (1982) Prospects for transplanting European models of regional integration to Southern Africa, *Politikon: South African Journal of Political Science*, 2, 32–41.

Vale, P. (1989a) Integration and disintegration in Southern Africa, *Reality: A Journal of Liberal and Radical Opinion*, May, 7–12.

Vale, P. (1989b) Whose world is it anyway? International relations in South Africa, in: H.C. Dyer and L. Mangaasarian (eds.) *The Study of International Relations: The State of the Art*, Macmillan, London.

Vale, P. (1990) *Starting Over: Some Early Questions on a Post-Apartheid Foreign Policy*, Southern African Perspectives, a Working Paper Series, No. 1, Centre for Southern African Studies, University of the Western Cape.

Van Brabant, J.M. (1974) On the origins and the tasks of the CMEA, *Osteuropa Wirtschaft*, 19, 192–193.

Van Ham, P. (1993) *The EC, Eastern Europe and European Unity: Discord, Collaboration and Integration Since 1947*, Pinter, London.

Vatikiotis, M. (1993) Market or mirage, *Far Eastern Economic Review*, 15 April, 48–50.

Vernon, R. (1971) *Sovereignty at Bay: The Multinational Spread of US Enterprises*, Basic Books, New York.

Vernon, R. (1977) *Storm Over the Multinationals: The Real Issues*, Harvard University Press, Cambridge, Mass.

Vernon, R. (1985) *Exploring the Global Economy: Emerging Issues in Trade and Investment*, Center for International Affairs, Harvard University.

Vesnik statistiki (1990a) Ekonomicheskiye vzaimosvyazi republik v narodnokhoy-aystvennom komplekse (Economic interrelations of the republics in the national economic complex), *Vestnik statistiki*, 3, 36–353.

Vestnik statistiki (1990b) Ob'em vvoza i vyvoza produktsii po soyiznym respublikam za 1988g vo vnutrennykh i mirovykh tsenakh (Volume of imports and exports of products by the Soviet republics in 1988 in internal and external prices), *Vestnik statistiki*, 4, 49–60.

Viner, J. (1950) *The Customs Union Issue*, Carnegie Endowment for World Peace, New York.

Wall Street Journal (1990) Foreign investment in Mexico grows, *Wall Street Journal*, 5 January.

Wallace, H., Wallace, W.V. and Webb, C. (eds.) (1983) *Policy-Making in the European Community*, Macmillan, London.

Wallace, W.V. (1992) From twelve to twenty-four? The challenges to the EC posed by the revolutions in Eastern Europe, in: C. Crouch and D. Marquand (eds.) *Towards Greater Europe? A Continent Without an Iron Curtain*. Blackwell, Oxford, 34–51.

Wallace, W.V. and Clarke, R.A. (1986) *COMECON, Trade and the West*, Pinter, London.

Walters, J. (1989) Renegotiating dependency: the case of the Southern African Customs Union, *Journal of Common Market Studies*, 38, 29–52.

Wang, Z.K. and Winters, L.A. (1991) *The Trading Potential of Eastern Europe*, Discussion Paper No. 610, Centre for Economic Policy Research London.

Webber, M.J. (1991) The contemporary transition, *Environment and Planning D: Society and Space*, 9, 165–182.

Weimer, B. (1991) The Southern African Development Coordination Conference (SADCC): Past and future, *Africa Insight*, 2, 78–88.

Weintraub, S. (1990) *Transforming the Mexican Economy: The Salinas Sexenio*, National Planning Association, Washington, D.C.

Whalley, J. (1992) CUSTA and NAFTA: Can WHFTA be far behind?, *Journal of Common Market Studies*, 30, 125–142.

Williamson, J. (1992) *Trade and Payments After Soviet Disintegration*, Institute for International Economics, Washington, D.C.

Winters, L.A. (1990) The road to Uruguay, *The Economic Journal*, 100, 1288–1303.
Winters, L.A. (1992) Goals and own goals in European trade policy, *The World Economy*, 15, 557–574.
Wise, M. (1984) *The Common Fisheries Policy of the European Community*, Methuen, London.
Wise, M. (1989) France and European Unity, in: R. Aldrich and J. Connel (eds.) *France in World Politics*, Routledge, London, 37–73.
Wise, M. (1991) War, peace and the European Community, in: N. Kliot and S. Waterman (eds.) *The Political Geography of Conflict and Peace*, Belhaven, London, 110–125.
Wise, M. and Chalkley, B. (1990) Unemployment: regional policy defeated?, in: D. Pinder (ed.) *Challenge and Change in Western Europe*, Belhaven, London, 179–194.
Wise, M. and Croxford, G. (1988) The European Regional Development Fund: Community ideals and national realities, *Political Geography Quarterly*, 7, 161–182.
Wise, M. and Gibb, R.A. (1993) *From Single Market to Social Europe: the European Community in the 1990s*, Longman, London.
Wonnacott, P. and Lutz, M. (1989) Is there a case for free trade areas? in: J. Schott (ed.) *Free Trade Areas and US Trade Policy*, Institute for International Economics, Washington, D.C., 59–65.
Wonnacott, R.J. (1990) *U.S. Hub-and-spoke bilaterals and the multilateral trading system: commentary*, C.D. Howe Institute, Paper No. 23, Ottawa.
Woolcock, S. (1991) *Market Access Issues in EC-US Relations: Trading Partners or Trading Blows?*, Royal Institute of International Affairs, Pinter, London.
World Bank (1992) *World Development Report 1992*, Oxford University Press, Oxford.
Yakovlev, A. (1991) Monopolizm v ekonomike SSSR (Monopoly in the economy of the USSR), *Vestnik statistiki*, 1, 3–6.
Yamazawa, I. (1992) On Pacific economic integration, *The Economic Journal*, 102, 1519–1529.
Yoffie, D.B. (1983) *Power and Protectionism: Strategies of Newly Industrialized Countries*, Columbia University Press, New York.
Yoffie, D.B. (1990) *International Trade and Competition: Cases and Notes in Strategy and Management*, McGraw-Hill, New York.
Yoffie, D.B. and Milner, H.V. (1989) An alternative to free trade or protectionism: why corporations seek strategic trade policy, *California Management Review*, 31, 111–131.
Zieba, R. (1992) 'Nowy regionalizm' w Europie a Polska (Poland and the 'New regionalism' in Europe), *Sprawy Międzynarodowe*, 43, 25–44.

Index